Principles of Polymer Engineering

A note on the cover photograph

The cover photograph shows a footbridge over the River Tay at Aberfeldy in the Scottish Highlands. The bridge, made from polymer-based composites, was commissioned by a golf club, in order to extend its course from 9 to 18 holes. The design brief specified that: (a) capital and future maintenance costs should be kept to a minimum; (b) construction should not involve heavy machinery that might damage the course; and (c) the appearance of the bridge should be compatible with its location in an area of outstanding natural beauty.

The club consulted staff of the Engineering Department at nearby Dundee University, who after consultation with the Maunsell Group recommended the use of composites, and collaborated in the design and construction. The bridge deck is made from glass-reinforced polyester (GRP), and was installed manually by final-year engineering students from the University, without the use of cranes. The A-shaped pylons are also made from GRP, and the supporting cables are of Kevlar fibre, which is sheathed in a protective coating of low-density polyethylene.

The structure was designed and specified by Maunsell Structural Plastics, of Beckenham, Kent, who kindly supplied the photograph. The company has won many awards for its composites-based designs, which provide good illustrations of the principles of polymer engineering.

Principles of Polymer Engineering

Second edition

N. G. McCrum

Hertford College
University of Oxford

C. P. Buckley

Department of Engineering Science
University of Oxford

C. B. Bucknall

Advanced Materials Department
Cranfield University

OXFORD • NEW YORK • TOKYO
OXFORD UNIVERSITY PRESS
1997

Oxford University Press, Great Clarendon Street, Oxford OX2 6DP
Oxford New York
Athens Auckland Bangkok Bogota Bombay Buenos Aires
Calcutta Cape Town Dar es Salaam Delhi Florence Hong Kong
Istanbul Karachi Kuala Lumpur Madras Madrid Melbourne
Mexico City Nairobi Paris Singapore Taipei Tokyo Toronto Warsaw
and associated companies in
Berlin Ibadan

Oxford is a trade mark of Oxford University Press

Published in the United States
by Oxford University Press Inc., New York

A catalogue record for this book is available from the British Library

Library of Congress Cataloging in Publication Data
McCrum, N. G.
Principles of polymer engineering / N.G. McCrum, C.P. Buckley,
C.B. Bucknall. – 2nd ed.
Includes bibliographical references and index.
ISBN 0 19 856527 5 (Hbk).   ISBN 0 19 856526 7 (Pbk)
1. Polymers. 2. Polymerization. I. Buckley, C. P.
II. Bucknall, C. B. III. Title.
TA455.P58M334 1997   668.9–dc21   97-12589

ISBN 0 19 856527 5 (Hbk)
ISBN 0 19 856526 7 (Pbk)

Typeset by Technical Typesetting Ireland
Printed in Great Britain by Bookcraft Ltd., Midsomer Norton, Avon

Preface to the second edition

Some eight years have passed since the first edition appeared. During this time we have used the first edition extensively in our own teaching at Oxford, UMIST, and Cranfield, and it has been widely used elsewhere. We have been delighted with the reception it has received. In particular, the distinctive style of the book—developing the principles of polymer engineering from a base of the underlying materials science, and aiming to treat topics at a self-contained quantitative level or not at all—seems to have worked well. In our experience it has engaged the interest and enthusiasm of student engineers. The large number of end-of-chapter problems has proved a particularly useful feature to instructors and students, as a means of exercising and testing an understanding of the text, expecially since the separate *Solutions Manual* was published (OUP, 1989).

Over the intervening years polymers have become even more firmly embedded in the university engineering curriculum, alongside traditional engineering materials. No engineering course can now afford to neglect the engineering properties of polymers or their important role in manufacturing. It is to serve this continuing teaching need that we have revised *Principles of Polymer Engineering*.

The original format is retained, but we believe we have strengthened the treatment in key areas. We find today's students much better attuned to the wider aspects of engineering decision-making, especially with respect to the environment, than used to be the case. We want to encourage this and have revised Chapter 0 substantially, to include a discussion of the environmental impact of polymers. Whether in providing lighter automobile components for fuel-saving, or in the ability of thermoplastics to be remelted and recycled, or in the recovery of heat by clean incineration, polymers seem set to piay a vital role in the more environmentally aware engineering of the future. We have tried to give a taste of the important issues involved. A continuing trend has been the growth in the variety of grades of polymer available, tailored for different applications: many of them are toughened grades. There is now increased understanding of the mechanisms by which toughening is achieved in polymers, and of how to accommodate it in fracture test methods, and we have extended Chapter 5 on yield and fracture to reflect the improved state of knowledge. Another significant revision is an enlargement of Chapter 7 on processing, especially to include a treatment of heat transfer. Although the underlying theory will be familiar already to many students, it is so central a topic in polymer forming processes that we believe readers will prefer having it presented together with other aspects of processing in the same book. We have also added to the end-of-chapter problems, particularly on manufacturing and design, Chapters 7 and 8. The book now contains some 195 problems, and a revised edition of the *Solutions Manual* will be published shortly. Finally we have taken the opportunity to clarify some points and to make other minor improvements.

There are several areas of polymer science and technology that have seen recent dramatic advances but are not included. We continue to aim to provide a treatment that is self-contained and quantitative within a book of reasonable length, and this means that hard choices must be made. Examples of topics not discussed are liquid crystal polymers and electrical properties of polymers. Some excellent monographs are available on these topics and are included in our extensive list of Further Reading for students who are interested.

We trust our readers will find the revised book a stimulating introduction to a fascinating subject.

Oxford   N.C.M.
Cranfield   C.P.B.
June 1997   C.B.B.

Preface to the first edition

Our purpose in writing this book has been to prepare a text which is

- an integrated, complete, and stimulating introduction to polymer engineering,
- suitable for the core course in mechanical or production engineering, and
- directed at undergraduates in their third or fourth year.

The integrated course includes elements of polymer chemistry and physics which today are classed as materials science. The level of treatment assumes a preliminary standard in this subject which would be reached after an introductory course based on, for example, Van Vlack's *Elements of Materials Science and Engineering*. Other prerequisite courses include introductory elasticity, strength of materials, thermodynamics, and fluid mechanics.

A materials science framework is essential because of the need to master a new vocabulary and to acquire a conceptual underpinning for the later chapters in polymer processing and design. This is the modern educational route in engineering metallurgy, in which physical metallurgy is taught in parallel with the macroscopic, phenomenological theories of metal plasticity and fracture. This is a far more attractive route than that in favour until forty years ago, in which the teaching was entirely practical and was based on tables of properties and selection rules for materials. In our view, undergraduates, having been taught engineering metallurgy by the modern route, will anticipate and respond to the same method in the teaching of polymer engineering.

The eight main chapters present a logical development from materials science to polymer technology. The purpose of this treatment, in a book of 400 pages, is not to train specialists but to present a short, integrated course which will stimulate and be attractive to all mechanical and production engineers, including, of course, the small number who will later go on to specialize as polymer engineers and who will require specialist courses such as will be found in the books by Middleman or by Tadmor and Gogos. Our overriding purpose is to implant an understanding of the scope and promise of the polymer revolution by describing polymer science and technology as it exists today together with insights into what is to come.

In writing the book we have laid emphasis on describing phenomena in depth, or not at all. For example, linear viscoelasticity is described in some depth because

- it brings into one framework the fundamental polymer phenomena of creep, stress relaxation, and mechanical damping and shows clearly the relationship between them;
- it is rapidly accepted by engineers, being intimately related to electrical network theory; and
- it yields valuable insights into non-linear viscoelasticity.

The theory of rubber elasticity is described in depth because

- the nature of the force between cross-links in rubber is of fundamental interest and gives to rubbers and polymer melts their unique and extraordinary properties;
- it leads to a 3-D closed-form relationship between principal stresses and strains; and
- it is a condensed-phase analogue of the kinetic theory of gases, which is normally part of introductory engineering courses in thermodynamics.

In this way we hope the book will be entirely satisfying in that the treatment is everywhere complete up to an appropriate level. This has meant that some topics—for example heat transfer in polymer forming—are not discussed at all. The problem is one of selection and we hope that our choice, particularly in the balance between polymer science and polymer technology, will be found to be correct.

A note system is used to bring the real commercial world into the book in an incisive way which does not break up the flow of each chapter. For example, when the polymerization of polypropylene, and its molecular structure, are described in Chapter 1 the relevant note—placed at the end of the chapter—details the strengths and deficiencies of the plastic, its relative position in the hierarchy, and examples of its application. The note system is occasionally used to insert theory or an extended footnote which—for one reason or another—would break up the flow of the chapter. The text can be read without using the notes, but preliminary trials have shown that undergraduates will find it a most useful system, particularly at a second reading. We have made use of worked examples and each chapter has over twenty study problems.

1986   N.G.M.
                                                                 C.P.B.
                                                                 C.B.B.

Acknowledgements

We are grateful to the publishers and authors for permission to publish the following figures: Fig. 0.1, *Chemicals Information Handbook 1992*, Shell International Chemical Company Ltd.; Fig. 0.3, *Plastics & Rubber Weekly*, 7 Aug. 1993; Fig. 0.4, *Plastics & Rubber Weekly*, 30 Nov. 1991; Fig. 0.5, *Plastics & Rubber Weekly*, 21 Nov. 1992; Fig. 0.6, Weber, A., *Plastics, Rubber & Composites Processing & Applications*, 1991, **16**, 143–6; Fig. 0.8, *Modern Plastics International*, August 1993, 32; Fig. 1.9, Society of Plastics Engineers, 14 Fairfield Drive, Brookfield Center, USA; Figs 1.12, 7.19, 7.20, 7.29, 7.36, 7.38, 7.39, 7.41, and 7.43, Modern Plastics Encyclopaedia, McGraw-Hill, Inc.; Fig. 2.10, McGraw-Hill Inc; Figs 4.9, 4.28, and Fig. 4.29, Elsevier Science Publishers; Figs 5.5 and 5.17, John Wiley and Sons, Inc.; Fig. 5.7, American Chemical Society; Fig. 5.16, Butterworth Scientific Ltd.; Figs 7.1, 7.5, 7.6, 7.13, 7.14, 7.15, 7.24, 7.26, 7.27 and 7.28, Longman Group, UK Ltd; Figs 7.3, 7.30, 7.31, 8.13, 8.14, 8.19, and 8.20, ICI plc, Petrochemicals and Plastics Division; Fig. 7.16, A. D. Peters and Co., Ltd.; Figs 7.22, 7.23, 8.7, and 8.21, General Electric Plastics; Figs 7.33, 7.34, 7.35, Barber-Colman Company; Fig. 8.1, Henry Crossley (Packings) Ltd.; Fig. 8.4, Choride Lorival Ltd. and Landrover UK Ltd.; Fig. 8.15, E.I. du Pont de Nemours; Figs 8.25 and 8.26, Bayer AG; Fig. 8.27, Westland Helicopters Ltd.

The frontispiece was kindly supplied by Maunsell Structural Plastics; a full description is given on p. ii.

Contents

0 *Introduction*

Despite the central role that plastics play in life today, there remains a trace of the old view that plastic products are cheap and nasty. This is reflected in a poll held recently in Italy. Of those interviewed, 15% were neutral, neither for nor against plastics, 25% had no view, 35% thought they were essential and approved, but a surprising 25% of those surveyed were quite opposed to them.

Polymers, in the form of plastics, rubbers and fibres, have for many years played essential but varied roles in everyday life: as electrical insulation, as tyres, and as packaging for food, to mention but three. There is no other class of material that could substitute for them. It might be thought that the public's view of plastics in food packaging would be favourable: after all, plastics packaging in the developed world leads to low wastage (less than 2%) whereas in the undeveloped world about 50% of the food produced becomes rotten. Plastic packaging brings with it also a great improvement in hygiene. Yet the public image of plastics as food packaging is poor, much lower than that of traditional materials such as glass, paper and tinplate. How is it, then, that the word 'plastic' is frequently used as a term of abuse in the sense of plastic bread or a plastic smile?

The root of this apparent contradiction is psychological. First and foremost is the feeling that plastics, having been conceived as substitutes, are inferior to the real thing: imitation marble laminates for the bathroom? polypropylene grass? mock onyx table lamps? Historically, there is no doubt that plastics were developed by entrepreneurs as imitation materials and that this form of replacement was intended. But what of the essential replacements that plastics also permit? False teeth are inferior to the real thing but are desirable if you have no other choice. And what of the artificial hip joint? Most people today have a close friend or relative whose life has been improved immeasurably by the polyethylene hip prosthesis. There are a vast number of other replacements, not quite so essential as these, which are highly advantageous. For instance, in automobile engineering, great improvements in safety, noise reduction, comfort and fuel economy are being derived from the increasing replacement of metal alloys with plastics.

The basic argument in favour of plastics is that they provide a choice, functionally, aesthetically or monetarily. If you need an artificial hip joint or false teeth, the choice is straightforward: to be specific, there isn't one. If you prefer leather suitcases to vinyl, or silk stockings to nylon, you are at liberty to buy them and preserve the quality of your life.

The second major cause of public antagonism towards plastics stems from a perceived threat to the environment: hedgerows decorated with carrier bags, pavements littered with burger boxes and grass verges strewn with plastic bottles have cast the plastics industry in the role of environmental villain. In the past, in the days of glass and paper, the environment had to look after itself. Today, pressure groups react to environmental abuse, and rightly so. Later in this chapter, and elsewhere in more detail, we examine the environmental issues. But we note at once that the part the plastics industry plays in reducing the world's reserves of oil is very low, Figure 0.1. The favoured environmental route for many in the plastics industry is to recycle plastics back to energy through incineration with energy recovery. This makes a lot of sense: the energy of the oil is recovered after the oil-derived plastic has been used as a bottle or a car

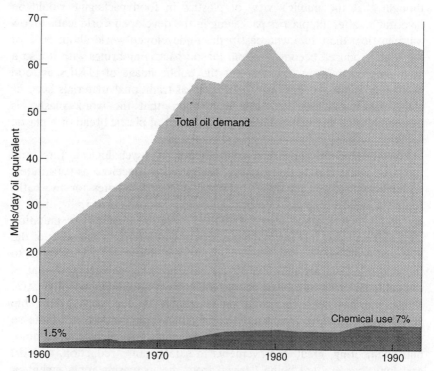

0.1 The chemical share of the total world oil demand. Somewhat more than half of the chemical share is turned into plastics. It will be seen therefore that the greater part of the world's recovered oil is burnt to produce electrical power or for automotive propulsion etc.

bumper. There are other possibilities such as materials recycling or chemical recycling which we describe later.

The third major cause of public antagonism to plastics is the undoubted incidence of abysmal design. This failure was, and to some extent still is, caused by ignorance: a lack of awareness on the part of designers as to which plastic to use for a specific application and how to use it. For instance, polystyrene, a low-cost but brittle plastic, should not be used in its untoughened state in stressed applications. Plastics differ significantly one from another. A design failure brings criticism, not of the designer, but of plastics as a group. This irrational treatment of plastics does not extend to metals. The failure of a metal part through gross design negligence (for instance the use of mild steel in a corrosive environment which requires stainless steel) is seen by all to be a failure in design and not in the use of metal alloys. At root, there is a lack of awareness by engineers and the public that the name 'plastics' is used to cover a large group of similarly based but significantly different materials.

It is interesting to note that the image of being inferior substitutes is not associated with rubbers, simply because natural rubber when it was introduced commercially to the industrial world in the mid-nineteenth century was the very first material of its kind. Its first uses were in waterproof clothing, balls, erasers, and belting for machinery. It then helped to transform the bicycle from being a dangerous and uncomfortable contraption into the first personal mechanical vehicle. Finally, the application of pneumatic tyres to the motor car by Michelin towards the end of the nineteenth century made possible the phenomenal growth of motor transport which has dominated the use of rubber every since.

The contrast between the practical uniqueness and utility of rubber and the tarnished image of plastics in their early years effectively disguised from the ordinary person the fact that **plastics and rubbers are sub-groups of the same class of materials—organic high polymers—differing only in detail in their molecular structures**.

0.1 The past

A century ago, the polymer industry was in its infancy. The first man-made plastic, a form of cellulose nitrate, was exhibited at the Great International Exhibition in London in 1862. The exhibits were arranged into 36 classes. Amongst the 14 000 exhibits of Section C of Class 4 (animal and vegetable substances used in manufacture) was a set of small mouldings made by Alexander Parkes from a material he named Parkesine. It was described in the exhibition leaflet as a replacement for natural materials such as ivory and tortoiseshell which were becoming rare and expensive.

Parkes was born in 1813 and was apprenticed as a brass-founder. He became active in the youthful rubber industry and in 1843 patented a waterproof fabric, a patent he sold later to Charles Macintosh. The waterproofing of fabrics was naturally a major early use of natural rubber, which was obtained at the time from trees in the South American jungle. The unpleasant tackiness of the early products was eliminated with the discovery by Goodyear and Hancock in the 1840s of the vulcanizing effect of sulfur. At first the sulfur was merely dusted on the surface, but this was soon to be followed by mechanical mixing of rubber and sulfur. The interest of Parkes in the infant rubber and plastics industry is symbolic of the scientific and industrial relationship which has always existed between them.

However, Parkesine was not commercially viable, because of its high cost. Parkes employed a large quantity of expensive solvent which was not recovered. The first truly commercial process for the production of plastic material from cellulose nitrate (Celluloid) was due to John Hyatt of Albany, New York State. In 1863, Hyatt, who was then a twenty-six-year-old printer, sought to win a $10 000 prize offered by the Phelan and Collander Company for a new synthetic material to make billiard balls. He studied the literature on cellulose nitrate and knew of the discovery of Parkes that the combination of pyroxylin and camphor produced a plastic resembling ivory. Hyatt conceived the idea of using a small quantity of solvent and supplementing it with heat and pressure. After early experiments conducted in the kitchen of his boarding house, he was expelled to a shed. Celluloid is one of the lower nitrates of cellulose—one of the higher ones is gun cotton! Hyatt was completely successful with his process, and the patent was issued in 1870. The importance of Celluloid is not only that it was the first plastic but that it was for forty years, until the development of Bakelite, the only one. Celluloid is still used today to make billiard balls and table tennis balls.

Parkes, Hancock, Goodyear and Hyatt were typical of the gifted and energetic practical men at whose hands the plastic and rubber industries developed in the middle of the nineteenth century. Their method of working was quite different from the methods which became widespread after 1920. Before that time there was no understanding of the molecular structure of polymers. For natural rubber, Faraday in 1826 had deduced the empirical formula C_5H_8. But there was no appreciation of the fact that natural rubber comprises enormously long molecules. The prevailing confusion was finally dispelled by Staudinger in 1920. His revolutionary idea met heated resistance but by the 1930s it was commonly accepted that all plastics and rubbers are polymers, or macromolecules as Staudinger termed them. In the case of rubber, for instance, identical C_5H_8 units are strung one after another in an immensely long chain. Everything started to fall

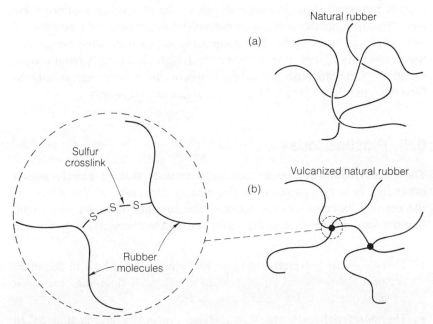

0.2 (a) Natural rubber is composed of immensely long threadlike molecules. In a raw gum from the rubber tree the molecules are separate, linear entities. (b) The effect of vulcanization is to cross-link the molecules one to another. If each molecule is attached to at least one other molecule somewhere along its length, then the whole specimen becomes one single molecular network.

into place: for instance, the effect of vulcanization is merely the attachment of one long molecule to another by bridges of sulfur atoms, $-S-S-S-$ (see Figure 0.2). The Staudinger hypothesis (for which he was awarded the Nobel Prize in 1953) had an enormous effect: it permitted the rational interpretation of experiment and so gave to industrial chemists a light to guide their labours.

The two great developments of the 1930s, nylon and polyethylene, arose from fundamental research programmes initiated by industrial chemical companies. In 1928, Du Pont de Nemours & Co. invited Wallace Carothers, a young organic chemist from Harvard University, to lead a research team whose object was to prepare polymeric materials of defined structures and to investigate how the properties of these structures depend on constitution. This programme was immensely successful and led to the commercialization of nylon in 1939. The first application of nylon was in ladies' hosiery, but by 1941 it was used in self-lubricating bearings, wire insulation, surgical sutures and catheters. In England, ICI initiated in 1932 a research programme on the effect of very high pressure on chemical reactions, including polymerization. Here also the work was crowned with success and resulted in a commercial process for the synthesis of polyethylene. The first development uses for polyethylene were as insulation in submarine and radar cables and comparable applications where negligible water absorption, low power factor, and high dielectric strength were essential. Simultaneously in other parts of the world, particularly in Germany, the development of other polymers moved rapidly.

0.2 Plastics today

The total volume of plastics produced exceeds that of metals and is expanding at a rate faster than the rate of expansion of the economy. Plastics may be split into two groups. The first and largest group is the thermoplastic polymers, and the second group the thermoset polymers.

- **Thermoplastic polymers** are those that when heated flow in the manner of a highly viscous liquid, and do so reversibly time and time again on subsequently being heated and cooled.

- **Thermoset polymers** are those whose precursors are heated to an appropriate temperature for a short time, so that they will flow as a viscous liquid; a chemical cross-linking reaction then causes the liquid to solidify to form an infusible mass. The precursor materials may be of low molecular weight; some precursors, after mixing, will flow and cross-link at room temperature.

It will be helpful to remark that the principal difference between these two groups is in the manner of forming. A thermoplastic is heated and then constrained by a mould or die (see Chapter 7) so that it flows into the required shape. It is then cooled and removed from the mould. The forming of thermosets is somewhat more complicated. The precursors are normally injected into a mould which is then heated so that the resulting polymeric mass, when set, has the required shape. It is then cooled and removed from the mould.

Of the 31 million tonnes of plastics sold in the USA in 1993, about 90% was thermoplastic. The worldwide consumption by market of plastics is shown in Figure 0.3. Packaging is the largest consumer of plastics materials worldwide, taking 31% of material, followed by building at 17% and transport at 7%; electrical applications also consume 7% and after this the other applications are very small. The technical significance is not readily measured by the volume of plastic consumed: for example, the volume of the epoxy family of polymers, which are thermosets, is small but their technical significance in electronic circuit boards and epoxy–carbon fibre

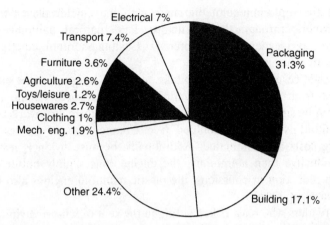

0.3 The worldwide consumption of plastics by market in 1992. Packaging is the dominant market, taking approximately one-third of all plastic produced. The hi-tech areas penetrated by the engineering plastics are mainly in mechanical engineering, transport (aerospace, automotive) and electrical engineering.

composites is enormous. Amongst the thermoplastics, the volume of poly-tetrafluoroethylene (Teflon) is very low but its use in low-friction and electrical applications makes it technically highly significant. In order to give the reader an insight into the application of plastics in engineering we will describe trends in one of the smaller and highly significant markets for plastics, the automotive, in some detail.

0.3 Automotive applications of plastics

In automobile construction, it is the lightness and simplicity of processing of plastics that attract the designer. A lighter vehicle implies less fuel and lower emissions. The two major constraints which dominate car design today stem from safety legislation—originating in the USA in the 1960s—and from the oil crisis. Both factors have led to a decrease in the amount of metals used and a corresponding increase in the average mass of plastics in cars, which began in the mid-1960s. Originally plastics were used only in the interior of cars, where they are now firmly established. Here they contribute to passive safety by covering edges, corners and surfaces in the passenger compartment with shock-absorbing materials. They also contribute to active safety, for instance in the improvement of handling characteristics due to aerodynamic fittings such as front air dams and rear spoilers, and the advances made in the use of plastic reinforcing fibres in tyre construction. The environmental compatibility of a car can also be enhanced by the application of plastics, noise emission being the outstanding example, for instance with plastic camshaft drive chain guides or the replacement of the chain with the quieter plastic toothed belt. Polymers, through their structure, have natural damping—noise abatement—properties (see Chapter 4). Customers often equate quiet engines with efficient engines!

The plastic (usually nylon) air intake manifold weighs half as much as one made from pressure-cast aluminium, and improves engine performance with its smoother inner surfaces, which minimize airflow resistance. The manifold is injection-moulded (see Chapter 7) and so secondary machining costs are eliminated. Additionally, because nylon is less thermally conductive than aluminium, the engine runs slightly hotter, which leads to a reduction in emissions: the plastic manifold engines also have a shorter warm-up time.

Those readers who have risked sitting in the seat of a moving vintage car will know that the modern automobile seat is a far more sophisticated product. It forms an integral part of the suspension system, being designed to support the driver/passenger during transit whilst cushioning the continuous motion generated by an irregular road surface. Without advances in polyurethane plastic foam technology this would not be possible. Moulded shapes for cushions provide driver stability, a real contribution to safety. This would not have been possible in the days of wood, leather, cotton flock and horsehair!

The increase in the mass of plastics in cars has occurred in the expensive as well as in the cheaper models. In BMW cars, for instance, the proportion of plastics has been above that of the average for German cars. The plastics used in the BMW 5 series are shown in Figure 0.4: they constitute 10% of the vehicle's mass: 149 kg comprising 13 major types of plastics, in order of application:

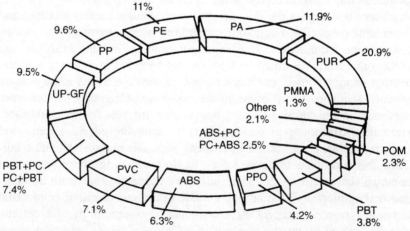

0.4 How the cookie crumbles: plastics use in the BMW 5 series. For the key to the symbols (PP = polypropylene, etc.) see 0.N.1.

- 58% in the interior
- 27% on the exterior
- 8% in the chassis
- 7% in the engine.

These plastics are mostly thermoplastic. They fall into three groups (for a key to the abbreviations see Note 0.N.1 at the end of this chapter):

- the 'commodity plastics' (PE, PP and PVC), which are manufactured in vast quantities, are essentially cheap and find wide application, for example in packaging as well as automotive engineering;
- the 'engineering thermoplastics' (often referred to as ETPs): these are high-quality, expensive polymers, for example PA, POM, ABS, PC, PBT, PPO and alloys of ABS–PC and PBT–PC;
- speciality plastics for foamed seating, PUR and optical plastics, PMMA for example (PC also has excellent optical properties).

From a technical point of view the engineering thermoplastics are particularly interesting. There are the plastics (there are some eight or nine of them) which combine lightness and corrosion resistance with a good balance of stiffness and toughness maintained over a wide temperature range. They are easily, precisely, and rapidly fabricated at modest temperatures (usually in the range 200 to 300°C). These are the plastics that are penetrating the traditional markets of metal alloys used in load-bearing situations. Of these, the most dominant, in terms of volume, Figure 0.5, are (in order) ABS, PA, PC, POM, PPO and PET, PBT. The properties of the smallest volume group (1% in 1985, growing to 4% in 1995) are of great technical significance. These niche polymers have formidable properties and are described later in the book.

Three specific mechanical examples of the use of plastics in BMW cars are: (1) the universal joint which is used to operate the water valve in the heating system; (2) the petrol tank; and (3) centrifugal fans for the heating system. The last, the centrifugal fan, is a good example of how the competition is not now so much between plastics and metal alloys, but between different types of plastic. The cheapest and most expensive candidate plastics considered by the BMW design engineers for the centrifugal fan were polypropylene (PP) and polyoxymethylene (POM). The price per kg of PP is one-third that of POM; PP is also less dense, its density being two-thirds that of POM. A superficial decision from these facts alone would favour PP, the cheaper and lighter. However, the elastic modulus of POM is three times that of PP, so that the use of POM permits a reduced wall thickness. This results in a lighter item and a moulding

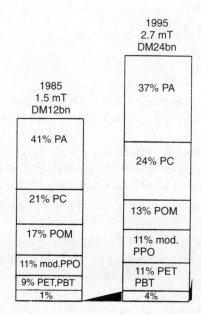

0.5 The worldwide consumption of engineering thermoplastics by polymer for 1985 and a prediction for 1995 (computed in 1992). For the key to the symbols (PA = polyamide, etc.) see 0.N.1.

cycle which is shorter (because the cooling rate is quicker: see Chapter 7). These advantages alone are not sufficient to favour POM. It is only when secondary factors are considered, in particular the simple pressing of the POM part on to a metal shaft, that the use of POM is justified. This case is quoted to show clearly that a comprehensive analysis of all factors is required to find the correct material, a point of design we take up again in Chapter 8.

Although the advance of plastics in automobile applications has been led by the engineering thermoplastics, the commodity plastics are making inroads into applications normally thought of as restricted to engineering thermoplastics. Pressure on costs and greater precision in defining end-use performance are major factors here. Thus as the high-performance end of the engineering thermoplastics market erodes the market for metal, the low-performance end of the engineering thermoplastic market is eroded by the commodity plastics: of these, PP is the undoubted star.

Another factor, which is of increasing dominance in material selection, is the environmental: specifically in what way, and at what environmental cost, can the plastic part be disposed of when the car is scrapped? This is materials selection in a Green Age, a revolutionary concept undreamt of twenty years ago!

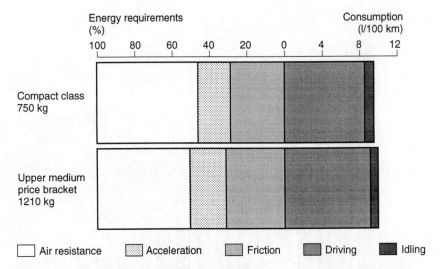

0.6 Percentage energy requirements and rate of gasoline consumption for cars in the compact class (0.75 tonne) and in the upper medium price bracket (1.21 tonne) for the standard '$\frac{1}{3}$ mix': that is, $\frac{1}{3}$ town driving, $\frac{1}{3}$ driving at 90 km h^{-1} on a main road and $\frac{1}{3}$ driving at 120 km h^{-1} on a motorway.

0.4 Environmental considerations

The thrust towards replacing heavier metal parts with lighter plastic parts is generated by an effort to lower emissions and to save fuel. Figure 0.6 shows the energy requirements for cars in the compact class and in the upper medium price bracket for the '$\frac{1}{3}$ mix': $\frac{1}{3}$ town driving, $\frac{1}{3}$ driving at 90 km h^{-1} on main roads and $\frac{1}{3}$ driving at 120 km h^{-1} on motorways.

It will be seen that the factors that depend on mass are dominant—acceleration and friction—and it is this fact which powers the drive towards the use of plastics to replace metal parts and hence lower mass. In the case of rotating parts, a rotational inertia has to be overcome in addition to the translational inertia.

Now the energy consumed in propelling a car, in an average life of 160 000 km, is 10 to 15 times the energy used to manufacture it. It is clear, therefore, that the achievement of a light vehicle is of paramount importance, even if the energy input for the production of a lighter plastic exceeds that of the heavier metal part it replaces.

Before examining the facts, it might be thought that since plastics are synthesized from hydrocarbons their energy content would be high. It turns out, however, that the energy content of plastics is less than that of competitive metals, when compared on a volume basis (see Figure 0.7). Frequently, engineering plastics permit part designs of about equal volume to the metal parts: in this case Figure 0.7 is the relevant guide. On a mass

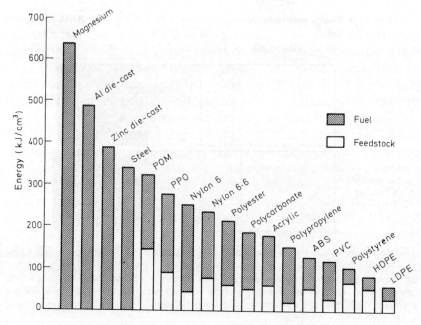

0.7 Energy requirements to produce metal alloys and plastics. For plastics the energy required to manufacture the plastic is shown separately from the fuel equivalent of the raw material.

basis they are also attractive when compared with competitive alloys (aluminium alloys in particular).

For example, a typical steel car bonnet requires as much as 27 kg of crude oil equivalent, taking into account the significant stamping waste that has to be recycled. The same bonnet in polymer composite requires 16 kg of crude oil equivalent because of its lower mass, negligible trimming waste, energy-conserving forming process and large proportion of mineral fillers and glass fibre reinforcement, which have a low crude oil equivalent. Thus the switch from steel to polymer leads to energy saving at two stages:

- at the end of the assembly line, before the engine is switched on;

- throughout its working life, owing to lower fuel consumption, until for the last time the engine is switched off.

For every 1 million vehicles the total savings achievable by replacing steel bonnets with plastic amounts to 23 000 tonnes of crude oil, considerably more than the energy content of the plastic bonnets. In Western Europe in the mid-1990s there are 130 million passenger cars with an average life cycle of 160 000 km.

There is a major political drive to reuse the plastic scrapped from old cars. The recycling of metal scrap from old cars is an enormous and well-established industry with a global turnover of about $100 billion. There are approximately 600 shredders operating in the world today, most of them capable of taking a whole car and granulating it into thousands of small pieces in seconds. The shredding industry plays a major economic and environmental role. In a motor car about 70% is iron and steel, 5% non-ferrous metals and 25% plastics including tyres, upholstery and carpets: the shredding industry is gravely concerned about this 25% since it is composed of about four rubbers and upwards of twenty mixed plastics for which there is no existing demand. The disposal of this waste is costly and has to be paid for from the profit on metal scrap. At present the greater part of it is disposed in landfill. The fraction of mixed plastics in the waste increases year by year, to the alarm of the shredding industry.

There are four ways of tackling the problems of plastic waste. In ascending order of perceived environmental benefit these are:

- bury: landfill
- recover: burn the plastic and recover the energy (energy or thermal recycling)
- recycle: transform the plastic back to the feedstock (chemical or feedstock recycling)
- reuse: grind up the plastic waste and then reform it into another plastic product (mechanical or material recycling).

The reason why the environmental benefit is in this order is apparent from Figure 0.7. For example, approximately 80% of the energy required to produce 1 cm^3 of polypropylene is expended on the polymerization and associated processes. So if a polypropylene bumper from a motor car can be recycled into another bumper (without the expenditure of a lot of energy), this is environmentally better than chemical recycling or energy recycling.

The problems with the mechanical recycling of plastic from used cars include:

- the plastics have to be stripped by hand from the motorcar, prior to shredding the metal;
- the twenty or so different plastics have then to be identified and separated;
- each plastic has then to be transported to the particular location where it is to be recycled.

This mechanical recycling procedure is hopelessly uneconomic compared with the recycling of steel, which, after shredding, is separated magnetically, washed or air-cleaned, and transported for conversion into a new product. There is no way in which this new product is inferior by virtue of its manufacture from recycled steel as opposed to virgin steel, whereas, for example, a recycled polypropylene bumper will be inferior to a bumper made from virgin polypropylene, because the injection moulding process causes a drop in relative molecular mass, and the recycled polypropylene is two, three or four times more expensive than the virgin polypropylene.

There is the possibility that if clear single polymer fractions can be economically and systematically collected and separated, they may be mechanically recycled into products with a less demanding role such as animal flooring, fencing, or pallets. So ideally, the same polymer might be used time and time again: this is the cascade model. The major problem is identification, separation, and cleaning. For example, scrap plastic fuel tanks are saturated with petrol: up to 100 g per tank. Another example: the two-component polyurethane paint system used on bumpers has a detrimental effect on the impact resistance of recycled bumpers, and must be removed before mechanical recycling. The Society of Plastics Engineers has proposed a labelling scheme to expedite identification and separation. Nevertheless, much plastic waste will either be difficult or impossible to identify and separate, or difficult to clean. There is also a severe logistical problem, and problems concerning purity and consistency of supply.

The logistical problems can be assessed from Figure 0.4, which shows the diversity of polymers used today in the BMW 5 Series. This is a typical example. It is not just that there are sixteen polymers specified but that within each chemical species—for different automotive components—there will be plastics of different relative molecular mass specified for optimum performance. Note also that at 7.4% we have mixed polymer alloys: PBT with added PC, and PC with added PBT. How is the operator stripping the plastic from a scrap BMW 5 to know which is which? One proposal to assist the stripping is to standardize polymeric types for automobile parts. This will assist the dismantling, but from a functional point of view it is a weak proposal, since it will retard initiative in design. And in any event, the mechanically recycled scrap is not only inferior in properties to virgin material but also more expensive.

Recycling procedures must be economic and show a net environmental gain. Recycling cannot be an end in itself: there is no point in expending a lot of petrol driving a few bottles to the bottle bank! It is irrational to argue that because mechanical recycling works for steel it must be made to work for plastics: they are very different materials. It is logical therefore to investigate the options to mechanical recycling (excluding the option of agreed last resort, which is landfill).

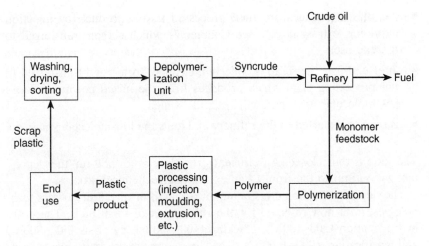

0.8 Schematic illustration of the chemical recycling routes for waste plastic. The depolymerization unit uses fossil-fuel-type technology (hydrogenation, cracking, pyrolysis, gasification) to produce syncrude (a mixture of hydrocarbons of low molecular mass) which enters the refinery, where it is recycled with incoming crude oil into fuel or monomer feedstock.

Chemical recycling

For polymers produced by condensation reactions (see Section 2.4), for example polyamides (PA) or poly(ethylene terephthalate) (PET), the mixed plastic waste may be converted back directly to the monomer by hydrolysis. There is much developmental and pilot plant work in this area. The monomers can then be separated and introduced into the virgin feedstock for polymerization in the normal way. There is no way in which, for example, a PA polymerized from chemically recycled monomer will be inferior to virgin material. This is the ideal procedure, but it is applicable in only a few cases to the remainder of the polymer kingdom, the addition polymers (see Section 2.4).

Generalized recycling routes applicable to all thermoplastic polymers (both addition and condensation) are illustrated schematically in Figure 0.8. This may be described as a crude, rough depolymerization as opposed to the precise surgical depolymerization achieved by the hydrolysis of a condensation polymer. Taking the procedure from the start:

- crude oil (a mixture of hydrocarbons) from the oil well enters the refinery and is mixed with syncrude (another mixture of hydrocarbons) from the chemically recycled plastic;

- the combined flow is then refined, one refined component becoming fuel and the other the feedstock for polymerization to form plastic granules;

- the plastic granules are then processed into a product (by injection moulding, extrusion, etc.; see Chapter 7), which is then transferred to the end-user;

- after end-use, the scrap is collected, cleaned, sorted and transferred to the processing plant which produces from the mixed polymer scrap a low molecular mass product, syncrude, which:

- is then transported to the refinery and joins the low-molecular-mass oil.

The loop is thus closed. An efficient procedure is to 'bolt-on' the depolymerization unit to the front of an existing petrochemical process.

The depolymerization units being investigated involve four process technologies, until now reserved for the refinement of fossil fuels. These are hydrogenation (300–500°C), cracking (400–600°C), pyrolysis (500–900°C) and gasification (900–1400°C). The first commercial-scale plant using hydrogenation opened in Germany in 1992. There is much pilot plant activity in this area elsewhere in Western Europe, America and Japan. This procedure is not attractive economically but highly attractive politically—in the current cultural atmosphere—compared with incineration with energy recycling, which is described next.

Energy recycling

The simplest and most obvious way to recycle the energy of waste plastic is to burn it and so generate steam and hence electricity or—with greater efficiency—use the recovered energy directly, in a district heating system for example. The calorific value of many plastics is comparable with that of heating oil and in excess of that of coal, Table 0.1 It might be thought that the plastic components of an automobile, after a long and virtuous life of 160 000 km, having played their vital part in mass reduction and so saved a considerable quantity of fuel, might be pensioned off and burnt to generate electricity. In *total life cycle* analysis, polymers win over steel, requiring about three times less energy than steel: this increases to five times if the energy content of the polymer is recycled.

Table 0.1   Calorific values of some polymers and other materials

Product	Calorific value $(MJ\ kg^{-1})$
PE, PP	46
Heating oil	44
PS	41
'Mixed plastics'	37
Coal	30
Wood	16
'Domestic rubbish'	11

In countries such as Japan, Denmark, and Switzerland, where over 70% of municipal solid waste (MSW) is burnt, the incineration solution appears less distasteful than in the rest of the world, where MSW goes mostly in landfill. In countries at present reliant on landfill, there is strong political objection to incineration. This is not to say that thermal recycling is not practised in these countries: in Western Europe 20% of plastic in MSW and 20% of plastic in the category Large Industry and Distribution (mainly packaging) is incinerated with thermal recycling. There is however strong political reaction against the expansion of incineration with thermal re-cycling, and a preference for mechanical and, to a lesser extent, chemical recycling.

The opponents of thermal recycling are concerned with the environ-ment: the presence of too high a proportion of poly(vinyl chloride) in the plastic waste, if the incineration is not carried out correctly, can lead to the presence of dioxins in the flue gases; there is also concern that the energy used in manufacturing the polymer is not recovered.

The supporters of thermal recovery argue that it has been proved that if

- combustion is well engineered and operated, and
- the latest emission control technology is used,

then modern incineration is safe. This is the view of Japan's Ministry of International Trade and Industry (MITI), whose long-term vision for plastics waste disposal places the main emphasis on thermal, rather than mechanical recycling (0.N.2). To this end, in Japan over 2000 incinerators have been built in the last 15 years, often in the heart of urban centres.

In conclusion, concerning recycling it must be said that we are at the crossroads. There is great unease about the squandering of the world's oil reserves. The different solutions lie in opposing directions. The most obvious solution is easily reached after glancing at Figure 0.1: because 90% of the world's supply of oil is burnt for power, meaningful economies are not to be obtained without major advances in fuel economy. Can the heating of homes be made more efficient? What of the use of petrol in private jet aircraft and pleasure craft? Could motorway maximum speeds be reduced and also enforced rigorously? These and many other possibili-ties bring in political considerations. The success or failure of the Japanese solution (incineration with energy recycling, and, in a minor role, chemical recycling) will be followed with great interest in the European Union. The success or failure of the European Union solution (mechanical recycling and, in a minor role, chemical recycling) will be followed with great interest in Japan. That the world has found its way to this trans-global experiment is most fortunate.

These selected comments on one area, transportation, illustrate a general trend in engineering, and one which will continue. The opportunity for innovative design with a new material is more easily seized than with an older one. In the years ahead, we can certainly look forward to great developments in the plastics and fibres available, which will present even greater opportunities to the enterprising and ambitious designer. For this reason we trust that the student engineer will find this book both profitable and, even more important, pleasurable.

Notes for Chapter 0

0.N.1

The following common abbreviations are used to indicate plastics:

PE	polyethylene
PP	polypropylene
PVC	poly(vinyl chloride)
PA	polyamide
POM	polyoxymethylene
ABS	acrylonitrile–butadiene–styrene
PC	polycarbonate
PET	poly(ethylene terephthalate)
PBT	poly(butylene terephthalate)
PPO	poly(phenylene oxide)
PUR	polyurethane
PMMA	poly(methylmethacrylate)

Details of the chemistry and statement of areas of application can be obtained by consulting the Index.

0.N.2

A most striking rejection of the Japanese view comes from the German Environment Ministry (report in *Modern Plastics International*, January 1994, 12) who reaffirm their commitment to mechanical and chemical recycling. A source at the Ministry stated that if thermal recycling was added to the ordinance 'then everybody will take the cheapest way, which means incineration with energy recovery'. The ordinance mentioned states the target to be 60% of all consumer plastics waste to be mechanically or chemically recycled by 1998: these are the environmental routes favoured in Western Europe.

1 *Structure of the molecule*

1.1 Introduction

A **polymer** (from the Greek **poly**, meaning many, and **meros**, meaning part) is a long molecule consisting of many small units (**monomers**) joined end to end. Polyethylene,

$$\phi_1 \!+\! \text{CH}_2 \!-\! \text{CH}_2 \!+\!_n \phi_2, \qquad (\text{I})$$

the simplest hydrocarbon polymer, will serve in this introduction as an example of a typical synthetic polymer. The number of ethylene monomers, n in (I), which join together to form the molecule is usually of order 10^4 but may be as high as 10^6 or as low as 10^3. The small end-groups ϕ_1 and ϕ_2 occur in very small concentration. They have no effect on the mechanical properties of the polymer except in so far as they influence chemical stability: for some polymers an unstable end-group when heated or irradiated with light can initiate the degradation of the molecule (Note 1.N.1). Apart from this factor, the end-groups are of no consequence and we shall confine attention in this chapter to the mechanically significant characteristics of the molecule.

There are two molecular factors that govern the mechanical properties of a polymer. The first is the length of the molecule, which is proportional to n and which is therefore proportional to the molecular size, or **relative molecular mass**. The second is the shape of the molecule. For example, the first polymerization of polyethylene (in 1935) produced a molecule with small side branches containing a few carbon atoms, most often two or four (Figure 1.1). The number of side branches may be varied by changing the polymerization conditions. Even small variations in the number of side branches can be of technological significance, and cause appreciable changes in, for instance, elastic modulus, creep resistance, and toughness. The commercial forms of **side-branched polyethylene** have approximately three side branches per 100 main-chain carbon atoms. These forms are produced by heating ethylene gas at $\sim 200°\text{C}$ at a high pressure of ~ 2000 atmospheres (1.N.2) in the presence of a suitable polymerization initiator.

An almost entirely unbranched form of polyethylene may be synthesized without heating or pressurizing the gas (1.N.3). This simpler process dates

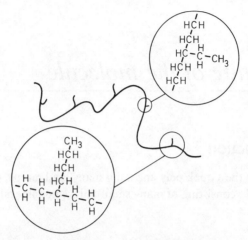

1.1 Side-branched polyethylene.

from the early 1950s. Ethylene gas is merely bubbled through a suitable liquid containing a Ziegler–Natta catalyst. The polymer forms on the surface of the catalyst. This form of polyethylene has less than one side branch per 100 carbon atoms and, for this reason, is frequently known as **linear polyethylene**.

Both linear and side-branched polyethylenes are of technological interest. As the number of side branches per 100 carbon atoms is increased from zero there are pronounced changes in physical properties (see Chapter 4), but the linear and side-branched polymers are identical in one respect, which is that they may be reversibly heated to melt and then cooled to crystallize time and time again. On melting they flow as does a liquid and are thus **thermoplastic** (or thermal-flow) polymers. In this they are distinguished from **cross-linked** polyethylene, which when heated will not flow.

In a cross-linked polymer the chains are joined chemically at tie points. There are many ways of preparing a cross-linked network. For polyethylene the simplest way is to irradiate the molten polymer with ionizing radiation, such as β- or γ-rays. The chains then cross-link so that the entire specimen is one giant molecular network (see Figure 1.2). In the liquid state the cross-links inhibit flow: they cause the polymer to be form-stable and to possess properties typical of rubbers. Once the specimen is cross-linked its natural shape is fixed: it can be deformed under stress when heated and forced into a different shape, which can be 'frozen in' by cooling to a sufficiently low temperature; but if the specimen is reheated it will revert to the shape it had when the cross-links were introduced. That shape can be changed only by degrading the network, for example by burning.

1.2 Illustration of the molecular structure of cross-linked polyethylene in the liquid state. The spaces between the sketched net are filled with other parts of the network.

Many cross-linked networks are produced by chemical reactions triggered by heating. After heating, the network (and consequently the specimen shape) are 'set': this is the origin of the term **thermoset**.

There are a vast number of monomers that may be polymerized. The resulting polymers fall broadly into two categories, **addition polymers** and **condensation polymers**, depending on the mechanism of polymerization. These two methods of polymerization will be illustrated in the following section, using as examples polymers of technological interest. It is important to keep in mind that the synthesized polymer—whether produced by an addition or a condensation reaction—may be linear (or linear but with small side-branches) or, on the other hand, cross-linked.

1.2 Addition polymers

Addition polymers are synthesized by the addition of unsaturated monomers to the growing chain. The synthesis of polyethylene from ethylene is formally represented by

$$
n\begin{bmatrix} H & H \\ | & | \\ C = C \\ | & | \\ H & H \end{bmatrix} \longrightarrow \begin{bmatrix} H & H \\ | & | \\ C - C \\ | & | \\ H & H \end{bmatrix}_n \tag{II}
$$

The number of atoms in the polymer equals the number of atoms in the n monomers from which it is synthesized: no small molecule, such as water, is split off in the reaction. The double bond of the monomer opens. The synthesis of a **vinyl** polymer is represented by

$$n \begin{bmatrix} & H & X \\ & | & | \\ & C = C \\ & | & | \\ & H & H \end{bmatrix} \longrightarrow \begin{bmatrix} & H & X \\ & | & | \\ -& C - C -\\ & | & | \\ & H & H \end{bmatrix}_n \tag{III}$$

in which for polypropylene X is CH_3 (1.N.4), polystyrene X is C_6H_5 (1.N.5), poly(vinyl chloride) X is Cl (1.N.6). Disubstituted polymers such as poly(methyl methacrylate) (1.N.7),

$$\begin{bmatrix} & H & CH_3 \\ & | & | \\ -& C - C -\\ & | & | \\ & H & \\ & & | \\ & & C \\ & O & O \\ & & CH_3 \end{bmatrix}_n \tag{IV}$$

are also synthesized by addition reactions.

The addition of the n monomers to form the polymer occurs in n separate but identical steps. For instance, suppose a polyethylene molecule to have been initiated at end-group ϕ_1, and to have reached the stage in its growth indicated in Figure 1.3. The growing end of the molecule has an unpaired electron (or free-radical structure), represented by a dot; this

1.3 One stage in the synthesis of a polyethylene molecule by a free-radical mechanism.

unpaired electron is unstable. When an ethylene monomer approaches the growing end of the molecule, the monomer is 'captured':

$$R-\underset{\underset{H}{|}}{\overset{\overset{H}{|}}{C}}-\underset{\underset{H}{|}}{\overset{\overset{H}{|}}{\overset{\bullet}{C}} + \underset{\underset{H}{|}}{\overset{\overset{H}{|}}{C}}=\underset{\underset{H}{|}}{\overset{\overset{H}{|}}{C}} \longrightarrow R-\underset{\underset{H}{|}}{\overset{\overset{H}{|}}{C}}-\underset{\underset{H}{|}}{\overset{\overset{H}{|}}{C}}-\underset{\underset{H}{|}}{\overset{\overset{H}{|}}{C}}-\underset{\underset{H}{|}}{\overset{\overset{H}{|}}{\overset{\bullet}{C}}} \qquad \text{(V)}$$

The double bond of the captured ethylene monomer opens, the monomer is added to the molecule, and the unpaired electron is transferred along the chain as shown in (V) and in Figure 1.3. The reaction can occur repeatedly until a chain termination reaction occurs, forming a stable end group ϕ_2:

$$\phi_1 \!\!\left[\begin{array}{c} \underset{\underset{H}{|}}{\overset{\overset{H}{|}}{C}}-\underset{\underset{H}{|}}{\overset{\overset{H}{|}}{C}} \end{array}\right]_n \!\!\phi_2$$

The n steps, sometimes several thousand, can occur in a time less than a second.

The anticipated polymer may be predicted with reasonable precision from the monomer structure, as for example in reaction (V). The rational prediction for the structure of polyethylene is clearly (I), and this was the structure first proposed in 1936 from the free-radical reaction ((V) and Figure 1.3). Five years later it was realized that this was not quite correct, and that the polymer in fact contained a few side branches. The simplest explanation of side branching is that every now and then, as the polymer propagates (V), the chain 'backbites', according to

$$\begin{array}{ccc} R-CH_2 & \overset{\bullet}{C}H_2 & R-\overset{\bullet}{C}H & CH_3 \\ \diagdown & \diagup & \longrightarrow & \diagdown \quad \diagup \\ (CH_2)_x & & (CH_2)_x \end{array} \qquad \text{(VI)}$$

By this 'backbiting' mechanism the free-radical on the end of the growing polymer removes a hydrogen atom from a CH_2 group in the already formed main chain. The polymer then continues to grow from this point in the main chain. This leaves a fully saturated side group of $x + 1$ carbon atoms. The free-radical polymerization is performed at high temperatures ($\sim 200°C$), which enhance the chain 'agility' so as to permit the twisting necessary to achieve 'backbiting' (1.N.8).

In the synthesis of a polymer on a Zeigler–Natta catalyst, the molecule grows in the manner of a hair (Figure 1.4). The growth point is at the catalyst surface. The monomers diffuse through the solvent to the catalyst

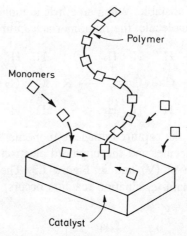

1.4 An illustration of polymerization by Ziegler–Natta catalyst. The monomers diffuse on to the catalyst surface. They then migrate across the surface to the growth point.

surface and thence to the growth point. They there react and the polymer grows out from the catalyst surface. The effect of the catalyst surface in shaping or 'tailoring' the molecule is of considerable importance. The monomer is inserted in a specific and reproducible manner. Since the growth point is firmly anchored to the catalyst surface there is no possibility of backbiting. This explains why polyethylenes synthesized on these catalysts have negligible side-branching.

1.3   Condensation polymers

In each step of the synthesis of a condensation polymer a small molecule, usually water, is split off. For example, the reaction of hexamethylene diamine

$$\begin{array}{ccc} H & & H \\ | & & | \\ N{-}(CH_2)_6{-}N & & \\ | & & | \\ H & & H \end{array} \qquad (VII)$$

and adipic acid

$$\begin{array}{ccc} OH & & OH \\ | & & | \\ C{-}(CH_2)_4{-}C & & \\ \| & & \| \\ O & & O \end{array} \qquad (VIII)$$

yields $(2n-1)H_2O$ plus

$$H{-}\left[\begin{array}{cc} H & H \\ | & | \\ N{-}(CH_2)_6{-}N{-}C{-}(CH_2)_4{-}C \\ & \| \qquad\qquad \| \\ & O \qquad\qquad O \end{array} \right]_n{-}OH \qquad (IX)$$

a polyamide known as nylon 6.6, there being 6 carbon atoms in each residue (1.N.9). The abbreviation PA 6.6 is frequently used. There are numerous related polyamides (nylon 6.10,[1] etc.).

A second example is the reaction of ethylene glycol

$$HO-CH_2-CH_2-OH \qquad\qquad (X)$$

and terephthalic acid

$$HO-\overset{\overset{O}{\parallel}}{C}-\overset{\displaystyle\bigcirc}{}-\overset{\overset{O}{\parallel}}{C}-OH \qquad\qquad (XI)$$

which yields $(2n-1)H_2O$ plus

$$HO\left[CH_2-CH_2-O-\overset{\overset{O}{\parallel}}{C}-\overset{\displaystyle\bigcirc}{}-\overset{\overset{O}{\parallel}}{C}-O\right]_n H \qquad\qquad (XII)$$

the polyester poly(ethylene terephthalate) (1.N.10).

Note that these condensation polymers are of the form

$$-A-B-A-B-A-B-. \qquad\qquad (XIII)$$

This absolute regularity in the order of placement of the monomers enhances the tendency of the molecule to form **crystals** when cooled from the liquid state: this molecular characteristic is consequently of the greatest mechanical significance, as will be shown repeatedly in the following chapters.

1.4 Copolymers

Copolymers are polymers composed of two or more different monomers. For example, if ethylene and propylene are polymerized simultaneously then the polymer will contain both ethylene and propylene units. If the mixture of gaseous monomers is mainly ethylene, then the copolymer will consist of linear ethylene sequences with, here and there, a propylene unit. Conversely, the polymerization of propylene with a small amount of ethylene produces linear sequences of polypropylene separated here and there by an ethylene unit. The reader will anticipate that the first copolymer will exhibit the properties of a modified polyethylene, the second of a modified polypropylene, and this is in fact the observed behaviour.

The combination of monomers to form copolymers can be compared with the mixing of metals to form solid solutions, which is the basis of alloy

---

1 For example, nylon 6.10 is the polyamide made from hexamethylene diamine and sebacic acid, $HOOC(CH_2)_8COOH$.

Random

Block

Graft

1.5 The arrangement of the monomers in a copolymer will be, as a rule, **random**. For special applications **block copolymers** or **graft copolymers** are synthesized.

formation. The chemical engineer by small variations in copolymer composition can synthesize polymers with subtly different properties. The properties which are controlled by changes in copolymer composition include elastic modulus, toughness, melt viscosity, and thermal stability (1.N.11). We return to this subject again in Chapters 4 and 5. Copolymers are also polymerized with **block** or **graft** structures (see Figure 1.5) for specific purposes (1.N.12).

1.5   Cross-linked polymers

The monomers discussed so far are all said to have a functionality of 2. Each monomer will react with two other monomers. The monomers in the condensation reactions have chemically active groups at both ends. In the addition polymers the opening of the double bond in a monomer leads to its attachment to two other monomers. It is obvious that a polymer produced from monomers with functionality equal to 2 will be linear: it is equally obvious that monomers with a functionality exceeding 2 present the opportunity for the polymerization of a cross-linked network (see Figure 1.6).

Suppose a monomer A (functionality 2) is mixed with a small proportion of monomer B (functionality 4); then, assuming the conditions are correct, a cross-linked network will be produced. Four linear chains of A will meet at each B cross-link. For example, suppose styrene (functionality 2)

$$CH_2 = CH$$

A ⟶   (XIV)

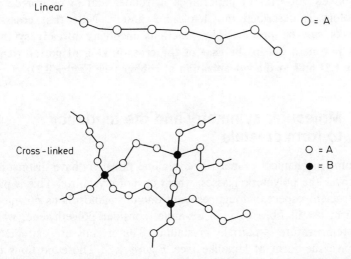

1.6 Monomer A (functionality 2) yields a linear homopolymer when polymerized, but when polymerized with a small amount of monomer B (functionality 4) it yields a cross-linked copolymer.

and divinyl benzene (functionality 4)

$$B \longrightarrow \quad \begin{array}{c} CH_2{=}CH \\ \bigcirc \\ CH{=}CH_2 \end{array} \qquad (XV)$$

are reacted. The resulting polymer will be formed by addition reactions. It will consist of chains of polystyrene meeting at a cross-linking divinyl benzene group

$$\begin{array}{c} H \\ | \\ -A-A-A-CH_2-C-A-A-A- \\ \bigcirc \leftarrow B \\ -A-A-A-C-CH_2-A-A-A- \\ | \\ H \end{array} \qquad (XVI)$$

The entire network will consist of chains of A units meeting at B cross-links as indicated in Figure 1.6. The proportion of B will, of course, determine the tightness of the cross-linked network. The polymerization of other cross-linked networks of importance in rubber and in reinforced plastic technology is described in Chapters 3 and 7. **Note that cross-linked networks can be produced by reactions involving already synthesized linear polymers (as in the case of the cross-linking of polyethylene (see Figure 1.2) and in the vulcanization of rubber (see Figure 0.2)).**

1.6   Molecular symmetry and the tendency to form crystals

For some mechanical properties, crystalline polymers have distinct advantages over the polymeric glasses, which do not crystallize. This is particularly true in respect of toughness and ductility: amorphous polymers will not form useful fibres. As an example, consider polyethylene, which at room temperature is partially crystalline. The crystals are extremely thin and have the form of lamellae (see Figure 1.7). Those portions of the molecule which have not crystallized are trapped between the lamellae. It is this two-phase structure, involving a high-modulus crystal fraction and a low-modulus rubbery amorphous fraction, intimately interconnected in a sandwich structure, that leads to good ductility and toughness.

1.7 Lamellae within adjacent bands of a banded spherulite of linear polyethylene crystallized at 125°C. The specimen was cut open after crystallization, and lamellae are revealed by permanganic etching of the cut surface. The electron micrograph is of a carbon replica of the etched cut surface (after D. C. Bassett). Scale bar = 10 μm.

What is the molecular characteristic which distinguishes the crystalline polymers from the non-crystalline? It is, to be brief, **molecular regularity**: an irregular molecular structure prevents crystallization.

- One form of regularity is the ABABA sequence which is mandatory for condensation polymers such as the nylons (Section 1.3).
- A source of irregularity is random side branching as in polyethylene; irregularity in the size and placement of the side branches inhibits the packing of the molecule into the crystal.
- Another form of irregularity can occur in the vinyl polymers:

$$\left[\begin{array}{cc} H & H \\ | & | \\ -C-C- \\ | & | \\ H & X \end{array} \right]_n$$

Crystallization can be disrupted by the asymmetry introduced by the atom X.

The three forms of symmetry (or **stereo-regularity**) of vinyl polymers are illustrated by the model in Figure 1.8. To facilitate comparison, each model is laid out for inspection in a planar zigzag, so that it is straight and not in a crumpled form. The placement of the X atom is then seen to take one of three patterns. In (a) the X atoms fall all on one side of the plane in which the carbon atoms lie; in (b) the X atoms alternate regularly from side to side; in (c) the X atoms are arranged randomly. These three configurations are termed **isotactic** (a), **syndiotactic** (b), and **atactic** (c). Of these three forms of stereo-regularity it can be stated that crystals are:

- always formed from the isotactic polymers;
- sometimes formed from the syndiotactic; and
- never formed from the atactic.

Note that there is **no** change in chemical composition, merely in the placement of the atom X. The pioneering work in the recognition and polymerization of isotactic, crystalline vinyl polymers is due to Ziegler and Natta, who were awarded the Nobel Prize for Chemistry in 1963.

The tremendous significance of the phenomenon of stereo-regularity is demonstrated best by polypropylene (X in CH_3). Propylene is a cheap and abundant monomer. It is of interest only when polymerized in the isotactic form using Ziegler–Natta catalysis. The surface structure of the catalyst (see Figure 1.4) exerts an influence on the presentation of the monomer to

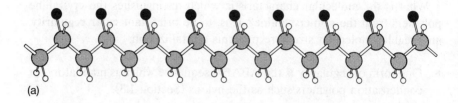

(a)

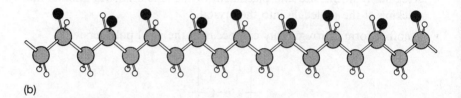

(b)

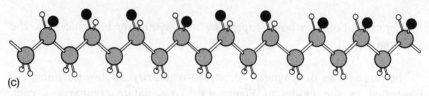

(c)

1.8 Stereoregularity in a vinyl polymer: three different molecular configurations (a) isotactic, (b) syndiotactic, and (c) atactic. The molecular shape may be changed from this planar zig-zag by rotation around C–C bonds. This will change the **conformation** of the molecule; it does not change the **configuration,** which is established at the instant of polymerization.

the growth point. This leads to an ordered reaction producing the isotactic polymer. The catalyst is never perfect, so a small fraction of atactic polymer is produced simultaneously.

1.7 Distribution of relative molecular mass

The size of a molecule is described most directly by n (structures II–IV, IX, XII), which is known as the **degree of polymerization**. Of more use operationally is the relative molecular mass, formerly termed the molecular weight. The term relative molecular mass[1] has been slow to gain acceptance, possibly because of its unwieldy length; we will abbreviate it to

---

1 The term relative molar mass is identical to relative molecular mass.

RMM. RMM has no units, being a pure number (it is a ratio between masses). Thus for molecule Y

$$RMM = \frac{\text{mass of molecule Y}}{(1/12) \text{ mass of an atom of carbon 12}}$$

$$= \frac{\text{mass of a mole of molecule Y}}{(1/12) \text{ mass of a mole of carbon 12}}.$$

A polyethylene molecule of degree of polymerization 10^4 has RMM

$$RMM = 10^4 \, (2 \times 12 + 4 \times 1)$$

$$= 280\,000,$$

since the RMM of carbon is 12 and that of hydrogen 1. This value of RMM neglects the insignificant effect of the end groups.

After polymerization, it is found that a polymer is composed invariably of molecules with many different sizes: the large and the small are completely and intimately mixed. For exceptional purposes—and at considerable expense—they may be separated into fractions, each fraction having a fairly uniform value of RMM. The mechanical strength of the fractions differs remarkably (see Figure 1.9). The rapid drop in tensile

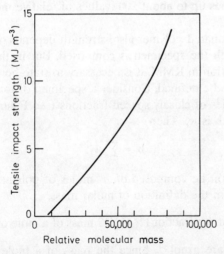

1.9 Effect of RMM on the tensile impact strength of linear polyethylene. A normal polymer with a distribution of RMM was split into a number of fractions: each fraction had a particular RMM. The tensile impact strength of each fraction was measured. The tensile impact strength increases with RMM. The experiment quantifies the common experience that paraffin wax (which can be looked upon as a polyethylene with less than ~ 200 CH_2 units in the molecule) is a brittle solid, with a toughness incomparably less than that of polyethylene. (After L. H. Tung, S. P. E. Conference Proceedings (1958) p. 959.)

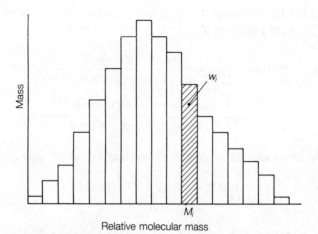

1.10 The distribution is thought of as a series of closely spaced fractions. The ith fraction, of which the specimen contains a mass w_i, is of relative molecular mass M_i.

impact strength as RMM tends to values typical of the paraffins is marked, but not unexpected. Figure 1.9 quantifies the common experience that paraffin wax is a brittle solid. The chemical structure of the paraffins is

$$CH_3 \text{--} (CH_2)_x CH_3,$$

with x taking values up to about 200 (values of relative molecular mass up to RMM $\approx 3\,000$).

For an unfractionated polymer also, strength depends on the size of the molecules of which the specimen is composed. But in this case, because there is a distribution in RMM, it is necessary to state how the average for the specimen is to be defined. Consider a specimen of total mass W to be made up of a series of closely spaced fractions (see Figure 1.10) of which the mass of the ith is w_i. Then

$$W = \sum w_i. \qquad (1.1)$$

Let the ith fraction be composed of n_i moles of polymer of molar mass M_i. Note that, from the definition of molar mass,

$M_i = $ (RMM of ith fraction) $\times$ (1/12 mass of a mole of carbon 12).

The units of M_i are g mol^{-1}. Since the mass of a mole of carbon 12 is 12.0011 g, we have

$$M_i = (\text{RMM of } i\text{th fraction}) \times (1.0001) \text{ g mol}^{-1}.$$

That is, M_i and RMM are essentially numerically identical and differ only in their units.

The mass of the ith fraction in the specimen is

$$w_i = n_i M_i. \tag{1.2}$$

The **number average** molar mass $\overline{M}_n$ is defined as

$$\overline{M}_n = \frac{\sum n_i M_i}{\sum n_i} = \frac{\sum w_i}{\sum n_i} = \frac{W}{\sum n_i}. \tag{1.3}$$

This is the simplest average: it is the mass of the specimen divided by the total number of moles present. Properties that depend on the total number of molecules in the specimen, independent of their size, correlate with $\overline{M}_n$.

The **weight average** molar mass $\overline{M}_w$ is defined as

$$\overline{M}_w = \frac{\sum w_i M_i}{\sum w_i} = \frac{\sum w_i M_i}{W}. \tag{1.4}$$

For some physical properties the contribution of the high M_i species dominates the behaviour. In these cases $\overline{M}_w$ (rather than $\overline{M}_n$) is the more appropriate average for correlation. This fact may be seen more easily by rewriting eqn 1.4 using eqn 1.2:

$$\overline{M}_w = \frac{\sum n_i M_i^2}{\sum n_i M_i}. \tag{1.5}$$

Because of the squared M_i term in the numerator, $\overline{M}_w$ is sensitive to the presence of high M_i species.

As an example, consider a blend of mass 2 g formed from 1 g of each of two paraffins: one $C_{95}H_{192}$ and the other $C_{105}H_{212}$. The two molar masses are then

$$M_{95} = 95 \times 12 + 192 = 1332 \text{ g mol}^{-1}$$

and $\qquad M_{105} = 105 \times 12 + 212 = 1472 \text{ g mol}^{-1}.$

It then follows that in the 2 g specimen the number of moles present is

$$n_{95} = \frac{1}{1332} = 7.51 \times 10^{-4} \text{ mol}$$

and $\qquad n_{105} = \frac{1}{1472} = 6.79 \times 10^{-4} \text{ mol}.$

The average molar masses are then, from eqn 1.3,

$$\overline{M}_n = \frac{1+1}{(7.51+6.79) \times 10^{-4}} = 1399 \text{ g mol}^{-1}$$

and, from eqn 1.4,

$$\overline{M}_w = \frac{1 \times 1332 + 1 \times 1472}{2} = 1402 \text{ g mol}^{-1}.$$

In this case the two averages are almost the same. Now, for a specimen formed from 1 g of $C_{10}H_{22}$ and 1 g of $C_{190}H_{382}$, the same equations give:

$$M_{10} = 142 \text{ g mol}^{-1}; \ n_{10} = 70.42 \times 10^{-4} \text{ mol} \qquad \left\{ \begin{array}{l} \overline{M}_n = 270 \text{ g mol}^{-1} \text{ and} \\ \overline{M}_w = 1402 \text{ g mol}^{-1} \end{array} \right.$$
$$M_{190} = 2662 \text{ g mol}^{-1}; \ n_{190} = 3.76 \times 10^{-4} \text{ mol}$$

and for a specimen formed from 1 g of $C_{10}H_{22}$ and 1 g of $C_{1000}H_{2002}$,

$$M_{10} = 142 \text{ g mol}^{-1}; \ n_{10} = 70.42 \times 10^{-4} \text{ mol} \qquad \left\{ \begin{array}{l} \overline{M}_n = 281 \text{ g mol}^{-1} \text{ and} \\ \overline{M}_w = 7072 \text{ g mol}^{-1}. \end{array} \right.$$
$$M_{1000} = 14\,002 \text{ g mol}^{-1}; \ n_{1000} = 0.71 \times 10^{-4} \text{ mol}$$

These elementary results illustrate the following points.

- $\overline{M}_n$ is sensitive to the admixture of molecules of low molecular mass; thus the $\overline{M}_n$ values for the 10/190 and 10/1000 blends are much lower than for the 95/105 blend.

- $\overline{M}_w$ is sensitive to the admixture of molecules of high molecular mass; the $\overline{M}_w$ for the 10/1000 blend greatly exceeds that for the other two.

- $\overline{M}_w$ always exceeds $\overline{M}_n$.

- the ratio $\overline{M}_w/\overline{M}_n$ gives a measure of the **range** of molecular sizes in the specimen: for these three blends the ratios are 1.00, 5.19, and 25.17. For polymer specimens the ratio $\overline{M}_w/\overline{M}_n$ normally falls in the range 2 to 100.

In a normal synthetic polymer the distribution of M is continuous, as indicated in Figure 1.11. The area shown shaded represents the mass dW of that fraction of the specimen with molar mass between M and $M + dM$. The plot is thus of dW/dM against M. Figure 1.11 is a typical plot, showing in this case a tail at high molecular masses.

The quickest and most precise method of measuring the distribution of M is gel permeation chromatography. The polymer is dissolved in a

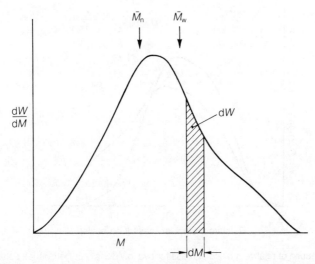

1.11 Illustration of a distribution of molecular mass. The area shaded represents the mass of the specimen with molecular mass between M and $M + dM$. A measure of the breadth of the distribution is provided by the two averages $\bar{M}_n$ and $\bar{M}_w$, which are defined by eqns 1.3 and 1.4.

solvent, for instance 1 g of polyethylene in 1 litre of xylene. A sample of the dilute solution is introduced into a solvent stream passing down a long tube packed with beads made of porous glass or porous cross-linked polystyrene. **The pores in the beads are of a size comparable with that of the macromolecules.** As the dissolved polymer molecules flow past the beads, the molecules are retarded as they diffuse in and out of the holes in the beads. The larger molecules can enter only a fraction of the holes, or are completely excluded. They are therefore retarded only slightly, or not at all, in their passage down the column. The smaller the molecule the greater the retardation in time. The largest molecules therefore pass most rapidly down the column, the smallest molecules most slowly. The amount of polymer appearing at the bottom of the column as a function of time is determined by optical or spectrophotometric methods to yield the distribution of molecular mass (see Figure 1.12).

Molecular masses can be measured by many different techniques, but all have one point in common: the measurements are made on the dissolved polymer. The reason for this is that the molecules when in dilute solution are isolated one from another (see Figure 1.13). When the molecules are aggregated together (in the crystal, glass, or liquid) they are closely intertwined. The process of dissolution into a solvent (normally stimulated by heating) occurs by the unravelling of each molecule from the aggregate. The separate molecules then diffuse off into the solvent as units. The molecule in solution is dynamic, its shape changing from instant to instant.

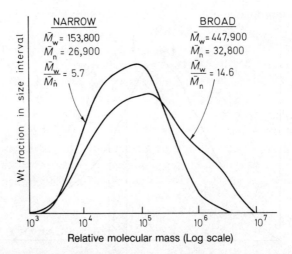

Relative molecular mass (Log scale)

1.12 Distribution of relative molecular mass for two low density polyethylenes determined by gel permeation chromatography: one with a narrow distribution, the other broad (after R. B. Staub and R. J. Turbett, *Modern Plastics Encyclopedia*, 1973–4).

The time-averaged shape is spherical, but at any instant the actual shape will be a more or less spherical, random tangle.

Note that, as indicated in Figure 1.13, the material within the tangle is mainly solvent: the molecule exerts its influence over a volume far in excess of its own molecular volume. This property of the polymer in dilute solution is typical also of the macromolecules when aggregated in glass, liquid, or crystalline solid (Chapter 2). Above all other distinctive characteristics, this is the one that distinguishes polymers from metals and ceramics. It is of particular relevance to the mechanical properties, particularly strength. In a tensile specimen broken under impact, the molecules under high stress at the tip of a propagating crack can be thought to act like reinforcing fibres. According to this simple view the long molecules reinforce the critical zone at the crack tip. The longer the molecule, the

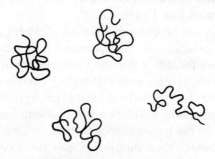

1.13 Schematic representation of polymer molecules in dilute solution.

greater the reinforcement at the crack tip: the strength thus increases with increasing molecular mass.

It is clear that an engineer concerned only with specimen toughness would always call for a polymer of the highest RMM. There are other important properties, however, which can be adversely affected by an increase in RMM. For instance, the viscosity of the molten polymer increases rapidly with RMM. It follows that the selection of the optimum RMM for a particular end-use involves a compromise, for example between the desire to increase impact strength and the requirement to form objects by injection moulding, inexpensively, and with conventional equipment. For any one polymer, manufacturers offer a large range of resins distinguished principally by differences in RMM. The normal catalogue description of the RMM of a specimen is an empirical quantity, the **melt flow index** or **melt index** (MFI), which is defined in Chapter 7.

Notes for Chapter 1

1.N.1

An excellent example of this is polyoxymethylene (POM)

$$\phi_1 \text{+\!CH}_2 \text{---} \text{O}\text{+}_n\, \phi_2, \qquad\qquad \text{(XVII)}$$

which is known sometimes as polyformaldehyde or acetal homopolymer. It was studied in detail by Staudinger in the 1920s. It has excellent mechanical properties, is tough, produces good mouldings, has reasonable high-temperature properties, but was not commercialized until as recently as 1959 since, unless the end-groups (or caps) are stable, the molecule 'unzips':

$$\text{+\!CH}_2 \text{---} \text{O}\text{+}_n \to n\text{CH}_2\text{O}, \qquad\qquad \text{(XVIII)}$$

the solid transforming back to formaldehyde. This is just the reverse of polymerization

$$n\text{CH}_2\text{O} \to \text{+\!CH}_2 \text{---} \text{O}\text{+}_n. \qquad\qquad \text{(XIX)}$$

With stable caps it is one of the best engineering plastics: a high-cost plastic produced in low volume. It is used for gears, appliance components, clock mechanisms, and heavy-duty conveyor belt links (see also 1.N.11).

1.N.2

For this reason this type of polyethylene is known as high-pressure polyethylene; other names include branched polyethylene or low-density polyethylene (LDPE) (**the latter is the most common name**). The largest application is film for packaging. The modulus is low and the polymer is not used in high-performance roles. It has excellent electrical properties, and is used therefore in cables; it is a low-cost plastic produced in high volume by many companies, and it does not absorb water.

1.N.3

This form of polyethylene is known as low-pressure polyethylene, linear polyethylene, or high density polyethylene (HDPE) (**the latter is the most common name**). It finds a wide application in bottles, crates, kitchenware, and pipe; it is a low-cost plastic produced in high volume by many companies, and it does not absorb water. The terms LDPE and HDPE are ambiguous since linear polyethylene, when of high relative molecular mass, can have a low density, comparable with that of branched polyethylene. The linear polyethylene used in hip-replacement is of this type. Linear low-density polyethylene (LLDPE) is the—somewhat unfortunate—name given to polyethylenes made by the copolymerization of ethylene with suitable comonomers using Ziegler–Natta type catalysis; a suitable comonomer is, for example, butene. The polymer is **not** linear but it is side-branched, and it has properties comparable with those of low-density polyethylene, with which it is in competition (see Problem 1.5).

1.N.4

Polypropylene (PP) is a high-volume, low-cost plastic first commercialized in 1959 and produced by many companies. It is unusual in that it is processed in large quantities by all four major fabrication processes: moulding, extrusion, film, and fibre technology. Major engineering applications include pipe and automobile parts. It does not absorb water.

1.N.5

Liquid styrene was observed by Simon to solidify after a month's storage as long ago as 1839; it was assumed at the time to form a glassy oxide. Large-scale industrial development was started first by Dow Chemical Company in 1935. Toughened (high-impact) polystyrenes (HIPS) were introduced in 1948 (see Chapter 5). PS is a low-cost, high-volume polymer; its major uses are packaging, kitchenware, and as expanded PS for thermal insulation.

1.N.6

Regnault observed the polymerization of vinyl chloride irradiated with sunlight in 1835. Poly(vinyl chloride) (PVC) was not produced commercially until 1930. It is frequently used in the plasticized state: this is done by mixing it with a low molecular weight liquid which softens it. PVC is a low-cost, high-volume plastic used (in the plasticized state) for hose, clothing, wire and cable insulation, and in wall and floor coverings. In the stiffer, unplasticized state (uPVC) it finds a large market in building (window frames and plumbing, for example).

1.N.7
Poly(methyl methacrylate) (PMMA) is the most important member of a large family of polymers known as **acrylic resins** which were commercialized first in Germany by Röhm and Haas in 1927. PMMA is a medium-cost, medium-volume polymer. Typical applications include glazing and illuminated outdoor signs.

1.N.8
This **intramolecular** backbiting mechanism is much more common than is the removal of a hydrogen atom by the growing chain from an adjacent molecule. The latter **intermolecular** process leads to long chain branching. Long chain branches have little effect on the properties of the solid polymer: they confer **melt elasticity** on the polymer liquid. This is most important in processing, particularly in extrusion, film forming, and thermoforming (see Chapter 7).

1.N.9
Nylon is the most widely used engineering thermoplastic and yields also excellent fibres. It is tough, strong, and abrasion resistant. Six types are used in engineering applications: nylon 6.6 (see Structure IX), nylon 6.10 and nylon 6.12 together with:

nylon 6,
$$H\left[N(CH_2)_5\!-\!\underset{\underset{O}{\|}}{C}\right]_n OH$$
with N–H above
(XX)

nylon 11,
$$H\left[N(CH_2)_{10}\!-\!\underset{\underset{O}{\|}}{C}\right]_n OH$$
(XXI)

nylon 12,
$$H\left[N(CH_2)_{11}\!-\!\underset{\underset{O}{\|}}{C}\right]_n OH$$
(XXII)

The presence of the NH and CO groups in the molecule is the reason why nylons absorb water. The water absorption characteristics are given in Table 1.1.

Table 1.1   Weight of water (%) absorbed by nylons at 23°C at 50% and 100% relative humidity

Nylon	6.6	6	6.10	6.12	11	12
50% RH	2.5	2.7	1.5	1.5	—	—
100% RH	9.0	9.5	3.5	3.0	1.9	1.4

It will be seen that as the CH_2 content of the molecule increases, the water absorption decreases. Water absorption leads to a lowering of modulus and an increase in dimensions. Nylon is frequently used in engineering applications reinforced with glass or carbon fibre or minerals.

1.N.10

Poly(ethylene terephthalate) (PET) is widely used as a fibre and is also an important engineering thermoplastic together with poly(butylene terephthalate) (PBT),

$$\text{HO}\!-\!\!\left[(CH_2)_4\!-\!O\!-\!\overset{\displaystyle O}{\overset{\|}{C}}\!-\!\!\left\langle\!\!\bigcirc\!\!\right\rangle\!\!-\!\overset{\displaystyle O}{\overset{\|}{C}}\!-\!O\right]_n\!\!-\!\!\text{H}$$ (XXIII)

They both come under the general heading of thermoplastic polyesters, as opposed to the thermoset polyesters (see Chapter 6). They are used in switch components, housings, bearings, and bottles.

1.N.11

Thermal stability is of great importance, not only for the end-use (for example a gear running hot), but also for the forming process. Thermoplastics are processed in the liquid state; it is of course essential that the polymer remain stable whilst it is liquid. A good example of enhanced thermal stability induced by copolymerization is in the copolymer

$$\dots\dots-(CH_2-O)-(CH_2-O)-(CH_2-CH_2-O)-(CH_2-O)-\dots\dots\ .$$

This, structurally, is a polyoxymethylene with, here and there, an ethylene oxide monomer. As noted in 1.N.1, if the chain commences to unzip, the formaldehyde molecules H_2CO are removed one by one. This unzipping is stopped at the CH_2CH_2O unit. Thus, a small percentage of ethylene oxide acts as a 'fail-safe' stabilizer in case unzipping occurs. The C—C bonds also stabilize the molecule against acidic and oxidative attack. This copolymer (with small ethylene oxide content) has properties only slightly inferior to those of the homopolymer, polyoxymethylene. It is a high-cost, low-volume copolymer, known also as acetal copolymer.

1.N.12

Technically important examples of block copolymers are the thermoplastic rubbers, such as the styrene–butadiene–styrene system. These molecules are essentially polybutadiene tipped with polystyrene end-blocks. Butadiene ($CH_2=CH-CH=CH_2$), on polymerization, yields a polymer consisting of three structures:

$$
\begin{array}{ccc}
\sim CH_2 \quad CH_2\sim & \sim CH_2 \quad H & \sim CH_2-CH\sim \\
\diagdown \quad \diagup & \diagdown \quad \diagup & | \\
C{=}C & C{=}C & CH \\
\diagup \quad \diagdown & \diagup \quad \diagdown & \| \\
H \quad\quad H & H \quad\quad CH_2\sim & CH_2 \\
\\
\text{I} & \text{II} & \text{III}
\end{array}
$$

The fraction of the molecule in each structural form is variable, but is usually in a proportion of order I/II/III = 3/6/1. Polybutadiene is a rubber above $\sim -90°C$. The ratio I/II/III determines the glass transition temperature (see Section 2.6).

Technically important examples of graft copolymers are the **high-impact** polystyrenes (HIPS). These are a mixture of PS and a rubber, such as polybutadiene. Maximum toughness is achieved if a small proportion of the PS chains is grafted on to some of the polybutadiene molecules. The high-impact PS then consists of PS, ungrafted polybutadiene, and the PS-polybutadiene grafted molecules: these last enhance the mixing of the PS and ungrafted polybutadiene.

Problems for Chapter 1

1.1 Write down chemical formulae for the addition polymers formed from the following monomers, and in each case calculate the relative molar mass of a polymer having a degree of polymerization of 1000:

(1) $CH_2=CH(C_2H_5)$
(2) $CH_2=CHF$
(3) $CH_2=CCl_2$
(4) $CH_2=C(CH_3)_2$
(5) $CH_2=CH(OH)$
(6) $CH_3-\underset{\underset{O}{\|}}{C}-O-CH=CH_2$

1.2 Which of the monomers listed in Problem 1.1 is capable of forming isotactic or syndiotactic sequences?

1.3 Write down chemical formulae for addition polymers formed from the following monomers, and indicate which are capable of forming stereo-isomers (i.e. molecules in which the same chemical groups are linked together, but alternative spatial configurations are possible; see Figure 1.8):

(1) $CH \equiv CH$

(2)
$$
\begin{array}{c}
CH_2 - CH_2 \\
CH_2 \qquad\qquad N-H \\
\qquad\qquad\qquad C=O \\
CH_2 - CH_2
\end{array} \quad ;
$$

(3) $CH_2 - CH_2$; with O bridging (epoxide)

(4)
$$
\begin{array}{c}
HC = CH \\
C \qquad C \\
O \quad O \quad O
\end{array}
$$

1.4 What alternative polymer structures can be formed from the **isoprene** molecule

$$
CH_2 = CH - \underset{\underset{CH_3}{|}}{C} = CH_2 \ ?
$$

1.5 Certain grades of commercial polyethylene made by the Ziegler–Natta process contain a small proportion of 1-hexene, $CH_2 = CH(C_4H_9)$, as co-monomer. How would you expect the properties of the resulting copolymer to differ from those of linear polyethylene homopolymer made by the same process?

1.6 What product is formed when hydrogen is added to the double bonds in polybutadiene if, using the nomenclature in 1.N.12, the molecule consists of:
(1) 100% structure (I)
(2) 50% each of structures (I) and (II)
(3) 46% each of structures (I) and (II) and 8% of structure (III)?

1.7 Polycarbonate is made by a condensation reaction in which HCl is given off (see 8.N.1 for the structure of polycarbonate). Which two monomers are required for this reaction?

1.8 What polymer is formed as a result of the condensation reaction between

$$NaO-\bigcirc-\underset{\underset{CH_3}{|}}{\overset{\overset{CH_3}{|}}{C}}-\bigcirc-ONa$$

and

$$Cl-\bigcirc-\underset{\underset{O}{\parallel}}{\overset{\overset{O}{\parallel}}{S}}-\bigcirc-Cl \ ?$$

1.9 Calculate the fraction of monomer units of the species named (in mole %) for copolymers having the following compositions by weight:
(1) styrene–butadiene copolymer (25 wt% styrene)
(2) ethylene–propylene copolymer (55 wt% ethylene)
(3) styrene–acrylonitrile copolymer (23.5 wt% acrylonitrile)
(4) butadiene–acrylonitrile copolymer (40 wt% acrylonitrile).
(Acrylonitrile is CH_2=$CH(CN)$; see 3.N.7.)

1.10 Suggest a reason for adding a small percentage of a third monomer, containing two $-\overset{|}{C}=\overset{|}{C}-$ units, when ethylene and propylene are copolymerized in roughly equal amounts. (The product is known as EPDM: ethylene–propylene diene monomer.)

1.11 What limits are imposed upon the RMM of the polyester formed from ethylene glycol and terephthalic acid when the ratio of monomer units is:
(1) 1.1:1
(2) 1.02:1
(3) 1:1?

1.12 The formation of nylon 6.6 from hexamethylene diamine and adipic acid depends upon the fact that both the diamine and the acid contain **two** reactive groups. If a small amount of acetic acid

$$CH_3-C\overset{\nearrow O}{\underset{\searrow OH}{}}$$

which has only one reactive group, is added to the mixture of monomers, it blocks the end of the chain, and prevents further growth. Calculate the number average RMM of the nylon 6.6 formed by reacting 100 moles of hexamethylene diamine with 99 moles of adipic acid and 2 moles of acetic acid.

1.13 Certain free radicals will remove a hydrogen atom from a polybuta-diene chain, leaving a free radical site on the chain. Show that this reaction leads to formation of graft copolymer, accompanied by cross-linking, when a solution of polybutadiene in styrene monomer is heated to polymerize the styrene.

1.14 The molecules of a block copolymer consist of a central block of butadiene units, with end blocks of polystyrene units. The average degree of polymerization is 2000 for the central block and 150 for each of the end blocks. In the solid material, the polystyrene chains form a separate phase, which is seen in the electron microscope as spheres approximately 25 nm in diameter. Calculate the weight fraction of styrene in the block copolymer, and estimate the number of polystyrene chains in each sphere. The density of polystyrene is 1050 kg m^{-3}.

1.15 Calculate the number average and weight average relative molecular mass for a sample of polystyrene made by blending eight separate fractions, each of which has a very narrow molecular mass distribution, as follows:

Relative molecular mass	15 000	27 000	39 000	56 000
Weight (g)	0.10	0.19	0.25	0.17
Relative molecular mass	78 000	104 000	120 000	153 000
Weight (g)	0.12	0.08	0.06	0.04

1.16 What is the effect upon $\bar{M}_n$ and $\bar{M}_w$ of adding 0.5 wt% of styrene monomer to the polystyrene sample defined in Problem 1.15?

1.17 Polyesters having the general formula

$$\text{HO}\text{-}(\text{R}\text{-}\underset{\underset{\text{O}}{\|}}{\text{C}}\text{-}\text{O})_x\text{H}$$

are formed by condensation polymerization of the monomer $\text{HO}\text{-}\text{M}\text{-}\text{H}$, where M represents the repeat unit (bracketed above). Show that when a fraction p of the $-\text{OH}$ groups has reacted, the number of molecules N_x in which there are x repeat units is given by

$$N_x = Np^{x-1}(1-p) = N_0(1-p)^2 p^{x-1},$$

where N is the total number of molecules present in the partly reacted system, and N_0 is the original number of monomer molecules. Hence show that the weight fraction of molecules containing x repeat units is given by

$$W_x = x(1-p)^2 p^{x-1}.$$

Plot N_x and W_x against x for the case where $p = 0.98$. (Hint: first show that N/N_0 = probability of finding an unreacted $-\text{OH}$ group $= 1-p$.)

2 *Structure of polymeric solids*

2.1 Introduction

This chapter opens with a description of the packing of polymer molecules in crystals. X-ray diffraction gives an extremely precise description of this, since the polymer lattice diffracts X-rays as does any other three-dimensional lattice. The shape and mutual arrangement of the minute crystals are then described. The crystals are separated one from another by amorphous regions whose dimensions are comparable with those of the crystals. The information on these matters stems from electron and light microscopy, techniques which are less precise than X-ray diffraction. **Crystalline synthetic polymers are invariably partly crystalline and partly amorphous: the term crystalline polymer always implies partially crystalline.**

Polymers, unlike metals, find wide application as completely amorphous solids. The two amorphous forms of greatest interest are the glasses and the elastomers or rubbers. Polymers found their earliest technical application as elastomers over 100 years ago, and it is in this form that they make a unique contribution to society. In examining the structure of the amorphous state we shall confine attention to the really significant point, which is the conformation of the molecule. It will be shown that the conformation resembles that of the molecule in dilute solution sketched at the end of the previous chapter. By means of a simple model a mathematical description of the conformation will be obtained which has fundamental and practical consequences for the mechanical properties of elastomers, the stress–strain curve in particular.

2.2 Structure of the crystal

Crystallization from the molten state may be thought of loosely as consisting of two stages. In the first, the molecule assumes its lowest-energy conformation. In the case of polyethylene this is the planar zigzag (see Figure 2.1(a)), in which the CH_2 chains have the same conformation as in the most common crystal structure of the normal paraffins, such as $C_{60}H_{122}$; the C—C bond length is 153 pm (1 pm = 10^{-12} m) and the angle between the bonds is close to 112°. In the second stage, the straight molecules pack together like parallel rods, as indicated in Figure 2.1(b)

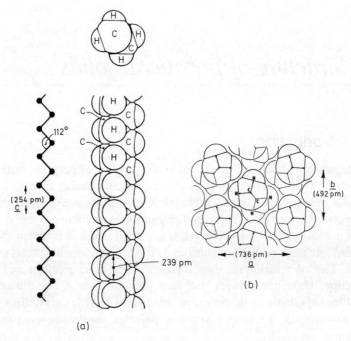

2.1 (a) Scale drawing of the conformation of a polyethylene molecule in the zigzag, crystalline conformation showing both the side and end view. (b) View of the crystal along the c axis in which the atoms have their correct external radii.

which shows a plan view. The lattice constants have the same meaning as for monatomic solids: the entire crystal can be generated by translations of the unit cell by lattice vectors equal to the quantities a, b, and c. For example, it is easily seen from Figure 2.1(a) that translations along the molecule parallel to and equal to the vector c generate the lattice in that direction: translations equal to vectors a and b will generate the crystal in two directions normal to c (see Figure 2.1(b)). The mutual arrangement of the molecules is such as to achieve the lowest energy state (lowest Gibbs free energy). The lateral packing is shown to scale in Figure 2.1(b). The zigzag planes lie at two orientations at 82° to each other.

The planar zigzag conformation of the C—C chain is not commonly observed in other polymers. Examination of Figure 2.1(a) shows that the hydrogen atoms on the next-but-one carbon atoms lie close together. From the length of the C—C bond (153 pm) and the angle between the bonds (112°), it follows that the centres of the hydrogen atoms are separated by $2 \times 153 \times \sin(112°/2) = 254$ pm, which is the $|c|$ component of the unit cell. The diameter of the hydrogen atom is only 15 pm smaller, a fact which is well displayed on the scale model shown in Figure 2.1(a). It is clear, therefore, that if the hydrogen atoms are replaced by atoms with diameters exceeding 254 pm the planar zigzag conformation of the carbon backbone cannot occur, since the larger atoms would then overlap.

For example, suppose all the hydrogen atoms are replaced by fluorine atoms with van der Waals diameter 270 pm. To accommodate the fluorine atoms a rotation around each $C-C$ bond of about 20° is induced. This is accompanied by a slight opening of the $C-C$ chain bond angle to about 116°. The result is that the molecular conformation of polytetrafluoroethylene (PTFE)

$$\left[\begin{array}{cc} F & F \\ | & | \\ -C-C- \\ | & | \\ F & F \end{array}\right]_n$$

in the crystal is a helix, as illustrated in Figure 2.2 (2.N.1). The helix repeats every 13 CF_2 units. The helical molecules then pack laterally to form the crystal.

In vinyl polymers (Structure III, Chapter 1) the molecules in the crystal form helices. The way in which this is achieved by rotation from the planar zigzag is illustrated for isotactic polypropylene in Figure 2.3. The interference between the large CH_3 groups is relieved by a rotation of 120° in

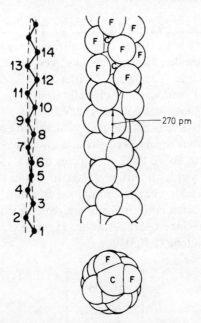

2.2 (a) Scale drawing of the helical conformation of a polytetrafluoroethylene molecule as it occurs in the crystal, showing both side and end views. There are 13 CF_2 units in one turn of the helix, which is 1.68 nm in length.

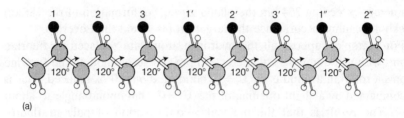

(a)

(b)

2.3 (a) Illustration of a vinyl molecule laid out in a polyethylene-type planar zigzag. It is not drawn to scale. If the large substituted atoms (numbered 1, 2, 3, etc.) exceed 254 pm in diameter, then this structure cannot occur since the substituted atoms would touch. In real vinyl polymers the substituted atoms (or groups of atoms) always exceed 254 pm in diameter and the molecule twists into a helix as illustrated in (b). This is the 3/1 helix which occurs in polypropylene. The groups, 1, 1', and 1" then lie in a line; and similarly for 2, 2', and 2" and 3, 3', etc. Hydrogen atoms are not shown in (b).

every second bond, as marked in Figure 2.3(a). The result is a helix (Figure 2.3(b)), termed a 3/1 helix (three chemical repeat units for one complete rotation of the molecule). Helices are observed in all isotactic polymers with bulky side groups. The helix is often more complex: for instance isotactic poly(methyl methacrylate) crystallizes in a 5/2 helix. The helical form is common in biological polymers (2.N.2).

It is sometimes observed that the helix which is in equilibrium at a high temperature transforms to another helix at a lower temperature. PTFE, on heating from below to above 10°C, transforms from a 13/1 to a 15/1 helix. The transformation yields an abrupt change in crystal volume of the order of 1%. This appreciable change in volume at about room temperature presents important, sometimes unfortunate, consequences for the dimensions of precisely machined parts. Another example of a helix transformation occurs in polybutene (2.N.3),

$$
\begin{bmatrix}
& H & H \\
& | & | \\
-& C & - C - \\
& | & | \\
& H & CH_2 \\
& & | \\
& & CH_3
\end{bmatrix}_n
$$

On crystallization from the melt an 11/3 helix forms. This helix slowly transforms to a 3/1 helix if the specimen is held for an extended period at room temperature. The change in crystalline and therefore in specimen dimensions is a complicating factor in applications of this polymer that require the maintenance of close dimensional tolerances.

Polymer crystals are extremely anisotropic. Along the molecule the bonds are covalent. In the two transverse directions the molecular packing is generated by much weaker secondary forces (van der Waals, dipole, and hydrogen bonds, depending on the chemical structure of the molecule). The most important consequence of this for the mechanical properties is that the moduli of the crystal depend critically on direction. In the direction of the molecule the modulus of polyethylene is of the order of 200 GPa (Problem 2.19); this value is typical of macromolecules, whose moduli (along the molecule) fall roughly in a band 100 to 400 GPa. In both transverse directions, the modulus is ~100 times lower.

Polymer fibres are mechanically transformed from the isotropic state by drawing (see Chapter 5). In the drawn state the molecular axes of the polymer molecules in the crystals lie along the fibre. This yields a high modulus along the fibre axis. At right angles the modulus of the fibre is less, but this is, in normal applications, of no consequence since it is along the fibre axis that high modulus and strength are required.

2.3 Crystal shape

The earliest and simplest model of a crystalline polymer is the fringed-micelle model (see Figure 2.4). The evidence upon which it was based

2.4 The fringed-micelle model of semicrystalline polymers. The solid consists of an intimate mixture of ordered crystals and randomly structured amorphous regions. The molecular length is considerably greater than the length of a crystal. A molecule thus passes through several crystals and several amorphous regions. The integrity of the two-phase solid is thus maintained by the long molecules.

came from X-ray diffraction. The diffraction pattern shows both sharp crystal lines and the diffuse lines typical of amorphous materials such as liquids, superposed one on the other. The fringed micelle model encapsulates the really important structural fact: this is that, although the molecules lie parallel to each other within the crystal, they do so over lengths far shorter than the total length of the molecule. As a useful order of magnitude guide, the molecule may contain 10 000 carbon atoms, but the length of the chain within the crystal may be only 100 carbon atoms. In the fringed micelle model a molecule meanders out of one crystal into the amorphous fraction and then into a neighbouring crystal. In this way it passes through many crystals. **All the crystals and the adjacent amorphous regions are thus firmly woven together by the long, thread-like macromolecules** (2.N.4).

The fringed-micelle model is today considered unrealistic except for polymers of exceptionally low crystallinity—for a polymer such as PVC, for example, with crystallinity which is hardly detectable. But the central fact of the model remains valid, which is that the two-phase crystal–amorphous solid is pinned together by the macromolecules which pass from one region to another and back again many times. It will be noted incidentally

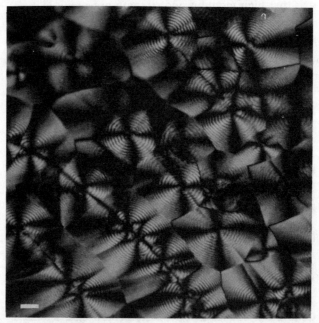

2.5 Banded spherulities of linear polyethylene within a commercial pellet. Optical micrograph of a thin section viewed between crossed polars. Note that sequences of spherulites have nucleated along lines, here in black contrast, which are probably a legacy of extrusion (after D. C. Bassett.). Scale bar = 10 μm.

that the longer the macromolecule the more effective the pinning. We will recall this point when considering the way in which relative molecular mass affects strength (see Chapter 1 and Figure 1.9).

The random arrangement of adjacent crystals in the fringed-micelle model (see Figure 2.4) does not accord with evidence obtained from microscopy. The electron microscope shows the crystals in polyethylene to be very thin twisted lamellae laid one upon another (Figure 1.7). The lamellae are too small to be observed with the light microscope. Light microscopic examination of thin films or sections between crossed polarizers reveals complex polyhedral objects known as spherulites (see Figure 2.5). They are in fact a complex ordered aggregation of the sub-microscopic crystals. In polyethylene the crystal lamellae are about 10 nm thick. They are separated one from another by thin lamellae of amorphous polymer of about the same thickness. The lateral dimensions of the crystals are far greater than 10 nm, so that the crystals have the simple proportions of a piece of paper. The spherulite is thus a polyhedral object with an internal texture akin to that of a telephone directory.

The diameter of a spherulite, although somewhat variable depending on crystallization conditions, is usually of the order of the wavelength of visible light. This is the origin of the white, milky appearance of crystalline polymers: the amorphous polymers are transparent. The light is scattered from the spherulites, not from the individual crystals, which are much smaller than the wavelength of light. In order to observe spherulites in a bulk crystallized specimen it is necessary to cut a specimen with a thickness of the order of the spherulite size. This is done quite easily using a microtome.

2.4 Crystallinity

The determination of the proportion of the solid that is crystalline (its **crystallinity**) is of considerable fundamental and often practical significance. The easiest method for measuring the crystallinity is the determination of specific volume (the specific volume is the inverse of the density). The simplest methods are Archimedes' method and the density gradient column. The measurement, by either method, is most conveniently performed at room temperature, say 20°C. At this temperature, let the specific volumes of the crystal and amorphous fractions be v_c and v_a. Then if x is the fraction of the mass that is crystalline and $(1 - x)$ the fraction that is amorphous, the specific volume v of the specimen is

$$v = xv_c + (1 - x)v_a. \tag{2.1}$$

From X-ray diffraction, for polyethylene, $v_c = 0.989 \times 10^{-3}$ m^3 kg^{-1} (see

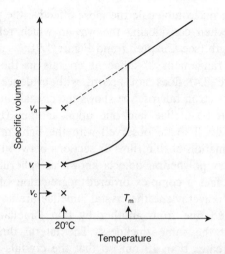

2.6 Specific volume versus temperature for semicrystalline linear polyethylene showing the effect of heating a specimen from 20°C to above the melting point T_m. The specific volume of the specimen at 20°C is v. The specific volume of the amorphous fraction v_a is obtained by extrapolating the v–T curve for the liquid down to 20°C. The specific volume of the crystalline fraction v_c is obtained from the lattice constants of the unit cell (see Problem 2.1). The crystal fraction can be obtained using these quantities in eqn 2.2.

Problem 2.1). It is not easy to obtain v_a. The preferred method is to extrapolate the volume–temperature line for the liquid polymer to 20°C (see Figure 2.6). This yields for polyethylene $v_a = 1.16 \times 10^{-3}$ m^3 kg^{-1}. Then, for example, if a measurement of specific volume of a specimen yields $v = 1.042 \times 10^{-3}$ m^3 kg^{-1}, from eqn 2.1,

$$x = \frac{v_a - v}{v_a - v_c},$$

(2.2)

so that $x = 0.69$. It would be normal practice to state that the crystallinity of this specimen was 69%.

There are very few polymers for which the specific volume method fails. One is PTFE. This polymer usually contains voids at the 1% level due to the method of fabrication (sintering). Fluctuations in void content around 1% completely mask the smaller changes in specific volume due to changes in crystallinity. The method also fails if the specific volumes of the amorphous and crystalline polymer are almost equal.

The modulus, strength, and most other mechanical properties depend critically on crystallinity. For a particular polymer the crystallinity will depend on the method of solidification. The crystal grows from the tangled melt by a process in which the chains straighten out and pack side by side. The growth of each crystal proceeds as far as the molecular tangle allows.

The longer the specimen is left at the crystallization temperature the greater the unravelling and the greater the crystallinity.

2.5 Crystallization and melting

When a polymer melt is cooled below T_m, crystallization is initiated at nuclei (often minute specks of impurity) at different points in the melt. The crystallization proceeds by the growth of the spherulites, which at this stage are spherical, each spherulite having a nucleus at its centre (see Figure 2.7). The spherulite expands at a constant rate if the temperature is

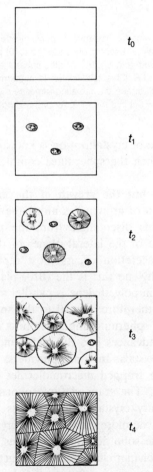

2.7 Illustration of five stages in crystallization showing spherulite growth. At time t_0—the supercooled melt. At later times t_1 to t_3—the growth of spherulites. Finally, at time t_4— the specimen is composed completely of spherulites.

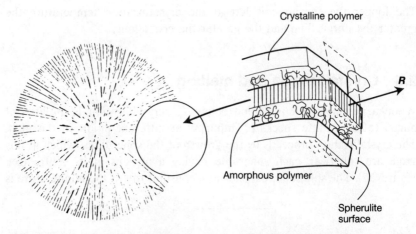

2.8 A polymer spherulite growing into the melt. In polyethylene the crystalline fibrils are thin lamellae. The molecules crystallize most rapidly on to the (010) plane: the *b* axis is therefore the direction of most rapid growth and is parallel to the spherulite radius *R*. The *a* and *c* axes are randomly distributed around *R*. If the solidification is isothermal, the lamellae are all of the same thickness. In order to fill space the radiating lamellae must branch and give birth to daughter lamellae as they grow out into the melt. Amorphous polymer is left trapped between the crystals.

held constant. This stage of crystallization is known as primary crystallization and is complete when the spherulites completely fill the space (see Figure 2.7).

Spherulite growth is but the growth of the many crystals which it comprises. The direction of growth lies along the spherulite radius. Segments of polymer chains disentangle from the melt and by accretion become attached to one of the lateral planes of the many small crystals (see Figure 2.8). The accretion occurs most rapidly on one particular crystal plane: in polyethylene this is the (010) plane. It follows that the vector *R*, in Figure 2.8, in polyethylene is parallel to *b*. The polymer chain axis *c* in the crystal is therefore normal to the spherulite radius. Radial growth ceases when the spherulite impinges on its neighbours.

As the growth front advances into the melt, amorphous polymer is left trapped between the crystals. In some polymers, linear polyethylene or PTFE for example, the trapped macromolecules are identical to those which have crystallized. The trapped (amorphous) macromolecules can crystallize in a secondary crystallization process after the cessation of primary crystallization. The more the secondary crystallization proceeds to completion the more the solid density approaches that of the crystal. On the other hand, if the polymer chain contains structural irregularities such as branch points, copolymerized units, cross-links, end-groups, or atactic segments, these will be rejected by the growing crystal and will be increasingly concentrated in the amorphous fraction. These structural irregularities play a role analogous to that of impurities in monatomic substances in

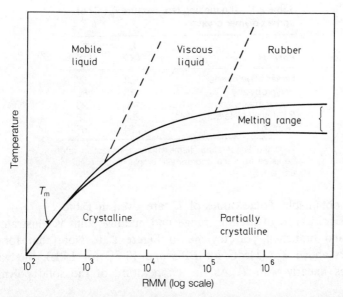

2.9 Dependence of melting temperatures T_m on relative molecular mass (RMM). At low RMM there is one T_m; at higher RMM there is a band of melting temperatures, which is due to the existence of a range of crystal thicknesses in the solid polymer.

their influence on melting and crystallization. The melting points are depressed and the melting range is broadened.

The dependence of melting point on relative molecular mass is indicated schematically in Figure 2.9. For values of RMM below 10^3 there is only one melting point, which decreases with decreasing chain length. Above 10^3 the melting point levels off, showing less and less increase with relative molecular mass; simultaneously the sharp melting point of the low-RMM molecules broadens into a melting range.

Even in the most linear (and perfect) polymers, a melting range is observed: this is due to a variation in crystal size. To be precise, the melting point of a polymer crystal depends on the length of the molecular segments arranged side by side within the crystal. As an example, consider a specimen composed of molecules with, on average, 10 000 carbon atoms in the polymer chain. Crystallization can be performed to achieve large crystals (say segments of 200 carbon atoms packed side by side) or small crystals (100 carbon atoms packed side by side): the thick crystals can be formed at a high crystallization temperature and the thin at a lower temperature. Now these two crystals will melt at markedly different temperatures, although both are formed from the same polymer. Polymers are rarely crystallized so that all the crystals are of precisely equal thickness. It follows that the specimen will exhibit a melting range even in specimens with no structural irregularities in the molecule. The melting temperature T_m is defined for the thickest, largest, and most perfect

Table 2.1   The melting temperature T_m(°C) of some polymer crystals

Polymer	T_m (°C)	T_1 (°C)
Linear polyethylene	138	56
Polypropylene	176	86
Polyoxymethylene	180	90
Poly(ethylene terephthalate)	266	158
Nylon 6.6	264	157

T_1 is the temperature of maximum crystallization rate deduced from the approximate empirical rule given in eqn 2.3.

crystals obtainable. Some values of T_m are given in Table 2.1.

The significance of melting range and melting point is illustrated for linear and branched polyethylene in Figure 2.10. Note that for both polymers above T_m (115°C for BPE and 138°C for LPE), the volume increases linearly with T. As the temperature of the solid polymer is

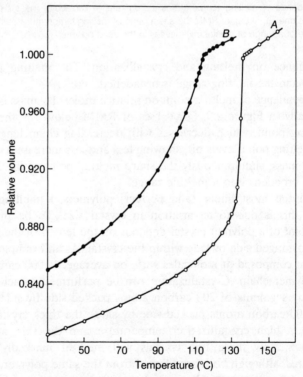

2.10   Plot of relative volume during heating against temperature T. The relative volume is the ratio of the specific volumes at T and T_m. Curve A is for linear polyethylene; curve B is for branched polyethylene (after Mandelkern).

increased, the volume of both polymers increases rapidly as T_m is approached, and equals the liquid volume at T_m. There are, however, significant differences in melting behaviour between the two polymers:

- LPE shows a much sharper melting point, 70% of the crystallinity disappearing in a 3–4°C interval;
- BPE melts over a wide temperature range, 60% of the crystallinity disappearing in a 40°C interval.

In fact, BPE will commence to premelt at temperatures as low as 20 or 30°C, and this is one reason why this polymer is not classed as an engineering polymer; other reasons are low melting point and low crystallinity—which yields a low modulus.

The rate of crystallization on cooling from the melt depends on two factors: **the rate of nucleation and the rate of crystal growth**. The maximum rate of crystallization occurs at a supercooling of approximately $0.2T_m$ (K). Values of the temperature T_1 of maximum rate of crystallization computed from this empirical rule,

$$T_1 = 0.8T_m,\tag{2.3}$$

are given in Table 2.1.

The relative size of the spherulite can be predicted from what is known of the crystallization of monatomic solids. Low-temperature solidification leads to small spherulites because the nucleation rate is high and the growth rate low. Conversely, high-temperature solidification leads to large spherulites because the nucleation rate is low relative to the growth rate.

2.6   The glass transition temperature

We first consider completely amorphous polymers, i.e. those polymers which do not crystallize even when cooled from the melt extremely slowly. The main effect of cooling the melt is to decrease the violence of the thermal agitation of the molecular segments. In the melt, segments of the molecules change place by thermally activated jumps. The number of jumps per second is very large indeed, well above 10^6 s^{-1}. If the cooling is continued, a temperature is reached at which the rate of segmental movement is extremely sluggish, and then, on further cooling, finally stops altogether. The polymeric specimen then consists of long molecules tangled in a liquid-like manner, but with a complete absence of the rapid molecular motion which is typical of a liquid. This is the glassy state. It is distinguished from the liquid state in one respect only: the immobility of the molecular backbones, which are frozen in crumpled conformations.

A simple manifestation of this cessation of molecular motion is seen in

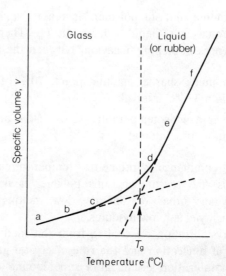

2.11 The glass transition temperature T_g is obtained from an experiment in which the specific volume *v* is measured whilst the specimen is cooled at a fixed rate, usually 1°C per minute. The construction for obtaining T_g from the data is shown. For polymer of low RMM, T_g marks the transition from glass to liquid; for high RMM, T_g marks the transition from glass to rubber.

the response of the specific volume to a change in temperature. In the glass (at a–b in Figure 2.11), thermal expansion is due to the increasing separation of the crumpled but immobile molecules. This, as in polymer crystals and in monatomic solids, has its origin in the anharmonicity of the secondary forces between molecules. The molecular conformation does not change: the molecules merely move further apart, one from another. **As the polymeric glass is heated through the temperature region in which segments of molecules commence to change position by discrete jumps, there appears an increase in thermal expansion coefficient (c–d in Figure 2.11).** On raising the temperature further, the expansion coefficient continues to increase until it finally attains the constant and high value of the liquid (e–f in Figure 2.11).

The glass transition temperature T_g is usually obtained from a volume–temperature plot of observations taken on cooling. The precise value of T_g depends slightly on the rate of cooling, being lower for lower rates of cooling. The usual cooling rate in observations of T_g is 1°C per minute. For example, for atactic polypropylene ($T_g = -19°C$) the temperature coefficients of volume are: for the glass just below T_g, $2.2 \times 10^{-4} \, \mathrm{K}^{-1}$; for the liquid $8.1 \times 10^{-4} \, \mathrm{K}^{-1}$. The coefficient for the glass is of the order of the crystal coefficient for the isotactic polymer at the same temperature. These coefficients are of the order of 20 times larger than the coefficients

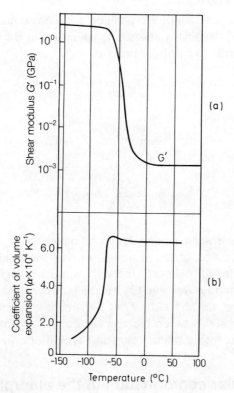

2.12 The glass–rubber transition in polyisobutylene:

$$\left[-CH_2 - \underset{\underset{CH_3}{|}}{\overset{\overset{CH_3}{|}}{C}} - \right]_n$$

(a) Dependence of shear modulus at frequencies ~1 Hz on temperature. (b) Dependence of coefficient of volume expansion on temperature. The glass transition measured in a volumetric experiment is −78°C.

for metals, a fact which has considerable significance in engineering design.

There are other manifestations of the glass transition. The most startling is a drop in modulus of the order of 10^3 (see Figure 2.12(a)). For this reason the glass transition temperature is sometimes loosely referred to as the softening point. The simplest way of observing the softening point is by means of a sharp, cone-shaped penetrometer. The penetrometer is forced into the polymeric solid with a constant load. Below T_g penetration is effected by the parting of immobile molecules, and the rate of penetration is low. However, above T_g the molecules flow and permit easy penetration.

The amorphous fraction of a crystalline polymer exhibits a glass transition. The ratio of the glass transition temperature to the melting point is observed empirically to be of the order of

$$\frac{T_g}{T_m} = 0.6. \tag{2.4}$$

For example, for isotactic polypropylene $T_m = 176°C$ and $T_g = -19°C$, so that

$$\frac{T_g}{T_m} = \frac{254\ K}{449\ K} = 0.57.$$

At temperatures between T_m and T_g the solid consists of rigid crystals and an amorphous fraction of low modulus, so that the solid is flexible and tough (see Chapter 3). Below T_g the solid consists of rigid crystals and a glassy, rigid amorphous fraction. Crystalline polymers find extensive application at temperatures above the T_g of the amorphous fraction: it is when the amorphous 'pads' between the crystals are rubbery that the polymeric solid exhibits that highly valued toughness typical of crystalline polymers.

2.7   Molecular conformation in the amorphous polymer

The reader will have noted that although we have made precise statements about the molecular conformation in the crystal, our description of molecular conformations in the amorphous fraction of crystalline polymers has been vague. There are two reasons for this. First, there is no precise physical technique available for determining the amorphous conformation.[1] Second, statistical methods are inapplicable for the short length of the molecule which is constrained between the crystals; for the long macromolecule in the completely amorphous state the position is quite different, as we now show.

Consider a single macromolecule, for example polyethylene (see Figure 2.13) in the planar zigzag, fully extended conformation. Now, rotation is possible around each main-chain C—C bond. It is as though each C—C bond were a flexible swivel. If the molecule is restrained within the crystal it will remain in the fully extended conformation, being caged in by its neighbours (see Figure 2.1(b)). But in the amorphous state (glass,

---

1 In recent years, however, neutron scattering has yielded evidence supporting the conclusions outlined in this section.

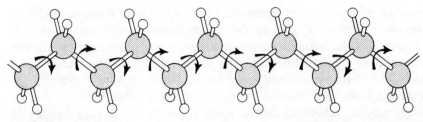

2.13 The rotation around C — C bonds which causes the polyethylene planar zigzag conformation to be changed into one of the enormous number of random conformations typical of the amorphous polymer.

rubber, or liquid) or in dilute solution the molecule assumes a random conformation.

Consider the first four carbon atoms C_1, C_2, C_3, and C_4 of a linear chain of n carbon atoms (see Figure 2.14). The valence angle is the same for each bond. Fix bond (1) at a particular orientation, say vertical. Then the position of carbon atom C_3 depends on the orientation of bond (2), which is specified by the angle ϕ_1 and which, for a random chain, can take any value between 0 and 2π. The position of atom C_4 depends not only on ϕ_2 but also on ϕ_1. Thus the position of the last carbon atom C_n of the macromolecule depends on angle ϕ_{n-2} and on all the other values of ϕ, of which there are $(n-3)$, each of which can take values between 0 and 2π. For values of n in the usual range 1000 to 100 000, the number of random molecular conformations becomes incomprehensibly large, increasing as an exponential function of n.

Nevertheless, despite the superficial complexity of the problem, the

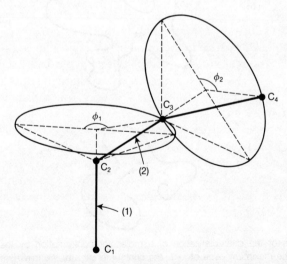

2.14 Rotational freedom in an idealized chain of four carbon atoms.

required information can be obtained statistically from a simple model. A specimen contains a vast number of molecules, which we denote N. Elementary calculation shows that in a cubic metre of amorphous polymer composed of identical molecules with $n = 10^4$ C atoms per molecule there are $N \approx 10^{24}$ molecules. Now at any instant in time it will be impossible to state the exact conformation of any one molecule. But, because N is so large, precise statements can be made about the relative probability of particular conformations. And since mechanical properties depend on averaging over all molecules, this is all the information required.

As a description or characterization of a molecular conformation we use the distance r between the ends of the molecule. If the molecule is in its most extended state then the extended or contour length L equals r (see Figure 2.15(a)). At the other extreme is the most crumpled conformation with the two ends coincident (see Figure 2.15(b)), where $r = 0$. These two extreme cases have a low probability of occurrence, a more likely state being that shown in Figure 2.15(c). Our intention is to obtain the average value of r.

The probability can be thought of physically in two ways, which appear at first sight different, but are in fact the same. Consider a cubic metre of

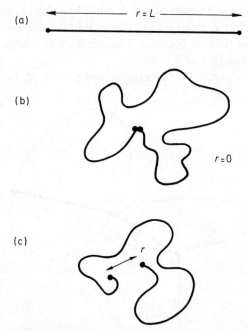

2.15 Illustration of the characterization of a molecular conformation by the end-to-end distance r: (a) the maximum value of $r = L$, the contour length; (b) the minimum value, $r = 0$, the ends coincident: (c) a more a priori likely value of r, $0 < r < L$.

liquid polymer comprising N linear molecules each of length L. In the liquid a particular macromolecule changes its conformation from instant to instant. Suppose we observe the values of r for each of the many conformations taken up be a particular macromolecule with the passage of time and compute a time average. In the second method we observe at one instant in time the average value of r for all N macromolecules in the specimen: we determine the instantaneous average for all macromolecules.

The result of both methods of averaging is the same for a polymer liquid. One macromolecule with the passage of time moves statistically through the whole gamut of conformations, which is also recorded by an instantaneous observation of all N macromolecules. Now for a glass, which is a frozen liquid, each molecule is set in a specific random conformation with a particular end-to-end distance r: its conformation and therefore its value of r do not change with time. In this case the only average we can form is the instantaneous average over all N macromolecules in the glass.

2.8 The freely jointed chain

In order to arrive at a physical estimate of r (the end-to-end separation) we use a simple mechanical model. The molecule is of length L and we consider it to be made up of n segments each of length l, joined flexibly one to another, so that

$$L = nl. \tag{2.5}$$

That is to say that the real polymer chain with its fixed valence angles is modelled by a hypothetical chain consisting of n rigid segments, each of length l, joined end-to-end in sequence but with absolutely no restriction on the angle between successive segments. The n segments are assembled commencing with segments 1 and 2 which are attached with angle θ_{12} between them (see 2.N.5). The angle θ_{12} lies between 0 and π and is selected randomly. Segment 3 is attached to segment 2, again with θ_{23} lying between 0 and π and selected randomly. This is continued until the nth segment is attached. The end-to-end vector r for this molecule is then

$$r = \sum_1^n l_i. \tag{2.6}$$

In order to obtain the scalar magnitude of r we form the dot product of each side with itself:

$$\begin{aligned} r^2 = r \cdot r &= (l_1 + l_2 + \cdots + l_n) \cdot (l_1 + l_2 + \cdots + l_n) \\ &= (l_1 \cdot l_1 + l_2 \cdot l_2 + \cdots + l_n \cdot l_n) \\ &\quad + 2(l_1 \cdot l_2 + l_1 \cdot l_3 + \cdots + l_n \cdot l_{n-1}). \end{aligned}$$

Since all the vectors $l_1, l_2, \ldots, l_n$ are of equal length l,

$$r^2 = nl^2 + 2l^2[\cos \theta_{12} + \cos \theta_{13} + \cdots + \cos \theta_{n,n-1}]. \tag{2.7}$$

Now this is the square of the end-to-end separation for one molecule. For all N molecules in the specimen (note $N \approx 10^{24}$, a much larger number than n, which for a macromolecule of normal size can be taken to be in the range 10^3 to 10^5) we can form the mean square of the end-to-end separation:

$$\langle r^2 \rangle = \frac{1}{N} \sum_1^N \{nl^2 + 2l^2[\cos \theta_{12} + \cos \theta_{13} + \cdots + \cos \theta_{n,n-1}]\}$$

$$= nl^2 + \frac{2l^2}{N} \sum_1^N [\cos \theta_{12} + \cos \theta_{13} + \cdots + \cos \theta_{n,n-1}]. \tag{2.8}$$

The following argument shows the second term on the right hand side to be zero. Consider first

$$[\cos \theta_{12} + \cos \theta_{13} + \cdots + \cos \theta_{n,n-1}],$$

which is one of the N terms and describes one particular molecule. The angles θ_{12}, θ_{13}, etc. are selected randomly in the range 0 to π. It follows that positive and negative values of $\cos \theta$ will occur with equal frequency. This summation (of a set of numbers in the range -1 to 1 distributed randomly positively and negatively) will yield a number which itself will be distributed equally positively and negatively around zero. It follows then, that for large N

$$\sum_1^N [\cos \theta_{12} + \cos \theta_{13} + \cdots + \cos \theta_{n,n-1}] = 0. \tag{2.9}$$

In which case the mean-square end-to-end distance is, from eqns 2.8 and 2.9,

$$\langle r^2 \rangle = nl^2. \tag{2.10}$$

The root-mean-square end-to-end distance, which we term R, is therefore

$$R = \langle r^2 \rangle^{1/2} = n^{1/2}l. \tag{2.11}$$

The ratio of R to the contour length is

$$\frac{R}{L} = \frac{\langle r^2 \rangle^{1/2}}{nl} = \frac{1}{n^{1/2}}. \tag{2.12}$$

Thus for $n = 10^4$, R is 1% of L. For a small molecule with $n = 10^2$, R is 10% of L. It is clear therefore that according to this model the molecule exists in a highly coiled state (2.N.6). This can be seen even more clearly if the root-mean-square distance of an element from the molecular centre of gravity is calculated; this quantity may be shown to equal $R/6$.

The coiled-up nature of the molecule may perhaps be illustrated best with a **macroscopic** example. Suppose the contour length to be 100 m and there are 10^4 segments each of length 1 cm. Then the root-mean-square end-to-end length is 1 m; the root-mean-square distance of an element from the centre of gravity of the model is 0.17 m. The molecules of course meander randomly and a few will be highly extended and others extended over a portion of their length, but the average molecule is coiled or crumpled over most of its length.

2.9 The Gaussian chain

The foregoing model gives a good physical picture of the flexible, randomly oriented molecule in the liquid or glass, but it has two weaknessess. First, there is some ambiguity concerning n: how many segments should be adopted in the model? Second, it doesn't lead to further analysis.

The Gaussian chain (or model) does not suffer from these weaknesses: the model assumes that the end-to-end separation of a macromolecule follows Gaussian statistics. It encompasses the freely jointed chain as a special case.

Consider a representative chain OA (see Figure 2.16) with a coordinate system attached at one end. Let the end-to-end vector of the chain be r and extend from the origin to the point (x, y, z), i.e.

$$r = ix + jy + kz \tag{2.13}$$

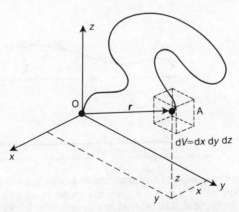

2.16 The statistical problem—what is the probability that for a chain, one end A lies in the volume element $dV = dx\,dy\,dz$ at r from the other end O?

The chain OA can take up an enormous number of different **conformations**, each characterized by a value of *r*. By conformation is meant a particular shape of the chain. The chain can be thought of in a simple manner, as a flexible string. When its ends are held at fixed *r* it can take up a certain number of conformations. Each value of *r* will have a specific probability: the greater the number of conformations for a particular *r*, the greater the probability of occurrence of that value of *r*. What is the probability that the chain displacement vector reaches from the origin to the point *r* and lies within the volume element $dV = dx\, dy\, dz$?

In order to see clearly in a simple way how the probability depends on *r*, consider first the behaviour expected of the chain if it is constrained artificially so that **end points O and A lie on the *x*-axis**. When the

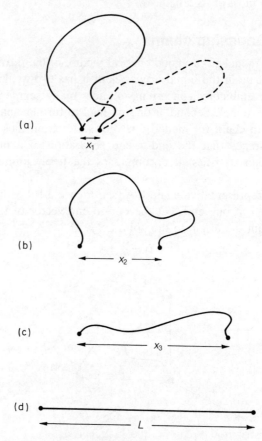

2.17 There are a certain number of conformations, or shapes, which the chain can take up with end-to-end separation x_1: two of the many are shown in (a). When the separation is increased to x_2 (b) the number of possibilities is less, and at x_3 (c) less again. In the limit, when the end-to-end distance equals the contour length *L*, there is only one conformation (d), the straight one.

end-to-end separation OA is x_1, much less than the contour length (see Figure 2.17), the chain can adopt a large number of conformations whilst keeping x_1 constant. For a larger end-to-end separation $x_2 \approx$ one half the contour length (see Figure 2.17), the number of conformations is less. For a value x_3, approaching the maximum length (which is the contour length of the chain), the number of possible conformations is lower again. Ultimately, when the separation between O and A equals the contour length, the chain is straight: **this conformation can be achieved in one way only, and has therefore an insignificant probability of occurrence.** At the other extreme, when O and A coincide, the number of conformations is at its greatest: the probability of occurrence of this value $x = 0$ is greater, therefore, than for any other value of x. This equating of the number of possible conformations and the probability rests on the entirely reasonable assumption that each conformation is a priori equally likely. If a particular value of x, say x_1, can be achieved by 10^3 times more conformations than x_2, and x_2 by 10^3 times more conformations than x_3, then, if the system is examined from time to time, it will be found that the frequency of occurrence of the values x_1, x_2, and x_3 will lie in the ratio $10^6/10^3/1$.

A function which models closely this behaviour in the region of interest is the Gaussian function

$$p(x) = \frac{\exp\left[-(x/\rho)^2\right]}{\sqrt{\pi}\,\rho}. \tag{2.14}$$

The quantity ρ is a representative length: a parameter of the model. The probability that the chain length lies between x and $x + \mathrm{d}x$ is, of course, linearly proportional to the magnitude of $\mathrm{d}x$. **So we define the probability of the end-to-end length lying between x and $x + \mathrm{d}x$ to be the product of $p(x)$ and $\mathrm{d}x$:**

$$p(x)\,\mathrm{d}x = \frac{\exp\left[-(x/\rho)^2\right]}{\sqrt{\pi}\,\rho}\,\mathrm{d}x. \tag{2.15}$$

The Gaussian function (eqn 2.14, see 2.N.7), is shown in Figure 2.23. It is a well-known bell-shaped curve, symmetrical about, and with a maximum at, $x = 0$. It has great application to any problem concerned with random processes—for example the distribution of rifle shots on a target. Even if the sight of the rifle is precise, errors due to the rifleperson will cause the shots to be distributed. Across a line such as AA' (see Figure 2.18(a)), the distribution of shots (number of shots per unit length) will be given by eqn 2.14. The number of shots in a length $\mathrm{d}x$ will be given by eqn 2.15. Note, however, that the number of shots in each of the rings is not a maximum in the bullseye (see Problem 2.17).

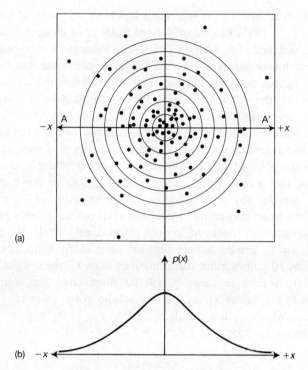

(a)

(b) −x ← ——————————— | ——————————— → +x

2.18 (a) The distribution of shots on a target. (b) The density of shots along a line such as AA' is proportional to $p(x)$, the Gaussian function (see eqn 2.14).

Returning to the three-dimensional problem posed in Figure 2.16, the Gaussian model gives the probability that the chain-end A lies between x and $x + dx$, y and $y + dy$, and z and $z + dz$ to be

$$P(x, y, z)\,dx\,dy\,dz = p(x)p(y)p(z)\,dx\,dy\,dz$$

$$= \frac{\exp - [(x^2 + y^2 + z^2)/\rho^2]}{(\sqrt{\pi}\,\rho)^3}\,dx\,dy\,dz \quad (2.16)$$

$$= \frac{\exp\left[-(r/\rho)^2\right]}{(\sqrt{\pi}\,\rho)^3}\,dx\,dy\,dz. \quad (2.17)$$

It will be seen that expressions similar to eqn 2.15 have been employed to describe the dependence of the probability on y and z as well as on x (see 2.N.7).

A major virtue of the Gaussian function is that it permits easy mathematical manipulation. For example, for a set of Gaussian chains the mean-square end-to-end distance is (see 2.N.7)

$$\langle r^2 \rangle = \tfrac{3}{2}\rho^2. \quad (2.18)$$

It will be recalled that for the flexible rod model of Section 2.8,

$$\langle r^2 \rangle = nl^2. \tag{2.19}$$

The constant ρ in the Gaussian function can thus be equated

$$\rho^2 = \tfrac{2}{3}nl^2, \tag{2.20}$$

$$\rho = \left(\frac{2n}{3} \right)^{1/2} l. \tag{2.21}$$

A further virtue of the Gaussian function is that it is frequently found to be a solution to problems involving random processes. The behavior of the flexible chain undergoing violent liquid-like motions is one such random process.

2.10 Molecular orientation

The Gaussian chain provides an excellent description of conformations in unperturbed polymer liquids and in the glasses obtained by cooling them.

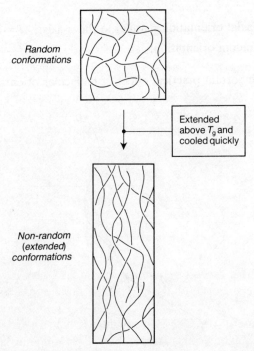

2.19 An amorphous polymer is extended above the glass transition and then quenched to the glassy state. The resulting conformations are no longer random: there is 'frozen-in' molecular orientation, which remains when stress is removed.

In real polymer products, however, the molecules often do not have the random shape of a Gaussian chain. Forces exerted on the liquid during forming cause the molecules partially to align. If this is followed rapidly by cooling, before the long molecules have time to wriggle back to random conformations, the cold, solid polymer is left with a frozen-in molecular orientation. Similar orientation can also be induced by plastic deformation in the solid state (Chapter 5). The case of an amorphous polymer is sketched in Figure 2.19.

It is useful to have a measure of the degree to which molecules are preferentially oriented. This is obtained by averaging the orientation of the elementary rigid units of which the molecules are constructed. Consider Figure 2.20. The segment AB of the molecule CD is inclined at angle ϕ to a reference axis—here the z-axis. The best measure of the mean degree of orientation is expressed in terms of $\overline{\cos^2\phi}$ (the value of $\cos^2\phi$ averaged over all segments of all molecules in the specimen). It is known as the **orientation factor** f and is defined by

$$f = \tfrac{1}{2}\left(3\,\overline{\cos^2\phi} - 1\right). \qquad (2.22)$$

This gives it the properties

- perfect uniaxial orientation parallel to the z-axis: $f = 1$; and
- perfectly random orientation: $f = 0$.

The really important practical effect of molecular orientation is that an

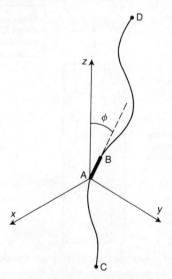

2.20 Section AB of molecule CD is inclined at angle ϕ to the z-axis.

oriented polymer is **anisotropic**. That is, if we exclude scalar properties (density, specific heat, etc.), we must expect all other physical properties to depend on the direction of testing. A simple general rule applies, as the degree of molecular orientation is increased:

- directions which molecular units are tending to rotate **towards** increasingly reflect the properties of covalent bonds (high stiffness, high strength, low thermal expansion, etc.).
- directions which molecular units are tending to rotate **away from** increasingly reflect the properties of van der Waals bonds (low stiffness, low strength, high thermal expansion...).

For example, consider the effects of orientation on the refractive index. Suppose molecules are preferentially aligned parallel to the z-axis (Figure 2.20). Then when plane-polarized light is passed through the polymer in the y-direction, different refractive indices are obtained when the plane of polarization lies parallel or perpendicular to the z-axis: $n_\parallel$ and $n_\perp$ respectively. For most polymers $n_\parallel > n_\perp$. The difference Δn is known as the birefringence:

$$\Delta n = n_\parallel - n_\perp \tag{2.23}$$

As might be expected, Δn and f are closely connected. With increase in f there is also an increase in birefringence. In fact they are linearly related:

$$\Delta n = \Delta n^0 f, \tag{2.24}$$

where Δn^0 is the limiting birefringence when all molecules are perfectly aligned, and is specific to any particular chemical structure. For this reason, and because polymers are often transparent to visible light, the birefringence is widely used in practice as a highly convenient measure of the degree of molecular orientation. The polymer moulder often wishes to minimize orientation in order to obtain superior properties, and this means minimizing birefringence. Low birefringence is required also in **high technology optical products**, of which the digital audio compact disc is the best example (see Chapter 8).

In an oriented **crystalline** polymer, crystals and amorphous regions each possess molecular orientation and have separate orientation factors f_c and f_a which, in general, differ. The measured birefringence is then the sum of contributions from both. It is usually found that crystals orientate faster than amorphous regions during extension, i.e. $f_c > f_a$. The structure of an oriented crystalline polymer is not spherulitic. Crystals are smaller in lateral extent, and are stacked into long, thin fibrillar units known as **fibrils**. A sketch is given in Figure 2.21. The fibrils are only weakly bonded together and may even be separated by occasional cracks.

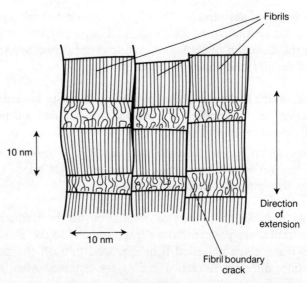

2.21 Fibrillar structure of an oriented crystalline polymer.

Great improvements in unidirectional mechanical properties are possible by converting crystalline polymers into oriented structures of this type. Many everyday polymer products owe their success to this fact (2.N.8). For example isotropic poly(ethylene terephthalate) (1.N.10) has a Young's modulus and tensile strength of 2.5 GPa and 70 MPa, respectively, at 20°C; oriented fibres of the same polymer used in tyre reinforcement have values of 14 GPa and 830 MPa. Orientation is introduced into a crystalline polymer by one of two routes:

- extension in the molten state followed by rapid crystallization of the oriented melt; or
- crystallization in spherulitic form followed by plastic extension (drawing) in the solid state.

In some products it is desirable to have enhanced strength in all directions in a **plane** instead of only in a single direction; the orientation must then be biaxial instead of uniaxial. This is achieved in a similar fashion as for uniaxial orientation, but the orientation process consists of simultaneous or sequential extension in two perpendicular directions (2.N.9).

Since molecular orientation is 'frozen-in', it follows that it can be 'unfrozen' by raising the temperature. An oriented amorphous polymer on being heated through its glass transition re-enters the rubbery state. The now mobile molecules recoil to random conformations, losing their orientation, and the polymer is seen to shrink back to approximately its original dimensions. Similar **thermal shrinkage** occurs on heating an oriented crystalline polymer, but in this case it is necessary to melt crystals before constraint on the amorphous fraction is totally released and complete shrinkage is obtained (2.N.10).

Notes for Chapter 2

2.N.1

PTFE was discovered in 1938 by R. J. Plunkett at the du Pont Jackson Laboratory, who insisted on cutting open a tetrafluoroethylene cylinder; the gas in the pressurized cylinder had apparently leaked, but no leak could be detected. There was in fact no leak: the pressure had dropped because impurities in the cylinder wall had catalyzed the polymerization of the gas. It is a most valuable plastic with a high melting point (327°C) and extremely low coefficient of friction: much used in highly corrosive circumstances in chemical plant, such as in gaskets, diaphragms, rings, tubing and taps, and in demanding electrical equipment such as insulators.

2.N.2

See, for example, J. D. Watson, *The Double Helix*. New American Library, New York (1969).

2.N.3

The isotactic crystalline form of polybutene is used in pipes and tubes. It has not found wide application as yet.

2.N.4

There is evidence that some molecules fold back directly into the crystal without entering the amorphous fraction. There is as yet no clear evidence on the extent of this phenomenon in normal specimens. It is a common phenomenon in crystals crystallized from polymers dissolved in dilute solution.

2.N.5

It may be helpful, in introducing this topic, to consider the ways in which a carpenter's rule may be opened. One such opening is illustrated in Figure 2.22 for a rule comprising four segments each of length l. Clearly, the end-to-end distance is r:

$$r = l_1 + l_2 + l_3 + l_4 = \sum_{i=1}^{4} l_i.$$

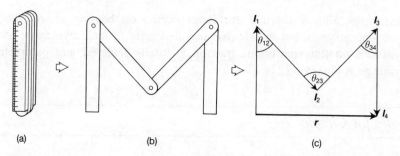

2.22 One way of opening a carpenter's rule. The end-to-end vector *r* is a description of the conformation.

If the rule is opened randomly many times and a record of |*r*| is kept it will be found that the average value of |*r*| will be *R*, which is given by (see eqn 2.11)

$$R = 4^{1/2}l = 2l.$$

An equivalent statement to 'opened randomly' is that θ_{12}, θ_{23}, and θ_{34} for each opening, are each selected randomly in the range 0 to π.

2.N.6

The problem tackled here is a celebrated one and applies to many phenomena; it comes under different names: random flight, random walk, drunkard's walk, etc. Suppose a drunken man in the centre of a large field takes *n* steps each of length *l*, and suppose that there is no correlation whatsoever between the direction of successive steps; how far does he travel on average from the point at which he started? It follows from eqn 2.10 that the mean square distance moved is

$$\langle r^2 \rangle = nl^2.$$

Suppose each step is $l = 1$ m and he takes 1 step per second, then after one hour $n = 3600$ and the total (contour) distance he has walked is 3600 m. But his average distance from the starting point is much less and is given

$$\langle r^2 \rangle = 3600 \times 1^2 \text{ m}^2.$$

So the rms distance moved is

$$R = \langle r^2 \rangle^{1/2} = 60 \text{ m}.$$

Note this is the average or expected distance: a single drunken man observed walking for one hour would not be likely to move precisely this distance.

2.N.7

The Gaussian (or normal) function has extremely wide application in engineering, and in statistics. First, it permits easy mathematical manipulation. For example, suppose we require to know for a Gaussian chain the most probable value of r irrespective of direction. The probability that the chain lies in the volume dV

$$dV = 4\pi r^2 \, dr,$$

between r and $r + dr$ is $P(r) \, dV$, where

$$P(r) = \frac{\exp - (r/\rho)^2}{(\sqrt{\pi}\rho)^3}. \qquad (2.N.7.1)$$

Hence this probability is

$$P(r) \, dV = \left[\frac{4\pi}{(\sqrt{\pi}\rho)^3} \right] r^2 \exp\left[-(r/\rho)^2 \right] dr. \qquad (2.N.7.2)$$

The function $4\pi r^2 P(r)$ is plotted against r in Figure 2.23. The most probable value of r is found by differentiating eqn 2.N.7.2, and this occurs at

$$r = \rho. \qquad (2.N.7.3)$$

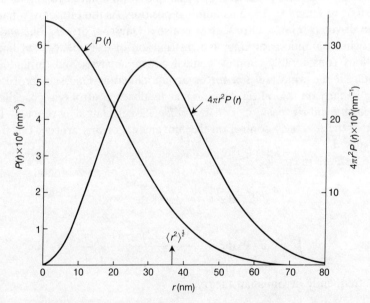

2.23 The Gaussian distribution functions of the end-to-end vector $\boldsymbol{r}$, $P(r) = (\sqrt{\pi}\rho)^{-3} \exp[-(r/\rho)^2]$ and $4\pi r^2 P(r)$ (see eqn 2.N.7.2) plotted against r for a polymer chain with $\rho = 30$ nm.

As a second illustration of the simplicity of the Gaussian function we calculate the mean-square value of r. It is obviously

$$\langle r^2 \rangle = \frac{\int_0^\infty r^2 P(r) 4\pi r^2 \, \mathrm{d}r}{\int_0^\infty P(r) 4\pi r^2 \, \mathrm{d}r}. \qquad (2.\mathrm{N}.7.4)$$

Now the denominator equals unity. Hence

$$\langle r^2 \rangle = \frac{4\pi}{(\sqrt{\pi}\rho)^3} \int_0^\infty r^4 \exp\left[-(r/\rho)^2\right] \mathrm{d}r$$

$$\langle r^2 \rangle = \tfrac{3}{2}\rho^2. \qquad (2.\mathrm{N}.7.5)$$

The straightforward mathematical manipulation is only one reason why the Gaussian distribution has found wide application.

An equally necessary reason is that the Gaussian distribution is frequently an exact solution to a **stochastic** problem. For example, in the tossing of a fair coin, as the number of tosses increases, the Gaussian distribution becomes an increasingly precise description. Or, for the polymer chain, when it is not highly extended and r is considerably less than the contour length, the Gaussian distribution is a useful solution.

Gaussian models are also of great utility in the *control* of systems perturbed by random processes. For example, wind and sea forces tend to move a deep-sea oil rig from its nominal position. As the output of a linear system driven by Gaussian processes is also a Gaussian process, the widely used twin assumptions of Gaussian distribution of disturbance and linear equations considerably simplify control system analysis and design. For example, if the wind and sea forces are modelled as Gaussian processes acting linearly on the oil rig, and a linear feedback control (via propellers) is used, the oil-rig position is a two-dimensional Gaussian process. The effect of the feedback control on the movements of this process can then be analysed.

The definite integrals

$$\int_0^\infty e^{-a^2 x^2} \, \mathrm{d}x = \frac{1}{2a} \sqrt{\pi}$$

$$\int_0^\infty x^{2n} e^{-ax^2} \, \mathrm{d}x = \frac{1.3.5\ldots(2n-1)}{2^{n+1}a^n} \sqrt{\frac{\pi}{a}}$$

occur frequently in Gaussian theory.

2.N.8

Many of the commodities we take for granted are uniaxially oriented crystalline polymers. Those we use most are the synthetic fibres—nylon, polyester, polypropylene, and others which are used in clothing, carpets, ropes, and tyre reinforcements. An interesting technical development is the current industrial trend towards extruding synthetic fibres (spinning) at incredibly high rates. Fibres are wound up at velocities of $100-300\ \mathrm{km\,h^{-1}}$ in high-speed spinning. This is an unusual coincidence between the requirements of product quality and process economics. High wind-up speeds cause the fibres to be highly extended at the moment of solidification, and hence to have a high degree of frozen-in molecular orientation. High speeds also mean the product is produced more cheaply: the costs of running plant for a given time are spread over a greater tonnage of fibre.

2.N.9

The great success of polymers as packaging films stems from their ability to be strengthened in a plane by biaxial extension during manufacture. In the majority of cases this is achieved by extruding the film as a tube and then inflating it in the molten state causing it to be biaxially extended (Chapter 7). Biaxially oriented polymer films also play critical roles in the electronics industry, as polyester audio and video cassette tapes and computer floppy disks. Here, an overriding requirement is dimensional stability, which implies high stiffness and low thermal expansion in the film plane. These properties are ensured by biaxially stretching the films following extrusion. The successful use of polyester bottles for carbonated drinks was made possible by the development of a special blow-moulding procedure which ensures that the polymer in the wall of the bottle is biaxially oriented. With uniaxial orientation alone, there is the danger of bottles splitting open—an explosive event because of the compressed gas inside.

2.N.10

Thermal shrinkage, accompanying release of moulded-in orientation, can be the cause of severe warping of polymer mouldings when heated, but desirable effects can also be achieved. Shrink-wrap polymer films have caused a minor revolution in methods of packaging. Heat-shrinkable polymer tubes and cable insulation are currently finding wide application.

Problems for Chapter 2

The following values of RMM should be used in these questions: $H = 1$; $C = 12$; $N = 14$; $O = 16$; $F = 19$; $Cl = 35.5$; Avogadro's constant $N = 6.023 \times 10^{23}\ \mathrm{mol^{-1}}$.

2.1   Values of $|a|$, $|b|$, and $|c|$ are listed below, in order, for a set of crystalline polymers which form orthorhombic unit cells, i.e. $\alpha = \beta = \gamma = 90°$, where α, β, and γ are the angles between the b and *c*, *a* and *c*, and *a* and *b* axes, respectively. Calculate the density of each crystal. The chain axis is marked[†].

(1) Polyethylene: 736, 492, 254[†] pm, $2[-CH_2-CH_2-]$ units per cell.

(2) Syndiotactic PVC: 1040, 530, 510[†] pm, $4[-CH_2-CHCl-]$ per cell.

(3) Poly(vinyl fluoride): 857, 495, 252[†] pm, $2[-CH_2-CHF-]$ per cell.

(4) Poly(vinylidene fluoride): 847, 470, 256[†] pm,

$$2[-CH_2-CF_2-]$$

per cell.

(5) Natural rubber: 1246, 886, 810[†] pm,

$$8[-CH_2-CH=C-CH_2-]$$
$$\qquad\qquad\quad |$$
$$\qquad\qquad\quad CH_3$$

per cell.

2.2 Polypropylene and nylon 6 form monoclinic crystals, in which $\alpha = \gamma = 90° \neq \beta$. Calculate the densities of these crystals. The chain axis is marked[†].

(1) Polypropylene: $|a| = 666$ pm; $|b| = 2078$ pm; $|c| = 650^{†}$ pm; $\beta = 99.6°$.

$$12[-CH_2-CH-]$$
$$\qquad\qquad |$$
$$\qquad\qquad CH_3$$

per cell.

(2) Nylon 6: $|a| = 956$ pm; $|b| = 1724^{†}$ pm; $|c| = 801$ pm; $\beta = 67.5°$.

$$8[-(CH_2)_5-C-N-]$$
$$\qquad\qquad || \quad |$$
$$\qquad\qquad O \ \ H$$

per cell.

2.3 Calculate the densities of nylon 6.6 and poly(ethylene terephthalate) (PET) crystals, which are triclinic (i.e. $\alpha \neq \beta \neq \gamma$).

(1) Nylon 6.6: $|a| = 490$ pm; $|b| = 540$ pm; $|c| = 1720^{†}$ pm; $\alpha = 48.5°$; $\beta = 77°$, $\gamma = 63.5°$. Unit cell contains a single

$$\left[-(CH_2)_6-N-C-(CH_2)_4-C-N-\right]$$
$$\qquad\qquad\quad | \ \ || \qquad\qquad\ || \ \ |$$
$$\qquad\qquad\quad H \ O \qquad\qquad\ O \ H$$

group.

(2) PET: $|a| = 456$ pm; $|b| = 596$ pm; $|c| = 1075^\dagger$ pm; $\alpha = 98.5°$; $\beta = 118°$, $\gamma = 112°$. Unit cell contains a single

$$\left[-CH_2CH_2-O-\underset{\underset{O}{\|}}{C}-\hspace{-2pt}\bigcirc\hspace{-2pt}-\underset{\underset{O}{\|}}{C}O- \right]$$

unit.

2.4 Bragg's law states that radiation of wavelength λ is diffracted through an angle 2θ when $n\lambda = 2d \sin \theta$, and the diffracting plane is at an angle θ to the incident beam. In this equation, n is an integer and d is the spacing between planes. Calculate the diffraction angle θ for first-order reflections ($n = 1$) from the (110) and (200) planes of polyethylene for copper K_α radiation ($\lambda = 154$ pm).

2.5 Syndiotactic PVC adopts a planar zigzag conformation when it crystallizes (crystallinity is usually small in PVC). Using van der Waals diameters of 239 pm for hydrogen and 356 pm for chlorine, and bond lengths of 107 pm for the C—H bond and 176 pm for C—Cl, make scale drawings of the PVC molecule in this conformation, in the style of Figure 2.1(a), showing clearly the positions of the chlorine atoms. Hence show that the repeat distance along the chain direction of the unit cell must be at least 508 pm. Is any valence angle distortion necessary to accommodate the large chlorine atoms?

2.6 The melting points of linear $C_x H_{2x+2}$ molecules can be fitted to the empirical equation

$$T_m \text{ (K)} = 1000/(2.4 + 17x^{-1}).$$

Plot graphs of T_m against x, and of $1000 T_m^{-1}$ against x^{-1}, using this equation, and compare the observed values of T_m listed below with both curves. Hence find T_m for high molecular weight polyethylene.

x	8	10	12	14	16	24	32	70
T_m (°C)	−56.8	−29.7	−9.7	2.5	14.7	47.6	67	104

2.7 (1) Calculate the mass fraction crystallinity of polyethylene samples of densities 926, 940, and 955 $kg\,m^{-3}$, using values of v_a and v_c given in Section 2.4. (2) Calculate the crystallinity of a polyoxymethylene (POM) sample of density 1410 $kg\,m^{-3}$, given that the densities of crystalline and amorphous phases are 1506 and 1250 kg/m^3, respectively.

2.8 The Clausius–Clapeyron equation states that the equilibrium temperature (expressed in K) at which a phase change takes place under pressure p is given by the equation

$$\frac{\mathrm{d}p}{\mathrm{d}T} = \frac{\Delta H}{T \Delta V},$$

where δH is the latent heat associated with the phase change, and ΔV is the corresponding volume change. Calculate T_m for polyethylene in an injection moulding machine under a hydrostatic pressure of 80 MPa. Take $\Delta H = 7.79$ kJ per mole of $-CH_2-CH_2-$ units, and $T_m = 143.5°C$ at 1 atmosphere.

2.9   The melting point of a polymer containing lamellae of thickness L is given by

$$T_m \ (\mathrm{K}) = T_m^\circ (1 - 2\gamma_s / L \Delta H_f),$$

where T_m is the melting point of the infinitely thick crystal, γ_s is the specific free energy of the lamellar surface, and ΔH_f is the enthalpy of fusion per unit volume. Calculate T_m for polyethylene containing lamellae 12.0 nm thick. Take $T_m^\circ = 416.7$ K, $\gamma_s = 0.0874$ J m^{-2}, and $\Delta H = 279$ MJ m^{-3}

2.10 Experiments were carried out to measure x_{max}, the maximum crystallinity, in a series of random copolymers of ethylene and propylene. Linear PE gave $x_{max} = 95\%$. Introduction of 4 CH_3- groups per 100 main-chain carbon atoms reduced x_{max} to 50%, and at 20 CH_3- groups per 100 main chain carbons $x_{max} = 0\%$. Calculate the weight fraction of propylene in the copolymer in order to obtain 50% and 0% crystallinity.

2.11 The glass transition T_g (K) of a random copolymer is given to a good approximation by

$$\frac{1}{T_g} = \frac{w_1}{T_{g_1}} + \frac{w_2}{T_{g_2}},$$

where w_1 and w_2 are weight fractions of the comonomers, and T_{g_1} and T_{g_2} are glass transition temperatures of the corresponding homopolymers. Calculate T_g for the non-crystalline ethylene–propylene copolymer described in Problem 2.10, taking $T_g = -120°C$ for PE and $T_g = -19°C$ for PP.

2.12 Draw sketches of log(Young's modulus) against temperature for each of the following polymers, including as much quantitative information as possible:
(1) polystyrene ($T_g = 100°C$)
(2) polybutadiene ($T_g = -100°C$)
(3) a random copolymer containing 25 wt% styrene and 75 wt% butadiene;
(4) a block copolymer of the same composition as (3). Note: the polystyrene and polybutadiene chains form separate phases.

2.13   Draw sketches of log(Young's modulus) against temperature for each of the following polymers, including as much quantitative information as possible:

(1) non-crystalline ethylene-propylene random copolymer (see Problem 2.11)

(2) nylon 6.6 ($T_g = 50°C$; $T_m = 265°C$; 35% crystalline)

(3) nylon 6.6 after prolonged immersion in water.

2.14   A linear polyethylene molecule has RMM = 84 000.

(a) Calculate the length of the fully extended chain, assuming no distortion of valence angles or bond lengths.

(b) Calculations based on a fixed valence angle of 109.5°, and completely free rotation about each bond give the result that a linear PE molecule containing y C—C bonds is equivalent to a freely jointed chain made up of $y/3$ segments. Using this result, find the root-mean-square end-to-end distance R of the PE molecule, and the rms distance of chain elements from the centre of gravity of the chain.

2.15   Rotation about C—C bonds does not occur freely. In PE, the carbon atoms are at their lowest energy in the planar zigzag conformation (all-*trans* structure). Rotation through 120° in either direction brings the carbon atom to another stable position, which was previously occupied by hydrogen (*gauche* position).

(1) Write down possible stable configurations of the following linear molecules, using the notation t, g^+, and g^-: C_4H_{10}, C_5H_{12}, and C_6H_{14}. Note that some rotations produce identical structures; these should not be counted twice.

(2) Calculate the number of possible conformations of a linear PE molecule of RMM = 70 000.

2.16   The Maxwell–Boltzmann equation states that the equilibrium population n_i occupying energy state ε_i at temperature T (K) is given by

$$\frac{n_i}{\sum n_i} = \frac{\exp(-\varepsilon_i/RT)}{\sum \exp(-\varepsilon_i/RT)},$$

where R is the gas constant. Taking $\varepsilon_i = 0$ for the *trans* conformation, and $\varepsilon_i = 2.1$ kJ mol^{-1} for gauche conformations, calculate the fraction of gauche structures in linear PE at 150°C.

2.17   A target of the type shown in Figure 2.18 is marked with 5 concentric rings of radius $N\rho/2$, where $N = 1, 2, 3, 4, 5$. The distribution of shots is given by eqn 2.14 both for the x direction and (after substitution of y for x) in the vertical direction. Calculate the fraction of shots striking each ring of the target.

2.18   Plot the radial distribution function $4\pi r^2 P(r)$ against r/ρ for a linear polyethylene molecule of RMM = 63 000, taking $l = 381$ pm, and $n = \frac{1}{3}$ (number of C—C bonds). Hence fins the value of r $(r > \rho)$ at which the distribution function falls to a tenth of its maximum value.

2.19 Consider the model of a polymer molecule consisting of n rods of length l joined flexibly as shown above. Let the total length be L and let force F applied at the ends extend the assembly to a length $L + \Delta L$. The extension is achieved by rod elongation and by an increase in the angle α. Show that

$$\frac{\Delta L}{F} = n\left[\frac{\cos^2\theta}{k_l} + \frac{l^2\sin^2\theta}{4k_\alpha}\right].$$

in which the force constants k_l and k_α are defined by

$$\Delta W_l = \tfrac{1}{2}k_l\,\Delta l^2,$$

$$\Delta W_\alpha = \tfrac{1}{2}k_\alpha\,\Delta\alpha^2.$$

ΔW_l is the energy needed to elongate a rod by Δl, and ΔW_α is the energy needed per bond to open the bond angle α by $\Delta\alpha$. If A is the effective cross-sectional area of the chain (in a plane perpendicular to F) show that the Young's modulus of the chain is,

$$E = \frac{l\cos\theta}{A\left[\dfrac{\cos^2\theta}{k_l} + \dfrac{l^2\sin^2\theta}{4k_\alpha}\right]}.$$

Calculate E using the following constants characteristic of the polyethylene molecule

$$A = 0.182\ \text{nm}^2,\ l = 153\ \text{pm},\ \theta = 34°,\ k_l = 436\ \text{N}\,\text{m}^{-1},$$

and

$$\left(\frac{k_\alpha}{l^2}\right) = 35\ \text{N}\,\text{m}^{-1}.$$

Comment on the result.

2.20 Draw a sketch showing the effect of increasing uniaxial drawing upon the X-ray diffraction pattern produced by the (200) plane in polyethylene.

2.21 The birefringence Δn, of an oriented polymer containing volume fraction x of crystals follows the equation

$$\Delta n = x f_c \, \Delta n_c^{\circ} + (1-x) f_a \, \Delta n_a^{\circ},$$

where f_a and f_c are orientation factors (determined from sonic modulus and X-ray diffreaction measurements) of the amorphous and crystalline phases, and Δn_a° and Δn_c° are birefringences for the fully oriented phases. Calculate Δn_a° and Δn_c° for poly-propylene from the data on uniaxially drawn samples given in Table 2.2.

Table 2.2 Information for Problem 2.21

Draw ratio	1.2	1.5	4	5	6
x	0.65	0.60	0.60	0.62	0.60
f_a	0.05	0.06	0.27	0.28	0.46
f_c	0.06	0.15	0.71	0.87	0.92
$1000 \, \Delta n$	2.2	4.0	19	22	27

2.22 When an isotropic rubber is extended uniaxially to a draw ratio λ ($= L/L_0$), chain units at an initial angle ϕ_1 to the draw direction rotate to ϕ_2. Assuming that deformation takes place at constant volume, and that the units rotate in the same way as lines joining pairs of points in the rubber, without change in the length of the unit, show that ϕ_1 and ϕ_2 are related by

$$\tan \phi_2 = \frac{\tan \phi_1}{\lambda^{3/2}}$$

2.23 Using the equation given in Problem 2.22, find the orientation factor f for a rubber stretched uniaxially to $\lambda = 3.0$.

3 *The elastic properties of rubber*

3.1 Introduction

Natural rubber is found in the latex of the tree *Hevea braziliensis*; its
principal constituent is polyisoprene:

$$\left[CH_2-\underset{\underset{CH_3}{|}}{C}=CH-CH_2 \right]_n.$$

Polyisoprene occurs in other plants, some of which have been of commercial significance, such as the Russian dandelion and the Mexican guayule
shrub. The well-known properties of natural rubber are displayed only at
temperatures above the glass transition temperature, which is −70°C;
below that temperature it is a brittle glass. Every amorphous polymer,
whether synthetic or natural, providing it is cross-linked,[1] displays comparable properties above its glass transition temperature. These properties
are, stated briefly:

- **extremely high extensibility, frequently up to ×10, generated by low
  mechanical stress**; and
- **complete recovery after mechanical deformation.**

To these must be added a third property, which is not so well known:

- **high extensibility and recovery are due to deformation induced changes
  in entropy.**

Since the third property is the basis of the other two unique properties of
rubber-like materials, we commence with a description of how the entropy
is changed when a specimen of rubber is deformed. From this understanding the stress–strain characteristics can be obtained. The chapter con-
cludes with an outline description of the rubbers of interest to technology.

---

1 If the polymer is not cross-linked it will, at temperatures above T_g, exhibit liquid-like
properties unless it is of extremely high relative molecular mass, in which case it will exhibit
properties comparable with those of a chemically cross-linked rubber.

3.2 Structure of an ideal rubber

The ideal, classical, rubbery properties are displayed by polymers cross-linked by valence bonds (main-chain bonds or sulphur bridges, see Introduction, Figure 0.1); these are termed chemical cross-links. Physical cross-links are also important in many useful rubbery materials. In physical cross-linking the chains are not chemically attached one to another, but are effectively pinned together in one of three ways:

- by the relative molecular mass being so high, the chains become grossly entangled—these entanglements act as physical cross-links;
- by the chains entering and leaving crystals—the chains are pinned together in the crystal (see Chapter 2); or
- by the chains entering rubbery domains from glassy domains (see 1.N.12).

The understanding of these physically cross-linked materials stem from the theory of rubber elasticity, which is based on a chemically cross-linked polymer, and it is to this simple idealized model that we confine attention.

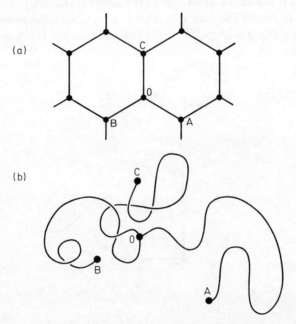

3.1 A rubber is a cross-linked net of polymer chains. (a) The network is laid out for illustrative purposes. The contour lengths of the chains are assumed to be equal (for example, OA = OB). (b) In the rubber the network is a random tangle. The space between the tie points OABC is filled with other sections of the net, including other tie points. The chain between tie points (e.g. OA) is termed a sub-molecule.

An ideal rubber consists of flexible cross-linked polymer chains under-going violent liquid-like motions. No matter what the type of cross-linking, physical or chemical, the elastomers all have this in common: the macro-molecules between cross-links undergo extremely rapid molecular move-ment.

We derive the relationship between stress and strain for a specimen consisting of ν chains, each chain being cross-linked into the network at both ends. (The specimen is the network and vice versa.) Figure 3.1(a) shows part of the molecular network laid out flat like a fishing net for illustrative purposes. In a rubber, the net is crumpled up; part of the crumpled network is shown in Figure 3.1(b). Three chains, OA, OB, and OC, are cross-linked at tie point O. The points A, B, and C are also tie points in the network. That three chains meet at a point, rather than, say, four or five, is of no significance to the theory. The essential fact is that the specimen contains ν chains of equal contour length, each of which is joined to the network at both ends.

The behaviour of the specimen as a whole follows from an understand-ing of the behaviour of a representative chain. Imagine it **detached from the network** with one end placed at the origin of a coordinate system (see Figure 3.2). It will be assumed that its randomness is exactly reproduced by the Gaussian theory (Section 2.9). That is to say, with the passage of time the molecule changes shape in a truly random manner. It follows from eqn 2.N.7.1 that

$$P(r) = \frac{\exp-(r/\rho)^2}{(\rho\sqrt{\pi})^3}. \qquad (3.1)$$

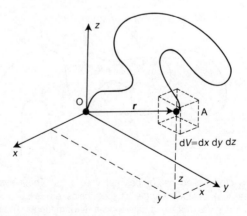

3.2 The representative chain OA detached from the network with a coordinate system at one end.

For this detached Gaussian chain, the mean-square end-to-end distance averaged over time is (see eqn 2.N.7.5)

$$\langle r^2 \rangle_o = \tfrac{3}{2}\rho^2. \tag{3.2}$$

The subscript signifies that the average over r^2 is taken with the chain **out** of the network. Since this chain is the same as all the other ν chains—if they too were examined detached from the network—the value of $\langle r^2 \rangle_o$ averaged over all chains is $(3/2)\rho^2$.

Imagine the chains now transferred back into the specimen and cross-linked, to reform the network (3.N.1). In order that the chains may be packed together to reform the specimen at volume V, the end-to-end distances have to be changed. The value of $\langle r^2 \rangle$, **when averaged over the assembled chains cross-linked into the network**, we signify by $\langle r^2 \rangle_i$ ('i' stands for 'in'). That is to say, when in our thought experiment, we move the un-cross-linked free chains (mean value of r^2, $\langle r^2 \rangle_o$) into the network, some constraint has to be imposed on the molecules, so that in the specimen the mean value of r^2 becomes $\langle r^2 \rangle_i$.

The value of $\langle r^2 \rangle_i$ is **not** an intrinsic property of the molecule but depends on the specimen volume V. If V is changed (by heating, for example, or by absorption of diluent (3.N.2)) then $\langle r^2 \rangle_i$ changes accordingly; $\langle r^2 \rangle_o$ is an intrinsic property of the macromolecule and does not change when the volume changes.

3.3 Entropy elasticity

In this section we explain how the conformational probabilities of the polymer chains contribute to the entropy of the rubber and, from this, how

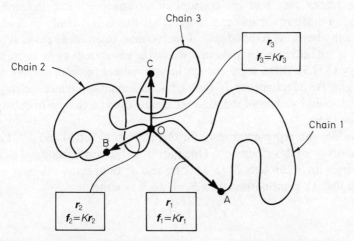

3.3 The point O is separated from the tie points A, B, and C by vectors r_1, r_2, and r_3. Three forces f_1, f_2, and f_3 act along r_1, r_2, and r_3. The forces are entropy-generated and linear in r, with force constant $K = (2kT/\rho^2)$.

the specimen is stabilized mechanically by a network of internal forces between adjacent tie points. The conclusions of this section are summarized in Figure 3.3; it may be opportune, at first reading, to accept Figure 3.3 and proceed directly to Section 3.4.

Consider the specimen, which is to be subject to an external tensile force F, to be a thermodynamic system. The differential of the Helmholtz free energy A with respect to the length L of the specimen is equal to F (Problem 3.2):

$$F = \left(\frac{\partial A}{\partial L} \right)_{T,V}. \tag{3.3}$$

Since

$$A = E - TS, \tag{3.4}$$

in which E is the internal energy of the system and S the entropy,

$$F = \left(\frac{\partial E}{\partial L} \right)_{T,V} - T \left(\frac{\partial S}{\partial L} \right)_{T,V}. \tag{3.5}$$

For solids other than rubbers—that is, for metals and ceramics, for example—E changes rapidly with L, and S not at all. We term these solids energy-elastic, and they are usually of high modulus.

For an ideal rubber a change in L causes no change whatsoever in E. There are, however, changes in the entropy S. $(\partial S/\partial L)_{T,V}$ is negative so that an increase in F is generated by an increase in L. **The decrease in S with increasing L is brought about by deformation-induced changes in the end-to-end vector of the network chains**.

The reader may find the concept of a retractive force induced by a change in entropy somewhat strange. That this is not really so is at once apparent when it is recalled that if the volume of an ideal gas is reduced, there is an increase in pressure which is generated by a decrease in entropy (3.N.3): there is no change in internal energy with volume for an ideal gas. An ideal rubber is analogous: deformation-induced changes in the end-to-end vectors of the network chains cause a change in entropy, as we now show.

Consider the representative chain OA in Figures 3.1(b) and 3.2. Let the total entropy of this chain be s. One part of s (the conformational entropy) will vary with r through $P(r)$. Making use of Boltzmann's equation[1] and noting that Ω is proportional to $P(r)$, we may write

$$s = s_0 + k \ln P(r), \tag{3.6}$$

---

1 Boltzmann's equation relates the entropy S of a system to Ω, the probability of that state, so that $S = k \ln \Omega$. See (3.N.3) for an application of the Boltzmann equation.

in which s_0 is a constant independent of r. For a Gaussian chain, from eqns 3.1 and 3.6,

$$s = s_0 - k\left[3\ln(\sqrt{\pi}\,\rho) + \left(\frac{r}{\rho}\right)^2\right], \qquad (3.7)$$

so that

$$\left(\frac{\partial s}{\partial r}\right) = -\left(\frac{2k}{\rho^2}\right)r. \qquad (3.8)$$

Hence for an ideal chain (for an ideal chain the internal energy is independent of the chain end-to-end vector r), this change in entropy s with length r implies a force (see eqn 3.5) f:

$$f = -T\left(\frac{\partial s}{\partial r}\right), \qquad (3.9)$$

which from eqn 3.8 is

$$f = \left[\frac{2kT}{\rho^2}\right]r. \qquad (3.10)$$

This force acts between the ends of the chain (see Figure 3.2), tending to draw the tie points O and A together and opposing an increase in r. Equation 3.10 may be written

$$f = Kr, \qquad (3.11)$$

in which K is a force constant:

$$K = \left[\frac{2kT}{\rho^2}\right]. \qquad (3.12)$$

This entropy-generated force

- is linear in T, i.e. the force constant **increases** with temperature, and
- is linearly elastic, i.e. the force is proportional to r.

This argument for the representative chain OA holds for all ν chains in the specimen.

Consider the three chains, which are labelled 1, 2, and 3, attached to tie point O in Figure 3.3. Let the end-to-end vectors of these chains be r_1, r_2, and r_3. According to eqn 3.11 there are forces

$$f_1 = Kr_1, \; f_2 = Kr_2, \; \text{and} \; f_3 = Kr_3 \qquad (3.13)$$

acting on tie point O.

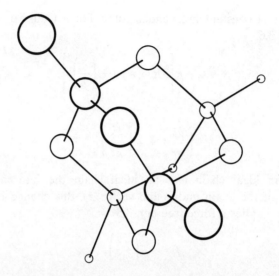

3.4 The spheres represent the positions of tie points of the molecular network; the network is not shown. The elastic properties of the network are represented by linearly elastic springs which act along the lines which interconnect the tie points. The macromolecules may be dismissed from further consideration.

The forces f, being proportional to r, can be represented by ν perfectly elastic springs (see Figure 3.4). The behaviour of the specimen can be understood in terms of the ν forces generated by these ν springs, each of spring constant K. The polymer molecules, which interconnect the tie points, **may be dismissed from further mechanical consideration**. It is fortunate that their effect is represented precisely by the springs.

The entropy forces are weak. They are strong enough, however, to cause the retraction of the specimen when the applied load is removed. This is because (we emphasize again) the polymer chains are in the liquid state. If the tie points did not exist—if the molecules were not pinned together at particular points—the assembly would flow in a liquid-like manner. It is this combination of long-range, weak entropy forces and liquid molecules which confers on rubbers their twin properties of high deformability and, paradoxically, complete retraction (3.N.4).

3.4 Elasticity of a network

Consider the specimen to be of initial dimensions X_i, Y_i, Z_i, and to be deformed by forces along the principal directions x, y, and z to the new dimensions X, Y, Z (see Figure 3.5(a)). When rubber is deformed it is always observed that there is essentially no change in volume. We may write, therefore,

$$XYZ = X_i Y_i Z_i.$$

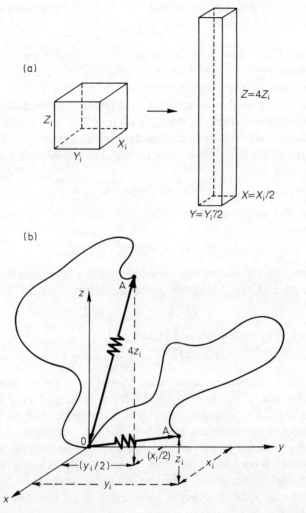

3.5   (a) A specimen of initial dimensions X_i, Y_i, and Z_i is deformed so that the dimensions become $X = \lambda_x X_i$, $Y = \lambda_y Y_i$, and $Z = \lambda_z Z_i$. (b) End A of the representative chain is at x_i, y_i, z_i before deformation; after deformation it is at $x = \lambda_x x_i$, $y = \lambda_y y_i$, $z = \lambda_z z_i$. The end-to-end vector changes in exactly the same way as the specimen dimensions. Scale drawing for $\lambda_x = \frac{1}{2}$, $\lambda_y = \frac{1}{2}$, and $\lambda_z = 4$. The deformation thus induces a change in the force between the two ends of the chain (see eqn 3.11). The end-to-end vectors of all ν chains in the specimen change in the same way.

The extension ratios in the three principal directions are

$$\lambda_x = \frac{X}{X_i}, \quad \lambda_y = \frac{Y}{Y_i}, \quad \lambda_z = \frac{Z}{Z_i}. \qquad (3.14)$$

It follows from this that the product of the three extension ratios is

$$\lambda_x \lambda_y \lambda_z = 1: \tag{3.15}$$

only two of the three may be varied arbitrarily and the third will then follow from eqn 3.15. Our aim is to calculate the external forces which produced the arbitrary deformation described by λ_x, λ_y, λ_z.

We assume that the deformation is **affine**, which means that every part of the specimen deforms as does the whole.[1] This will include the end-to-end vectors of all ν chains including the representative chain OA in Figure 3.5(b). It will be convenient to write the end-to-end vector in the undeformed state

$$r_i = ix_i + jy_i + kz_i. \tag{3.16}$$

Because the deformation is affine, this vector in the deformed state becomes

$$r = i\lambda_x x_i + j\lambda_y y_i + k\lambda_z z_i. \tag{3.17}$$

The change in the end-to-end vector produces a change in the end-to-end force (eqns 3.11 and 3.16) from f_i before deformation,

$$f_i = K[ix_i + jy_i + kz_i], \tag{3.18}$$

to f after deformation (eqns 3.11 and 3.17):

$$f = K[i\lambda_x x_i + j\lambda_y y_i + k\lambda_z z_i]. \tag{3.19}$$

Figure 3.5 is drawn to scale with $\lambda_z = 4$, $\lambda_x = \frac{1}{2} = \lambda_y$ (see eqn 3.15). (Note that the z component of the force (Figure 3.5(b)) **increases** by $Kz_i(\lambda_z - 1) = 3Kz_i$, whereas the x and y components both **decrease** by $0.5Kx_i$ for the x component and $0.5Ky_i$ for the y component.)

The stress–strain properties are calculated by **equating the work done by the internal forces with the work done by the external forces** which generate the deformation. Consider first the work done by f_z, the z-component of f, as z changes from z_i to $\lambda_z z_i$. Let this work for the representative chain be w_z:

$$
\begin{aligned}
w_z &= \int_{z_i}^{\lambda_z z_i} f_z \, dz \\[2mm]
&= \left(\frac{2kT}{\rho^2}\right) \int_{z_i}^{\lambda_z z_i} z \, dz \\[2mm]
&= \left(\frac{kT}{\rho^2}\right) [\lambda_z^2 - 1] z_i^2, \tag{3.20}
\end{aligned}
$$

1　It is reasonable to assume that a solid will deform in an affine manner: this is the elastic assumption. Note that in metal crystals plastic deformation by dislocation glide on slip planes is not affine. The crystal between the slip planes is not plastically distorted: all the deformation occurs at the slip plane.

making use of eqns 3.12, 3.18, and 3.19. Now this is the work done by the f_z force of one chain: it depends on z_i^2, the square of the initial value of the z component of the end-to-end vector of **that** chain. The work done by the f_z forces of all ν chains in the specimen is obtained by averaging over all values of z_i^2:

$$\sum_1^\nu w_z = \sum_1^\nu \left(\frac{kT}{\rho^2} \right) [\lambda_z^2 - 1] z_i^2.$$

All chains have the same ρ and imposed λ_z, hence

$$\sum_1^\nu w_z = \left(\frac{kT}{\rho^2} \right) [\lambda_z^2 - 1] \sum_1^\nu z_i^2. \tag{3.21}$$

By definition

$$\sum_1^\nu z_i^2 = \nu \langle z^2 \rangle_i, \tag{3.22}$$

where $\langle z^2 \rangle_i$ is the mean-square value of z_i in the undeformed state. The undeformed state is isotropic, so clearly

$$\langle z^2 \rangle_i = \langle y^2 \rangle_i = \langle x^2 \rangle_i = \frac{\langle r^2 \rangle_i}{3}. \tag{3.23}$$

It follows from eqns 3.21–3.23 that

$$\sum_1^\nu w_z = \left(\frac{kT}{\rho^2} \right) [\lambda_z^2 - 1] \frac{\nu \langle r^2 \rangle_i}{3}. \tag{3.24}$$

For a Gaussian chain ρ^2 controls the mean-square end-to-end distance $\langle r^2 \rangle_0$ of the chain when it is unperturbed by the cross-links (eqn 3.2). It follows, therefore, from eqns 3.2 and 3.24 that

$$\sum_1^\nu w_z = \frac{\nu kT}{2} \cdot \frac{\langle r^2 \rangle_i}{\langle r^2 \rangle_0} [\lambda_z^2 - 1]. \tag{3.25}$$

The equations for the work done by all the f_x and f_y forces follow by an identical argument, so that the total work done during the deformation is

$$W = \sum_1^\nu [w_x + w_y + w_z] \tag{3.26}$$

$$= \frac{\nu kT}{2} \frac{\langle r^2 \rangle_i}{\langle r^2 \rangle_0} [\lambda_x^2 + \lambda_y^2 + \lambda_z^2 - 3]. \tag{3.27}$$

This work is identical to the work done by the external forces applied to the surface of the specimen and is equal to the change in Helmholtz free energy.

Example 3.1

A bar of ideal rubber containing 5×10^{20} chains between cross-links is extended uniaxially at 20°C until its length is double the initial length. Calculate the heat gained or lost. Assume a Gaussian network and $\langle r^2 \rangle_i = \langle r^2 \rangle_0$.

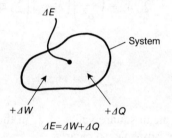

$$\Delta E = \Delta W + \Delta Q$$

Procedure

We need the work put into the system, which is given by eqn 3.27 (this work is designated ΔW following the usage of thermodynamics). Work put **in** is positive; heat put **in** is positive; $\Delta E = 0$ (ideal rubber).

Solution

From the first law of thermodynamics

$$\Delta E = \Delta W + \Delta Q,$$

which for $\Delta E = 0$ yields

$$\Delta Q = -\Delta W.$$

Therefore

$$\Delta Q = -\frac{\nu k T}{2} \left[\lambda_x^2 + \lambda_y^2 + \lambda_z^2 - 3 \right].$$

From eqn 3.15 the transverse extension ratios (which must be equal) are

$$\lambda_x = \lambda_y = 1/\lambda_z^{1/2}.$$

Therefore, for $\lambda_z = 2$

$$\Delta Q = -\frac{5 \times 10^{20} \times 1.38 \times 10^{-23} \times 293}{2} \left[\tfrac{1}{2} + \tfrac{1}{2} + 4 - 3 \right]$$

$$= -2.02 \, \text{J}.$$

The negative sign implies that the heat leaves the bar: all the work put in appears as heat. The entropy change is

$$\Delta S = \frac{\Delta Q}{T} = -\frac{2.02}{293} = -6.89 \times 10^{-3}\,\text{J}\,\text{K}^{-1}.$$

Comment

The entropy decreases as the specimen extends.

It is convenient to rewrite eqn 3.27 in terms of N, the number of sub-molecules (or chains) per unit volume,

$$N = \frac{\nu}{V}; \tag{3.28}$$

it will be recalled that ν is the total number of submolecules in the specimen which is of volume V. It follows that from eqns 3.27 and 3.28 that

$$W = \frac{VG}{2}\left[\lambda_x^2 + \lambda_y^2 + \lambda_z^2 - 3\right] \tag{3.29}$$

in which

$$G = NkT \cdot \frac{\langle r^2 \rangle_i}{\langle r^2 \rangle_o}. \tag{3.30}$$

It is left as an exercise for the reader to show that G is in fact the shear modulus of the specimen (Problem 3.6). Equation 3.29 is of central significance in the theory of rubbers and it is a tribute to our understanding that it may be derived from first principles (3.N.5).

Example 3.2

A rubber consists of a cross-linked network of chains each of RMM = 2×10^4; the density of the specimen is 900 $\text{kg}\,\text{m}^{-3}$. Calculate the shear modulus at 0°C. Assume a Gaussian network and $\langle r^2 \rangle_i = \langle r^2 \rangle_o$.

Procedure

The shear modulus is $G = NkT$ (from eqn 3.30). We require to find N, the number of chains per m^3.

Solution

A mole of chains (6.023×10^{23} chains) of RMM = 2×10^4 is of mass (see Section 1.7)

$$M = 2 \times 10^4 \times \frac{12.0011}{12} = 2.0 \times 10^4\,\text{g}.$$

Hence in a mass of 900×10^3 g (of volume $1\,\mathrm{m}^3$) the number of chains is

$$N = \frac{900 \times 10^3}{2.0 \times 10^4} \times 6.023 \times 10^{23} = 2.71 \times 10^{25}\,\mathrm{m}^{-3}.$$

Therefore

$$
\begin{aligned}
G &= 2.71 \times 10^{25} \times 1.38 \times 10^{-23} \times 273 \\
  &= 1.02 \times 10^5\,\mathrm{Pa}.
\end{aligned}
$$

Comment

Measurement of G for a rubber is the simplest method of obtaining N, the number of molecular chains per m^3.

3.5   Stress–strain relationship

The use of eqn 3.29 will be demonstrated for a specimen deformed in tension—the simplest case. The forces for any more complex deformation may be derived by a parallel argument (Problems 3.14, 15, 16).

Let the specimen be deformed along the z axis from initial length L_i to a final length L, by an applied force F (see Figure 3.6). The usual assumption of constant volume deformation will be made so that the volume V is

$$V = A_i L_i = AL, \tag{3.31}$$

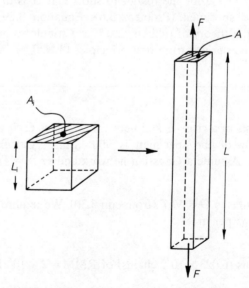

3.6 A specimen of initial length L_i is deformed by a tensile force F to length L. The cross-sectional area changes from A_i to A.

in which A_i and A are the cross-sectional areas before and after deformation. The theory is formulated simultaneously for compression, in which case the force F is negative, so $L < L_i$, and $A > A_i$, with eqn 3.31 still valid. Let

$$\lambda_z \equiv \lambda = \frac{L}{L_i}. \tag{3.32}$$

Rubber deforms at constant volume, so that

$$\lambda_x \lambda_y \lambda_z = 1, \tag{3.33}$$

and since in the transverse directions the extension ratios must be equal, we have

$$\lambda_y = \lambda_x = \frac{1}{\lambda^{\frac{1}{2}}}.$$

It then follows that the total work done (eqn 3.29) is

$$W = \frac{VG}{2}\left[\lambda^2 + \frac{2}{\lambda} - 3\right]. \tag{3.34}$$

But since the force F producing the deformation is

$$F = \frac{dW}{dL}$$

$$= \frac{dW}{d\lambda}\frac{d\lambda}{dL}, \tag{3.35}$$

we have, from eqns 3.34 and 3.35 and using $d\lambda/dL = 1/L_i$ from eqn 3.32,

$$F = \frac{VG}{L_i}\left[\lambda - \frac{1}{\lambda^2}\right], \tag{3.36}$$

therefore (see eqn 3.31)

$$F = A_i G\left[\lambda - \frac{1}{\lambda^2}\right]. \tag{3.37}$$

Example 3.3

The specimen of Example 3.1 is of initial length $L_i = 0.10$ m. What is the force required to double its length?

Procedure

Use eqn 3.37: $G = NkT$, eqn 3.30: A_i must be eliminated since it is not given.

Solution

$$N = \frac{\nu}{V} = \frac{\nu}{A_i L_i}$$

$$F = A_i NkT \left[\lambda - \frac{1}{\lambda^2} \right]$$

$$= A_i kT \left(\frac{\nu}{A_i L_i} \right) \left[\lambda - \frac{1}{\lambda^2} \right] = \frac{kT\nu}{L_i} \left[\lambda - \frac{1}{\lambda^2} \right]$$

$$= \frac{1.38 \times 10^{-23} \times 293 \times 5 \times 10^{20}}{0.10} \left[2 - \frac{1}{4} \right].$$

$$F = 35.4 \text{ N}.$$

If the nominal stress σ is required, then, using eqn 3.37,

$$\sigma = \frac{F}{A_i} = G \left[\lambda - \frac{1}{\lambda^2} \right]. \tag{3.38}$$

The dependence of the true stress on λ may also be obtained. The true stress is the applied load F divided by the instantaneous area of the cross-section A:

$$\sigma_t = \frac{F}{A} = \frac{FL}{V},$$

which, from eqns 3.36 and 3.32, yields

$$\sigma_t = G \left[\lambda^2 - \frac{1}{\lambda} \right]. \tag{3.39}$$

The measured dependence of σ on λ is shown in Figure 3.7 for natural rubber. The fit is good up to $\lambda = 1.2$. The fit for compression ($\lambda < 1.0$) is excellent. The lack of fit above $\lambda = 1.2$ is due to several factors which include: (i) rather simple assumptions in the model; (ii) that the chains cannot be Gaussian at high extensions since they cannot extend further than their own contour length (see (3.N.6)); and (iii) at high extensions vulcanized natural rubber crystallizes.

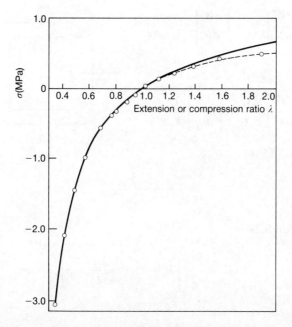

3.7 A comparison of theory and experiment for the tensile elongation and compression of a rubber prism. The full line is eqn 3.38; points show the result of experiment; σ is the force divided by the initial, unstrained, area of cross-section (after Treloar).

Nevertheless, in outline, the theory is well supported by experiment at values $\lambda < 1.2$. For example, the predicted dependence of the shear modulus (eqn 3.30) on both T and N is closely obeyed. The theory is also independent of chemical composition (as is the kinetic theory of gases) and this is supported by experiment. Note that although we have used eqn 3.29 to predict the results of a tensile experiment it may also be used for more complex situations, such as a pressurized tube (Problem 3.16) or a sheet subject to biaxial deformation (Problem 3.15).

Example 3.4

A rubber parallelepiped is deformed by forces along the x and y axes (the shear modulus of the rubber is $G = 1.0$ MPa). What are the nominal stresses σ_x and σ_y required to deform it to $\lambda_x = 3/2$, $\lambda_y = 2/3$. Assume a Gaussian network.

Procedure

The specimen is deformed from initial dimensions X_i, Y_i, Z_i to X, Y, Z. The forces required are f_x and f_y. At this point the strain energy is given by eqn 3.29, and the two forces are

$$f_x = \left[\frac{\partial W}{\partial X}\right]; f_y = \left[\frac{\partial W}{\partial Y}\right].$$

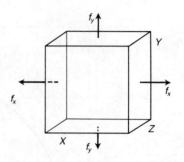

Solution

$$f_x = \left[\frac{\partial W}{\partial \lambda_x}\right]\frac{\mathrm{d}\lambda_x}{\mathrm{d}X} = \left[\frac{\partial W}{\partial \lambda_x}\right]\frac{1}{X_i}.$$

From eqn 3.15

$$\lambda_z = \frac{1}{\lambda_x \lambda_y},$$

and from eqn 3.29

$$W = \frac{VG}{2}\left[\lambda_x^2 + \lambda_y^2 + \frac{1}{\lambda_x^2 \lambda_y^2} - 3\right].$$

Therefore

$$f_x = VG\left[\lambda_x - \frac{1}{\lambda_x^3 \lambda_y^2}\right]\frac{1}{X_i}.$$

Therefore

$$\sigma_x = \frac{f_x}{Y_i Z_i} = \cancel{V}G\left[\lambda_x - \frac{1}{\lambda_x^3 \lambda_y^2}\right]\frac{1}{\cancel{X_i} Y_i Z_i}$$

$$= G\left[\lambda_x - \frac{1}{\lambda_x^3 \lambda_y^2}\right].$$

Similarly (interchange x and y)

$$\sigma_y = G\left[\lambda_y - \frac{1}{\lambda_x^2 \lambda_y^3}\right].$$

Inserting $G = 10^6$ Pa, $\lambda_x = 3/2$, $\lambda_y = 2/3$,

$$\sigma_x = 0.833 \text{ MPa}; \quad \sigma_y = -0.833 \text{ MPa}.$$

Comment

This is a biaxial stress system producing a pure shear deformation ($\lambda_z = 1.0$); see (3.N.5) for a solution to this problem using the Mooney equation.

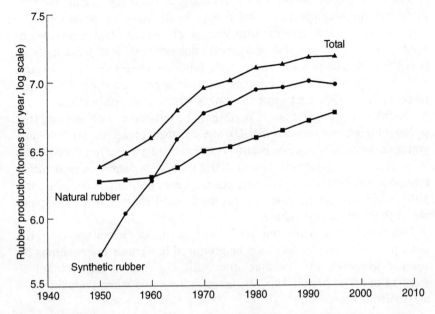

3.8 Yearly production of natural and synthetic rubber since 1950. The major increase in synthetic rubber production occurred in the 1950s and 1960s.

3.6 Engineering rubbers

In this section we outline the properties of the rubbers of interest in engineering. Commercially available natural rubber, unvulcanized, is mechanically weak, is deficient in elastic recovery, and is greatly swollen—if not completely dissolved—by many organic liquids. All the properties of natural rubber are improved by the application of the first great discovery of rubber technology: vulcanization or cross-linking. The most common vulcanization is through sulfur (see Figure 0.2): in normal soft vulcanizates there is one cross-link per 500 to 1000 isoprene units. The density of cross-links controls stiffness (eqns 3.30 and 3.39). Other forms of cross-linking are used: these achieve $C-C$ linkages as described for cross-linked polyethylene in Chapter 1, Figure 1.2.

The second great discovery in rubber technology was of the substantial benefits conferred on rubber by mixing it with carbon black. The carbon acts as a mechanical reinforcement (see Chapter 6), but it does bring other advantages: it reduces the degradation caused by sunlight and ozone, and reduces liquid absorption.

Figure 3.8 shows the increase in rubber production since 1950 and the proportion of this which is natural rubber. It will be seen that synthetic rubber production had risen to equal that of natural rubber by 1960 and is now considerably greater. The primary drive for synthetic rubber was generated by the two world wars, particularly in Germany. During the first world war the shortage of natural rubber in Germany led to curious tyre contraptions—wood covered with canvas or leather, and steel springs inside steel shoes. At that time both polyisoprene and polybutadiene (1.N.12) were synthesized. These early products were inferior to natural rubber, and it was not until advances in stereospecific catalysis, which took place in the 1950s, that synthetic polyisoprene and polybutadiene were produced in large quantities. Copolymers of butadiene with styrene, (the styrene–butadiene rubbers (SBR)) are another group of large-volume synthetic rubbers. Butadiene is also copolymerized with acrylonitrile (3.N.7) to produce the copolymer rubber NBR (one of a family termed nitrile rubbers). These synthetic systems, polyisoprene, polybutadiene, SBR, and NBR, when vulcanized, have general mechanical characteristics comparable to those of natural rubber.

The copolymers SBR and NBR have a major advantage over the homopolymers: that is, **they may be produced in various compositions for specific purposes**. For example, the NBR copolymers are resistant to swelling in oil or petrol. This resistance is increased when the content of acrylonitrile

$$-(CH_2-\underset{\underset{\underset{N}{\overset{|||}{}}}{\overset{|}{C}}}{CH})-$$

is increased. The acrylonitrile units with their nitrogen atoms repel hydrocarbon liquids, whereas the butadiene component attracts the chemically similar hydrocarbons (3.N.8). Thus, for applications for which a high hydrocarbon resistance is important the acrylonitrile content is raised. The usual range of acrylonitrile contents is, by weight, from 18% to 40%. If the acrylonitrile content is increased above 40% the glass transition of the copolymer is raised so much that the rubber becomes brittle at temperatures encountered in cold countries.

The reason for this change in T_g with composition is that the glass transitions of the two components, as homopolymers, are $-87°C$ for polybutadiene and $106°C$ for polyacrylonitrile. The glass transitions of the copolymers lie in between these two extremes. This is both the anticipated and the observed behaviour. The glass transition of a copolymer poly AB will lie at a temperature between the glass temperatures of the two homopolymers poly A and poly B. For example, the dependence of T_g on composition for the styrene–butadiene copolymers is shown in Figure 3.9: the styrene contents of the commercial SBR rubbers fall in the range 10 to 40%.

Polybutadiene is also used as the rubbery (soft) component in the new

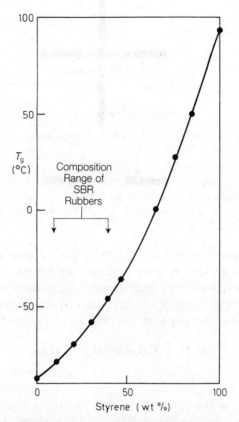

3.9 Glass transition temperatures of random copolymers of styrene and butadiene as a function of percentage styrene in the copolymer. The glass transitions of the homopolymers are: polystyrene 95°C; polybutadiene −87°C. The styrene contents of commercial SBR rubbers fall in the range 10 to 40% (after Illers).

thermoplastic rubbers: the glassy (hard) component is polystyrene. Thermoplastic rubbers are of great interest because they may be processed in conventional equipment for thermoplastics, unlike thermoset rubbers. The thermoplastic (TR) polymers are composed of long polybutadiene chains tipped at each end with polystyrene end-blocks (see 1.N.12). The molecules may be linear (see Figure 3.10(a)) or branched (see Figure 3.10(b)). The strength and effective cross-linking are provided by the polystyrene end-blocks, which at ambient temperatures agglomerate to form domains of glassy polystyrene (see Figure 3.11). The glassy domains are separated by a rubbery matrix of polybutadiene: the domains are pinned together by chains of polybutadiene. The domains liquefy when heated above the T_g of polystyrene, and the TR can then be made to flow under pressure like any other thermoplastic. On cooling through the T_g of polystyrene, the domains reform. TRs are now used increasingly to toughen polypropylene:

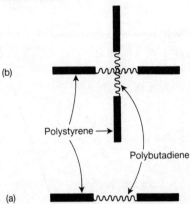

(b)

Polystyrene →

Polybutadiene

(a)

3.10 Styrene–butadiene block copolymers. (a) A linear molecule composed of polybutadiene with end-blocks of polystyrene. (b) A branched polybutadiene molecule with four polystyrene end-blocks (after Bull).

they are finely mixed in with the polypropylene to form alloys. The TRs have inherent advantages in alloys because they can be easily mixed with polypropylene in conventional thermoplastic equipment.

Valuable synthetic rubbers are produced by the copolymerization of ethylene and propylene. The structure may be represented as

$$\text{+CH}_2\text{—CH}_2\text{)}_{n'}\text{(CH}_2\text{—CH)}_{n''}$$
$$\underset{\displaystyle \text{CH}_3}{|}$$

with n' and n'' in the range 10 to 20. The mole fractions, ethylene/propylene, range from 40/60 to 60/40. In this range the copolymer is completely amorphous (3.N.9).

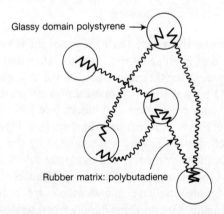

Glassy domain polystyrene →

Rubber matrix: polybutadiene

3.11 Illustration of phase structure of a thermoplastic rubber composed of styrene–butadiene block copolymers. The polystyrene molecules aggregate in glassy styrene domains which are linked together by the rubbery matrix in which the polybutadiene blocks have congregated (after Bull).

Another speciality system with a resistance higher than natural rubber to oils and petrol is the group of chloroprene rubbers. The structure is essentially polybutadiene (1.N.12) with a chlorine replacement for a hydrogen atom. Over 80% of the structure is of the form

$$-CH_2 \diagdown \qquad \diagup CH_2-$$
$$\diagdown \qquad \diagup$$
$$CCl=CH$$

This was the first of the high-cost speciality rubbers (neoprene rubber, first manufactured in 1931). In addition to its hydrocarbon resistance, it has other superior properties compared with natural rubber, in respect of heat resistance, permeability to gases, resistance to ozone oxidation, and crack susceptibility when exposed to direct sunlight. Other rubbers for high temperature stability and high resistance to hydrocarbon liquids are the speciality rubbers containing fluorine. There is a great need for improvement in elastomers for high-temperature (close to the engine) automotive applications.

Notes for Chapter 3

3.N.1
The reader will recall that the representative chain, originally part of the network, has been examined in isolation (detached from the other chains). The Gaussian function gives the time average of the end-to-end vector r for the chain when left undisturbed and permitted to move from one conformation to another by thermally induced, liquid-like movements. If ν such detached chains are examined at an instant in time, it will be found that their end-to-end vectors also conform to a Gaussian distribution. The formation of a real network can be thought of as the bringing together of these ν detached chains and the cross-linking of them together. The assembled chains in the network will then have a Gaussian distribution of end-to-end vectors. With the passage of time the value of r for a particular chain in the network will fluctuate, but these fluctuations will occur also in a random manner and, because ν is an extremely large number, the distribution will remain Gaussian.

3.N.2
A major factor in the selection of elastomers for automotive applications is the extent to which they absorb petrol or oil. When an elastomer absorbs diluent (vapour or liquid) it is said to swell (this is illustrated in Figure 3.12). The increase in volume shown is ×4. The increase in length of any

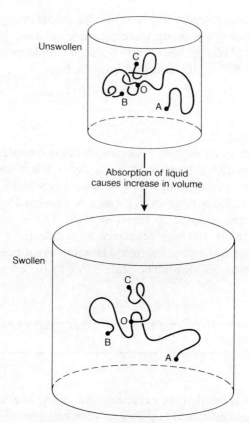

3.12 When an elastomer absorbs diluent (liquid or vapour) it expands. The end-to-end lengths (e.g. OA, OB, OC, etc.) all increase. If the diluent is removed, the elastomer contracts to its original volume.

linear dimension is $4^{\frac{1}{3}} = 1.6$. The increase in the value of $\langle r^2 \rangle_i$ (mean-square value of all lengths such as OA, OB, OC, etc.) is $4^{\frac{2}{3}} = 2.5$. The swelling is reversible: if the diluent is removed the elastomer recovers its original volume. The network can be looked upon as a sponge. The chemical nature of the diluent and of the elastomer determines the degree of swelling (i.e. the magnitude of the absorption).

3.N.3

Let an ideal gas consisting of ν molecules occupy a container of volume V_0. Now the probability that one molecule may be observed at any instant in the volume V_0 is of course unity; that it may be observed in a smaller part of the container of volume V has a probability V/V_0. The probability that all ν molecules will spontaneously move to the volume V is then

$$\Omega = (V/V_0)^\nu.$$

If the gas is compressed from V_0 to V the induced **change** in entropy is

$$S = k \ln \Omega$$

(using Boltzmann's equation) so that

$$S = \nu k \ln\left(\frac{V}{V_0}\right). \tag{3.N.3.1}$$

Recalling that the pressure P of a gas is

$$P = -(\partial A/\partial V)_T,$$

we have from eqn 3.4

$$P = -\left(\frac{\partial E}{\partial V}\right)_T + T\left(\frac{\partial S}{\partial V}\right)_T.$$

For an ideal gas $(\partial E/\partial V)_T = 0$, so that since the change in entropy with V is given by eqn 3.N.3.1:

$$P = \frac{\nu k T}{V}.$$

This derivation of the ideal gas law is very close in all its steps to the theory developed in Sections 3.3 and 3.4 for the elasticity of rubber. In the case of rubber, which is essentially incompressible, the deforming forces cause the specimen to change shape; the deformation is shear deformation with no change in volume.

3.N.4
Entropy forces between physical cross-links play important roles in other circumstances, notably in causing:

- elastic effects in flowing polymer liquids (Chapter 7); and
- recovery of polymeric solids following large strain deformation (Chapter 5).

The latter effect is promoted by warming, which both softens the polymer and increases the entropy forces (eqn 3.12).

3.N.5
The work done appears as strain energy, so that eqn 3.29 is often known as the strain energy function when referring to a unit volume of rubber. In this case it may be written

$$W = \frac{G}{2}\left[\lambda_x^2 + \lambda_y^2 + \lambda_z^2 - 3\right]. \tag{3.N.5.1}$$

In 1940, well before the derivation of the Gaussian network theory which leads to eqn 3.N.5.1, Mooney derived, by mathematical arguments involving considerations of symmetry,

$$W = C_1\left(\lambda_x^2 + \lambda_y^2 + \lambda_z^2 - 3\right) + C_2\left(1/\lambda_x^2 + 1/\lambda_y^2 + 1/\lambda_z^2 - 3\right), \quad (3.N.5.2)$$

in which C_1 and C_2 are constants. Clearly the Gaussian network theory coincides with the Mooney theory when

$$C_2 = 0,$$

and

$$C_1 = \frac{G}{2}. \quad (3.N.5.3)$$

The Gaussian network theory fits the data less well than the Mooney theory. This could be partly due to the Mooney theory having two adjustable constants.

Example 3.5

Repeat Example 3.4 for a rubber obeying the Mooney equation (eqn 3.N.5.2) with $C_1 = 0.20$ MPa and $C_2 = 0.10$ MPa.

Solution

$$\sigma_x = \frac{f_x}{Y_i Z_i} = \frac{1}{Y_i Z_i}\left(\frac{\partial W}{\partial \lambda_x}\right)\frac{1}{X_i} = \frac{1}{V}\left(\frac{\partial W}{\partial \lambda_x}\right).$$

Similarly,

$$\sigma_y = \frac{1}{V}\left(\frac{\partial W}{\partial \lambda_y}\right).$$

$$\frac{W}{V} = C_1\left[\lambda_x^2 + \lambda_y^2 + \frac{1}{\lambda_x^2 \lambda_y^2} - 3\right] + C_2\left[\frac{1}{\lambda_x^2} + \frac{1}{\lambda_y^2} + \lambda_x^2 \lambda_y^2 - 3\right]$$

$$\therefore \quad \sigma_x = 2C_1\left[\lambda_x - \frac{1}{\lambda_x^3 \lambda_y^2}\right] + 2C_2\left[\lambda_x \lambda_y^2 - \frac{1}{\lambda_x^3}\right].$$

For $\lambda_x = 3/2$, $\lambda_y = 2/3$,

$$\sigma_x = 2 \times 0.2 \left[\frac{3}{2} - \frac{1}{3/2} \right] + 2 \times 0.1 \left[\frac{2}{3} - \frac{1}{(3/2)^3} \right]$$

$$= 0.407 \text{ MPa}$$

$$\sigma_y = 2 \times 0.2 \left[\frac{2}{3} - \frac{1}{2/3} \right] + 2 \times 0.1 \left[\frac{3}{2} - \frac{1}{(2/3)^3} \right]$$

$$= -0.708 \text{ MPa}$$

Comment

As mentioned in 3.N.5, the Mooney equation will provide at large strains a better fit to the data than the Gaussian network theory.

3.N.6

The reader will note that in this theory of rubber elasticity we disregard the architecture of the C—C—C chain, and the van der Waals forces between chains. The molecules are assumed to be infinitely supple chains; they are sometimes described as phantom chains, in that, as they diffuse to and fro through space, they pass through one another in a way that real chains could not. Even the hooking of molecules around one another—to form entanglements—does not cause a deviation from the theory, to a first approximation. Entanglements act as physical cross-links, and must be added to the chemical cross-links in calculating the total effective cross-link density, which determines the number of sub-chains per unit volume N and hence the shear modulus; see eqn 3.30. On closer inspection, however, the data do reveal discrepancies. Deviation of the stress–strain relation from the Gaussian theory at extensions $\lambda > 1.2$ (see Figure 3.7) is best explained in terms of the physical cross-links allowing some relative movement of the molecules that are apparently 'joined', whereas true chemical cross-links allow none. This reduces the stress below the Gaussian prediction, and is accounted for empirically by the Mooney theory (3.N.5).

At much larger extensions, the Gaussian theory is seriously in error. It does not take account of the finite lengths of the sub-chains between crosslinks. Whereas for a Gaussian chain

$$P(r) \to 0 \quad \text{as} \quad r \to \infty$$

(see eqn 3.1), for a real chain of n links each of length l the end-to-end distance cannot exceed the contour length nl and hence

$$P(r) \to 0 \quad \text{as} \quad r \to nl$$

(we ignore here the very small elastic stretching of interatomic bonds and the possibility of bond rupture). It follows from eqns 3.6 and 3.9 that the entropy of a real chain decreases rapidly at this point, leading to

$$f \to \infty \quad \text{as} \quad r \to nl.$$

The chain is taut and cannot be stretched further. It is a straightforward matter to estimate how far the network can be stretched before this occurs. Clearly, when the sub-chains are stretched taut, all their end-to-end lengths r in the direction of stretching must equal the fully extended length $r = L = nl$. Before stretching commenced, they had a range of values: let us assign to r the root-mean-square value $r = R = n^{1/2}l$ (see Section 2.8). It follows that the limiting extension ratio in the direction of stretching is

$$\lambda = \frac{L}{R} = \frac{nl}{n^{1/2}l} = n^{1/2}. \tag{3.N.6.1}$$

This agrees with experience. Rubbery polymers are well known to become stiffer as they approach a limiting extension: the reader can check this easily by stretching a rubber band. Figure 3.13 shows the pronounced departure from the Gaussian prediction that results (see also Problem 3.19). In some rubbery polymers however, including natural rubber, stress-induced crystallization intervenes at large extensions. This may contribute to the stiffening.

3.N.7

The monomer acrylonitrile polymerizes according to

$$n(CH_2{=}CH) \longrightarrow {-}(CH_2{-}CH)_n{-}$$

with pendant groups C≡N.

This polymer, polyacrylonitrile (PAN), can be processed into a valuable fibre. The best high modulus **carbon fibres** are produced from the pyrolysis of PAN. The copolymer of acrylonitrile and styrene is a random linear molecule consisting of styrene units and acrylonitrile units: the structure is

$${-}(CH_2{-}CH)_n{-}\ {-}(CH_2{-}CH)_{n'}{-}$$

The excellent resistance to swelling in organic solvents is conferred by the N atom in the acrylonitrile segments.

3.N.8

The general rule that 'like dissolves like' holds well with polymers; for example, polymers absorb considerable quantities of their own monomers. If the polymer is linear it may well completely dissolve in its own monomer or other good solvent. If the solubility is lower, then the polymer, when immersed in solvent, will absorb **in equilibrium** a fraction of its own mass and will not dissolve: the lower the solubility the lower the fraction. In 1.N.9 the equilibrium water contents of nylons are given for 50% and 100% RH. This is the behaviour observed for vapours; equilibrium absorption increases with vapour pressure. The absorbed molecules are absorbed only into the amorphous fraction of crystalline polymers such as the nylons. Cross-linking in rubbers lowers the absorption: the swelling of the network is impeded by the cross-links. As the network expands because of the absorption of diluent, the entropy forces between tie points increase (Figures 3.3 and 3.4) and this limits the swelling.

3.N.9

We have seen (in Chapter 2) that both of the homopolymers polyethylene (PE) and polypropylene (PP) are partially crystalline. The glass transitions of the amorphous fraction of PE is $-120°C$ and of PP $-19°C$. Thus the glass transitions of the copolymers are expected to be in the range $-120°C$ up to $-19°C$: the glass transitions of the copolymers of commercial interest lie in the range $-50°C$ to $-58°C$. In this range of composition there is no crystallinity.

Problems for Chapter 3

Where needed, take Avogadro's constant to be $N_0 = 6.023 \times 10^{23}$ mol^{-1}, and Boltzmann's constant to be $k = 1.38 \times 10^{-23}$ J K^{-1}.

3.1 State the three conditions under which a polymer displays rubbery behaviour.

3.2 A bar of elastic material is stretched isothermally to a length L (see Figure 3.6). Show that the force F required to hold the ends a distance L apart is given, in terms of the free energy of the bar A, by

$$F = \left(\frac{\partial A}{\partial L} \right)_{T,V},$$

when the deformation takes place at constant volume. (Hint: apply the first and second laws of thermodynamics to an infinitesimal displacement of the length from L to $L + dL$. Note that isothermal deformation of any **elastic** material is a **reversible** thermodynamic process.)

3.3 If the bar of Problem 3.2 is held at constant length L while temperature T is raised, show that the force changes at the rate given by

$$\left(\frac{\partial F}{\partial T}\right)_L = -\left(\frac{\partial S}{\partial L}\right)_T,$$

where S is the entropy of the bar, if it is assumed that the stress-free length is independent of temperature (i.e. free thermal expansion is ignored). In practice, when this experiment is conducted on polymers in the **rubbery** state, it is found that $(\partial F/\partial T)_L$ is negative at small extensions but is positive at large extensions. Explain this. Hint: recall that for any continuous function z of two variables x and y the following relation holds:

$$\frac{\partial}{\partial y}\left(\frac{\partial z}{\partial x}\right)_y = \frac{\partial}{\partial x}\left(\frac{\partial z}{\partial y}\right)_x.$$

3.4 The bar of Problem 3.2 is extended adiabatically (e.g. quickly). Show that the temperature T changes with increase in L at the rate given by

$$\left(\frac{\partial T}{\partial L}\right)_S = \frac{-T}{Mc}\left(\frac{\partial S}{\partial L}\right)_T,$$

where T is absolute temperature, c is the specific heat capacity, and M is the mass of the bar. In practice, when this experiment is conducted on polymers in the **rubbery** state the temperature is found to rise. Explain this. Hint: recall that for a continuous function z of two variables x and y the differentials are related as follows

$$\left(\frac{\partial y}{\partial x}\right)_z = -\left(\frac{\partial z}{\partial x}\right)_y \bigg/ \left(\frac{\partial z}{\partial y}\right)_x.$$

3.5 Continue Problem 3.4 to show that, in the case of an ideal rubber of density ρ and specific heat capacity c under a nominal tensile stress σ, temperature rises with extension ratio λ during adiabatic stretching according to

$$\left(\frac{\partial T}{\partial \lambda}\right)_S = \frac{\sigma}{\rho c}.$$

(Hint: recall eqn 3.5.) Hence find the temperature rise on rapid (adiabatic) stretching of a cross-linked natural rubber from the stress-free state to 200% extension (i.e. $\lambda = 3$). Take the properties of the rubber to be $\rho = 909$ kg m^{-3}, $c = 1.92$ kJ kg^{-1}K^{-1}, and G (shear modulus) $= 500$ kPa, and assume it behaves as an ideal rubber and obeys Gaussian statistics (i.e. the analysis given in this chapter is applicable). In practice, the temperature rise is usually found to exceed the predicted value when extension ratios become large. Why should this be? (Refer to Section 6.5 as well as this chapter.)

3.6 A rectangular block of ideal rubber is deformed in simple shear by the action of a shear stress τ, as shown below.

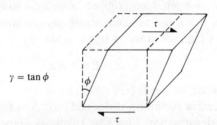

$$\gamma = \tan \phi$$

The shear γ is related to the two principal extension ratios λ and $1/\lambda$ in the plane of shear as follows:

$$\gamma = \lambda - \frac{1}{\lambda}.$$

Show that, if the rubber obeys Gaussian statistics (and hence eqn 3.27 applies), Hooke's law ($\tau = G\gamma$) is obeyed in simple shear and the shear modulus is given by

$$G = NkT \frac{\langle r^2 \rangle_i}{\langle r^2 \rangle_o}.$$

3.7 A sample of cross-linked natural rubber (polyisoprene) is found to have shear modulus $G = 217$ kPa at 20°C. Deduce N, the number of sub-chains between cross-links per m^3 and hence the number average degree of polymerization between cross-links. The chemical structure of polyisoprene is given in Section 3.1. You may take for relative atomic mass C $= 12$ and H $= 1$, and for density $\rho = 909$ kg m^{-3}. Assume $\langle r^2 \rangle_i = \langle r^2 \rangle_o$.

3.8 A sample of a certain rubbery polymer has tetrafunctional cross-links (that is, four chains meet at each cross-link). The rubber is found to have shear modulus $G = 800$ kPa at 40°C. Estimate the number of cross-links per m^3.

3.9 A rubbery polymer is immersed in a liquid and swells, absorbing the liquid until it reaches equilibrium, at which there is a volume fraction ϕ of polymer in the mixture of polymer and liquid (see

3.N.2). Find the ratio G_s/G_u of the shear moduli in swollen and unswollen states in terms of ϕ. Hence calculate the shear modulus of the rubber sample of Problem 3.7 when it has absorbed an equal volume of liquid.

3.10   Show that, according to eqn 3.38, the Young's modulus E of a rubbery polymer is related to the shear modulus G by

$$E = 3G.$$

This results from just one characteristic feature of a rubbery polymer. What is it?

3.11   Show that, for an ideal rubber obeying Gaussian statistics (and hence eqn 3.38), the nominal tensile stress–strain curve is approximated to a high accuracy at small strains by

$$\sigma = E\varepsilon(1 - \varepsilon).$$

Find the error at $\varepsilon = 0.1$.

3.12   In Chapter 5 a condition is derived (eqn 5.7) for determining when localized deformation ('necking') will spontaneously occur during uniaxial extension of a bar of material. Show that this condition is never satisfied during uniaxial extension of an ideal rubber obeying Gaussian statistics. Such a material therefore extends uniformly.

3.13   A sample of a certain ideal rubber has a number average RMM between cross-links of 5000 and a density of 900 kg m^{-3}. A block of this rubber, a cube of side 100 mm, is tested at a temperature of 300 K. What is its tensile modulus? What is its shear modulus? Axes X, Y, and Z are chosen parallel to the edges of the cube. A compressive force F_x is applied in the X-direction, to reduce the X-dimension from 100 mm to 75 mm. Calculate F_x. What do the Y and Z dimensions become? A further compressive force F_y is now applied in the Y-direction to reduce the Y-dimension to 75 mm, the X-dimension remaining at 75 mm. What now are the magnitudes of the forces F_x and F_y? What has the Z-dimension now become? How much strain energy is now stored in the block? Assume the rubber obeys Gaussian statistics and use the approximation $\langle r^2 \rangle_i = \langle r^2 \rangle_o$.

3.14   A spherical rubber balloon is of initial wall thickness 0.5 mm and diameter 100 mm. It is inflated to a final diameter of 500 mm. Calculate (1) the final thickness, (2) the true stress in the plane of the balloon wall, and hence (3) the internal air pressure required. Assume the rubber is ideal and obeys the Gaussian statistics, and take the shear modulus to be 1 MPa.

3.15   A sheet of ideal rubber is 200 mm square and 10 mm thick. Its edges are aligned parallel to axes X and Y. Forces F_x and F_y are applied in the X and Y directions respectively to stretch the sheet homogeneously, to bring the X-dimension to 400 mm and the Y-dimension to 300 mm. Calculate F_x and F_y, assuming the rubber to obey Gaussian statistics and to have a tensile modulus of 1.5 MPa.

3.16 A flexible rubber pipe is formed from an ideal rubber with tensile modulus $E = 1.5$ MPa. The external pipe diameter initially is 30 mm and the wall thickness is 3 mm. It is filled with a fluid pressurized to 50 kPa. Find the new pipe diameter and wall thickness. Assume the axial component of internal pressure is carried by rigid pipe connectors: the pipe itself therefore carries no axial stress. (You will find this problem has no analytical solution. Use either a graphical method or numerical method (with microcomputer or programmable calculator) to obtain the solution.)

3.17 Show that a rubber whose strain energy function is given by the Mooney equation 3.N.5.2 is linearly elastic in simple shear deformation, with a shear modulus G given by

$$G = 2(C_1 + C_2).$$

Use the relation between shear strain and principal extension ratios given in Problem 3.6.

3.18 For a certain rubber, it was found by experiment that in uniaxial extension by up to 100% the strain energy function was accurately given by the Mooney equation 3.N.5.2, with $C_1 = 300$ kPa and $C_2 = 100$ kPa. Find the tensile stress, based on the original cross-sectional area, required to extend a bar of this rubber by 100%. If an approximate prediction of this stress is obtained by applying the Gaussian approximation (eqn 3.N.5.1) to this material, find the magnitude of the error which results.

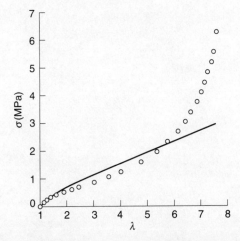

3.13 Nominal stress versus extension ratio, for uniaxial stretching of a sample of crosslinked natural rubber. Full line shows the Gaussian prediction (eqn 3.38).

3.19 The following are the first seven data points on the graph of Figure 3.13.

λ	1.00	1.136	1.241	1.377	1.596	1.882	2.168
σ (kPa)	0	145	229	308	409	502	592

(1) Show that the Mooney theory would predict a linear relation between $\sigma/(\lambda - 1/\lambda^2)$ and $1/\lambda$. Hence verify that these data satisfy the theory and find the constants C_1 and C_2, and hence the shear modulus G; see Problem 3.17.

(2) Apply the equation $G = NkT$ to find the number average degree of polymerization between cross-links for this sample of natural rubber (see Problem 3.7).

(3) Look again at Figure 3.13 and estimate the number n of chain links in the sub-molecules between cross-links (refer to 3.N.6), and hence estimate the number of carbon–carbon backbone bonds that constitute one link.

3.20 List three examples of load-bearing engineering components where the properties of polymers in the rubbery state are exploited. Add brief notes on each to indicate what you believe were the primary reasons for the decision to employ a rubbery polymer.

4   *Viscoelasticity*

4.1   Introduction

If a weight is suspended from a polymeric filament the strain will not be constant but will increase slowly with time. The effect is due to a molecular rearrangement in the solid induced by the stress. On release of the stress, the molecules slowly recover their former spatial arrangement and the strain simultaneously returns to zero. This effect is termed creep and is a manifestation of a general property of polymeric solids known as **viscoelasticity**: the solid is elastic in that it recovers, but is viscous in that it creeps.

Polymers are viscoelastic at all temperatures—they are never simple elastic solids—so that in considering the strains induced in service it is always required to take into account not only the stress, but the time for which it is applied. The viscoelastic properties are also highly temperature-dependent so that the maximum temperature must be clearly specified, and taken into consideration.

The major difference between elastic solids, such as metal alloys, and polymers can be stated briefly. Polymers, because of the great length of the molecular chain, cohere as solids even when discrete sections of the chain (or side branches) are undergoing Brownian motion[1] moving by diffusional jump processes from place to place. An applied mechanical stress leads to time effects in the strain as the mobile sections of the macromolecules flow. This flow builds up a back stress so that when the original stress is removed the back stress causes, in due time, a complete recovery. If the applied stress is left on indefinitely, the back stress continues to increase; when it equals the applied stress the net stress is zero and the solid ceases to creep. It is a matter of considerable significance that although the polymers used in engineering creep, **they do not creep indefinitely and they recover completely when the stress is removed**.

The theoretical and experimental framework of viscoelasticity was established in the nineteenth century by eminent physicists of the day, including

---

1   In the liquid the whole polymer chain undergoes vigorous Brownian motion; the molecules move as a whole by 'snake-like' motions, as envisaged in the theory of rubber elasticity. In the glass it is clear that although the chain is essentially immobile, **limited** Brownian motion is possible before the onset of the liquid-like Brownian motion at the glass transition.

Maxwell, Boltzmann, and Kelvin. Their interest was aroused partly by the creep and recovery of materials used as suspensions in electrical measuring instruments, such as glass, metals, silk, and natural rubber. In the latter part of the twentieth century viscoelasticity has become a subject of significance to engineers largely because of the increasing use of synthetic polymers as materials in engineering. The reason is clear: viscoelastic effects must be allowed for in quantitative design. At the end of this chapter we will indicate briefly how this is done and, in greater detail and in a broader context, in Chapter 8.

It will be necessary to describe the definition and measurement of the parameters used to quantify viscoelastic effects, in the sense that elastic moduli allow one to quantify elastic effects. For example: how do you design the size of a torsional shaft when the polymer creeps; for a plastic bolt after it is tightened, how do you compute the rate of stress decay; or, for the damping of mechanical vibrations what parameters would you require to design the polyurethane foam used to control and dampen vibrations and resonances in a hi-fi tone arm?

4.2   The nature of viscoelasticity

4.2.1   Creep

The results of typical creep experiments are shown in Figure 4.1. We discuss here specimens subjected to shear: related effects are, of course, observed in tension. If a constant shear stress σ_1 is applied to a viscoelastic specimen (4.N.1) the strain is observed to be time-dependent (see Figure 4.1(a)). Suppose the specimen to be allowed to recover and a larger constant stress σ_2 applied. The time dependence of the strain is shown in Figure 4.1(b). Now, if the strains at a particular time, say t_a after the application of the stress are plotted against the stress, it then is observed that these strains $\gamma(t_a)$ are linear in the stress (see Figure 4.1(c)). For a later time, say t_b after the application of the stress, the strains $\gamma(t_b)$ are again linear in the stress. Thus, if at an entirely **arbitrary** time t the strains at the two stresses are $\gamma_1(t)$ and $\gamma_2(t)$, then

$$\frac{\gamma_1(t)}{\sigma_1} = \frac{\gamma_2(t)}{\sigma_2}. \tag{4.1}$$

The strains in the two experiments at the same time t are proportional to the imposed stresses. This fact leads to the definition of the creep compliance at time t:

$$J(t) = \frac{\gamma_1(t)}{\sigma_1} = \frac{\gamma_2(t)}{\sigma_2}. \tag{4.2}$$

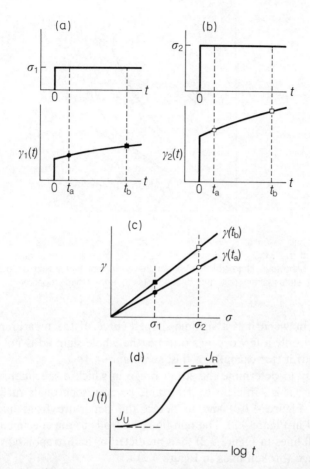

4.1 Linear viscoelastic creep: (a) constant stress σ_1 applied at $t=0$ leads to time-dependent strain $\gamma_1(t)$; (b) a higher stress σ_2 applied at $t=0$ leads to time dependent strain $\gamma_2(t)$; (c) from (a) and (b) the strains at t_a, $\gamma(t_a)$, and at time t_b, $\gamma(t_b)$, are linear in the stress; (d) the observed dependence of $J(t)$ (eqn 4.3) on log t through one complete relaxation. J_U and J_R are the unrelaxed and relaxed compliances, respectively.

In general for stress σ,

$$J(t) = \frac{\gamma(t)}{\sigma}. \tag{4.3}$$

Polymers exhibit this property of **linear viscoelastic creep** at low stresses (stresses sufficiently low that the strains are below ~0.005). In a creep experiment, the plot of strain against stress at a specific time is known as an **isochronal**.

If $J(t)$ is measured over a number of decades of time and plotted against log t, it has, in general, the form indicated in Figure 4.1(d). At very short and very long times $J(t)$ shows little or no time dependence; in the

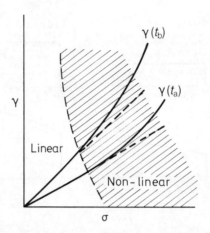

4.2 Isochronals taken at t_a (see Figure 4.1) after the initiation of the creep experiment, $\gamma(t_a)$, and at t_b, $\gamma(t_b)$. The diagram illustrates the transition from linear to non-linear viscoelastic behaviour. Note that this is not the $\gamma-\sigma$ plot that would be obtained in a conventional stress–strain test; the data are taken from creep experiments at different stresses.

period in between it is a sigmoid-shaped curve. If the measurements are taken over only a few decades of time, the whole sigmoid is not revealed: only part of it (for example of this, see Figure 4.4).

It is easy to determine the strain range in which a specimen is linearly viscoelastic: it is sufficient to determine several isochronals such as those shown in Figure 4.1(c) and to detect the departure from linearity (as illustrated in Figure 4.2). The non-linearity leads to the specimen creeping faster (full lines in Figure 4.2) than predicted by an extrapolation from the linear range (broken lines in Figure 4.2).

The reader will note that in the linear range the compliance $J(t)$ is independent of stress; the value of the stress used to determine $J(t)$ is of no significance; the compliance observed at time t is exactly the same whether the stress used in the creep experiment is σ_1, σ_2 or any arbitrary stress σ. Conversely, once $J(t)$ is known (for example in the range 1 s to 1 Ms) the strain for any stress and time is known through eqn 4.3 (in the range 1 s to 1 Ms). For non-linear deformation this is not so: each value of the stress leads to a time-dependent strain which can be obtained only by an experiment conducted at that stress. In practice, interpolation methods are used in the non-linear range (Chapter 8); this procedure is sufficiently precise for quantitative design.

The simplest method of determining $J(t)$ is to apply a constant torque to a thin-walled tube. Let the tube be of wall thickness s, radius r ($s \ll r$) and length l (see Figure 4.3). The torque applied is then

$$\Gamma = (2\pi rs)\sigma r, \tag{4.4}$$

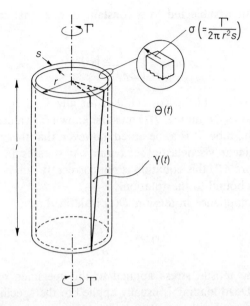

4.3 A thin-walled tube twisted by a torque Γ: the shear stress in the tube is σ. The rotation of one end with respect to the other is θ and the shear strain is γ. If the tube is viscoelastic, then both quantities depend on time, $\theta(t)$ and $\gamma(t)$.

in which σ is the shearing stress acting on a section. Hence

$$\sigma = \frac{\Gamma}{2\pi r^2 s}. \tag{4.5}$$

For an elastic tube the relationship between the angle of twist θ and the shear strain γ is

$$l\gamma = r\theta. \tag{4.6}$$

When the specimen is viscoelastic, nothing is changed except that both the shear strain and the angle of twist are functions of time. Hence we can write

$$l\gamma(t) = r\theta(t). \tag{4.7}$$

The quantity required is the compliance $J(t)$, which from eqns 4.3, 4.5, and 4.7 is

$$J(t) = \left[\frac{2\pi r^3 s}{l}\right] \frac{\theta(t)}{\Gamma}. \tag{4.8}$$

Thus $J(t)$ can be obtained by observing $\theta(t)$ for a fixed value of Γ.

Conversely, if in a design problem it is required to know the rotation of a thin walled tube subjected to a constant torque, this can be obtained from eqn 4.8:

$$\theta(t) = \left[\frac{l}{2\pi r^3 s}\right]\Gamma J(t). \qquad (4.9)$$

This equation will yield the correct $\theta(t)$ for any Γ; it is merely required that $J(t)$ be known. Naturally, $J(t)$ must be known for values of t up to the design life of the tube. It is to be noted, however, that this will hold only in the range of linear viscoelastic behaviour. At stresses in the non-linear range (see Figure 4.2) this equation is not correct. It can, however, be used to give a lower bound to the rotation.

The creep compliance in tension $D(t)$ is defined as

$$D(t) = \frac{\varepsilon(t)}{\sigma}, \qquad (4.10)$$

where σ is the tensile stress applied to the specimen and $\varepsilon(t)$ is the tensile strain. Dead loading is usually applied to the specimen. The strain is normally observed with an extensometer.

Example 4.1

A bar of polypropylene is of length 200 mm and has a rectangular cross-section of dimensions 25 mm × 3 mm. It is subjected to a constant tensile load of 250 N acting along its length. 100 s after the load was applied the length is measured and is found to have increased by 0.5 mm. Determine the 100 s tensile creep compliance.

Solution

Creep stress is the tensile load divided by cross-sectional area normal to the load, therefore

$$\sigma = 250/(25 \times 10^{-3} \times 3 \times 10^{-3}) = 3.33 \times 10^6 \text{ Pa}.$$

Creep strain at 100 s is the change in length at 100 s divided by original length, therefore

$$\varepsilon(100 \text{ s}) = 0.5 \times 10^{-3}/(200 \times 10^{-3}) = 2.50 \times 10^{-3}.$$

Hence the 100 s tensile creep compliance (eqn 4.10) is given by

$$D(100 \text{ s}) = \frac{\varepsilon(100 \text{ s})}{\sigma} = \frac{2.5 \times 10^{-3}}{3.33 \times 10^6} = 7.51 \times 10^{-10} \text{ Pa}^{-1}$$

$$= 0.751 \text{ GPa}^{-1}.$$

Example 4.2

A spherical vessel is moulded from a polymer whose one-month tensile creep compliance $D(1 \text{ month})$ is 2 GPa^{-1}. The vessel is of diameter 400 mm and wall thickness 5 mm. A constant internal pressure is applied to the vessel, giving rise to a tensile stress 1.6 MPa acting uniformly in all directions in the plane of the vessel wall. Find (a) the change in diameter, and (b) the change in wall thickness, after 1 month of pressurization, assuming the polymer to be linearly viscoelastic with constant Poisson ratio $\nu = 0.41$.

Solution

In this problem, the stress acts in more than one direction. Choose Ox and Oy to be two orthogonal axes in the plane of the vessel wall and Oz to be normal to the wall. A stress σ which is uniform in all directions in the plane of the vessel wall is equivalent to two tensile stresses σ_x and σ_y which are equal in magnitude and act parallel to axes Ox and Oy, respectively. For a linearly viscoelastic polymer the strains resulting from σ_x and σ_y will be additive. After 1 month of pressurization, the strains resulting from σ_x will be

$$\varepsilon_x(1 \text{ month}) = D(1 \text{ month})\sigma_x$$

$$\varepsilon_y(1 \text{ month}) = \varepsilon_z(1 \text{ month}) = -\nu\varepsilon_x(1 \text{ month}) = -\nu D(1 \text{ month})\sigma_x.$$

The strains resulting from σ_y will be

$$\varepsilon_y(1 \text{ month}) = D(1 \text{ month})\sigma_y$$

$$\varepsilon_x(1 \text{ month}) = \varepsilon_z(1 \text{ month}) = -\nu\varepsilon_y(1 \text{ month}) = -\nu D(1 \text{ month})\sigma_y.$$

Adding yields the total strains

$$\varepsilon_x(1 \text{ month}) = \varepsilon_y(1 \text{ month}) = [1 - \nu]D(1 \text{ month})\sigma$$

$$\varepsilon_z(1 \text{ month}) = -2\nu D(1 \text{ month})\sigma.$$

Substituting numerical values yields

$$\varepsilon_x(1 \text{ month}) = \varepsilon_y(1 \text{ month}) = [1 - 0.41] \times 2 \times 10^{-9} \times 1.6 \times 10^6$$

$$= 1.89 \times 10^{-3}$$

$$\varepsilon_z(1 \text{ month}) = -2 \times 0.41 \times 2 \times 10^{-9} \times 1.6 \times 10^6 = -2.62 \times 10^{-3}.$$

The change in diameter is obtained from the in-plane (circumferential) strains and the initial diameter:

Change in diameter after 1 month = $1.89 \times 10^{-3} \times 400 = 0.756$ mm.

The change in wall thickness is obtained from the normal strain and initial wall thickness:

$$\text{Change in wall thickness after 1 month} = -2.62 \times 10^{-3} \times 5$$

$$= -1.31 \times 10^{-2} \text{ mm.}$$

Figure 4.4 shows values of $J(t)$ for a polymer, in this case linear polyethylene, observed at nine temperatures between 15°C and 75°C. It will be noted from Figure 4.4 that this creep process (termed the α-creep

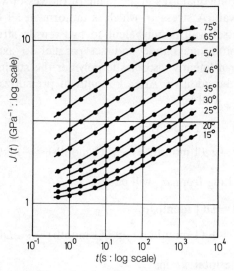

4.4 Shear compliance $J(t)$ of linear polyethylene at different temperatures in the region of the α-relaxation. Measurements are taken at times between 0.8 and 2000 s. The data are plotted against log t, as a plot against t will not reveal the significant differences in the shape of the curves at different temperatures (after McCrum and Morris).

process,[1] see also Figure 4.12) is apparently centred at 46°C; that is, for the observed time scale (10^{-1} to 10^4 s), the slope of the log $J(t)$ versus log t plot is greater at 46°C than at lower or higher temperatures. The origin of this typical viscoelastic behaviour is outlined in Section 4.3.

Creep occurs in metal alloys also. There are three principal differences between polymeric and metallic creep. In metals:

---

1 A polymer will normally have several viscoelastic processes, each one being caused by the onset of a particular form of molecular motion. These occur normally at widely spaced temperatures.

- the creep is not linearly viscoelastic;
- the creep is not recoverable; and
- the creep is significant only at high temperatures.

The creep of polymers is significant at all temperatures above $\sim -200°C$.

4.2.2 Stress relaxation

The alternative step-function experiment is stress-relaxation. A constant shear strain, say γ_1, is applied at $t = 0$ and the stress $\sigma(t)$ required to maintain γ_1 constant is observed. $\sigma(t)$ is found to decrease with time, as shown in Figure 4.5(a) (4.N.2). Suppose the specimen to be allowed to

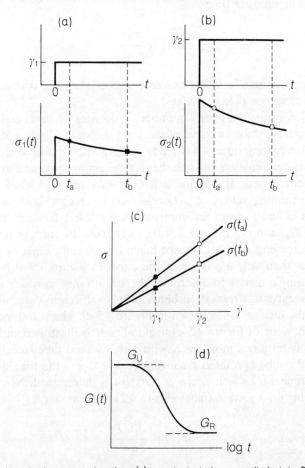

4.5 Linear viscoelastic stress relaxation: (a) constant strain γ_1 applied at $t = 0$ leads to a time-dependent stress $\sigma_1(t)$; (b) a higher strain γ_2 applied at $t = 0$ leads to time-dependent stress $\sigma_2(t)$; (c) from (a) and (b) the stresses at time t_a, $\sigma(t_a)$, and at time t_b, $\sigma(t_b)$, are linear in the strain; (d) the observed dependence of $G(t)$ on log t through one complete relaxation. G_U and G_R are the unrelaxed and relaxed moduli, respectively.

recover and a larger strain γ_2 is then applied. The time dependence of the stress is shown in Figure 4.5(b). At low strains (as in creep) it is observed that the isochronals are linear, as illustrated in Figure 4.5(c). Thus, if at an arbitrary time t the stresses in the two experiments are $\sigma_1(t)$ and $\sigma_2(t)$, then

$$\frac{\sigma_1(t)}{\gamma_1} = \frac{\sigma_2(t)}{\gamma_2}.$$

The stresses in the two experiments at the same time t are proportional to the imposed strains. This fact leads to the definition of the stress-relaxation modulus at time t:

$$G(t) = \frac{\sigma(t)}{\gamma}. \tag{4.11}$$

This is another manifestation of linear viscoelasticity; it is observed in all polymers at strains below ~ 0.005.

If $G(t)$ is measured over a number of decades of time, and is plotted against $\log t$, it exhibits a curve of the form shown in Figure 4.5(d). At very short and very long times $G(t)$ tends towards the limiting values G_U and G_R; in these extreme regions the shear is becoming more and more elastic, independent of time. It is found that $G_U^{-1} = J_U$ and $G_R^{-1} = J_R$, J_U and J_R being the limiting values of $J(t)$ observed in creep (Figure 4.1(d)). The subscripts U and R stand for unrelaxed and relaxed. Between the limiting values of G_U and G_R (Figure 4.5(d)), $G(t)$ takes the form of a sigmoid. If the measurements of $G(t)$ are not taken over a sufficiently large number of decades then only a portion of the curve in Figure 4.5(d) is displayed.

It is a simple matter to determine the strain range in which a specimen exhibits linear stress-relaxation behaviour; it suffices to determine several isochronals, such as those shown in Figure 4.5(c). The transition from the linear behaviour of Figure 4.5(c) to non-linear is illustrated in Figure 4.6.

$G(t)$ is determined most easily with a thin-walled tube (see Figure 4.2). At $t = 0$ the tube is rotated through an angle θ and the time dependence of the torque $\Gamma(t)$ which keeps θ constant is determined. Now the shear stress acting on a section (from Figure 4.2) is, from eqn 4.5,

$$\sigma(t) = \frac{\Gamma(t)}{2\pi r^2 s}, \tag{4.12}$$

and the constant strain is

$$\gamma = \frac{r\theta}{l}.$$

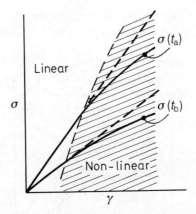

4.6 Isochronals taken at t_a (see Figure 4.5) after the initiation of the stress relaxation experiment, $\sigma(t_a)$, and at t_b, $\sigma(t_b)$. The diagram illustrates the transition from linear to non-linear behaviour. Note that this $\sigma-\gamma$ plot cannot be obtained in a conventional stress–strain experiment: it must be obtained by a series of stress relaxation experiments, as illustrated in Figure 4.5.

It follows from eqns 4.11 and 4.12 that

$$G(t) = \left[\frac{l}{2\pi r^3 s} \right] \frac{\Gamma(t)}{\theta}. \tag{4.13}$$

$G(t)$ is measured by determining $\Gamma(t)$ for fixed θ. Conversely, in a design calculation, the time dependence of the torque in a tube twisted through an angle θ may be obtained if $G(t)$ is known.

The stress-relaxation modulus in tension $E(t)$ is defined as

$$E(t) = \frac{\sigma(t)}{\varepsilon}, \tag{4.14}$$

where ε is the constant tensile strain applied to the specimen and $\sigma(t)$ is the tensile stress. The measurements are made most conveniently with conventional testing equipment in which a strain is imposed at $t = 0$ and the force generated is measured as a function of time.

Typical values for $E(t)$ for an amorphous polymer in the region of the glass–rubber transition are shown in Figure 4.7. The polymer is poly-isobutylene,

$$\left[\begin{array}{c} H \\ | \\ C \\ | \\ H \end{array} - \begin{array}{c} CH_3 \\ | \\ C \\ | \\ CH_3 \end{array} \right]_n .$$

The T_g is $-78°C$ (see Figure 2.12). Note that the relaxation in the time

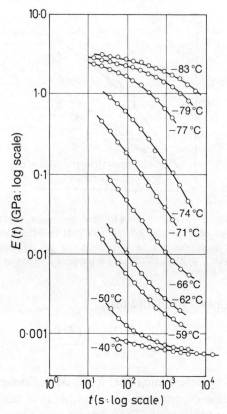

4.7 Stress relaxation modulus observed in tension $E(t)$ of polyisobutylene at different temperatures in the region of the glass–rubber relaxation ($T_g = -80°C$). At $-83°C$ at short time, $E(t)$ approaches asymptotically the modulus of the glass; at $-40°C$ at long time, $E(t)$ approaches asymptotically the modulus of the rubber. The relaxation is centred in the region of $-66°C$. Note the immense reduction in $E(t)$ of over 3 decades in a temperature rise of 43°C: this behaviour is typical of amorphous polymers at the glass–rubber relaxation.

scale of the experiment (10 to 10^4 s) is centred at $\sim -66°C$; the $\log E(t)$ versus $\log t$ plot is of greatest slope in this region; at lower and higher temperatures the slope is less.

4.2.3 Dynamic properties

It is part of our common experience that a plastic beaker, when struck, emits a dull note of short duration, which is quite different from the ringing note emitted by a bell or a crystal wine glass. This property of high mechanical damping is another manifestation of viscoelasticity. It is a property that is frequently of value, for instance in shock absorbers. In plastic structures subject to forced oscillation, mechanical vibrations at the natural frequencies of the structure do not easily build up, due to the high damping capacity of the plastic. A good example of this is the application of plastic materials in sailing craft, particularly in hull construction. Vibrations of the hull stimulated by the elements are rapidly damped. Further, noise created by ships' machinery is less readily transmitted to and radiated into the sea where it could cause interference with acoustic depth sounders and similar equipment. Where components of machinery are themselves made of plastic—for example gears, shafts, or even boats' propellers—high damping characteristics mean that vibration generated is not sustained and transmitted so that it becomes a nuisance. Another advantage is that the sound of the sea is diminished due to the damping of the acoustic waves as they pass through the hull. Now in these examples the frequencies of vibration are somewhat different: **how do polymer damping characteristics vary with frequency**?

We describe next the characteristics, or parameters, used to quantify dynamic viscoelastic properties and introduce briefly some of the methods of measurement. The close relationship between the dynamic parameters (for which the variable is frequency) and the parameters from step-function experiments (with time as variable) is described later.

Suppose an oscillatory shear strain of angular frequency ω,

$$\gamma = \gamma_0 \sin \omega t, \tag{4.15}$$

is generated in a specimen. For a linear viscoelastic material the stress response is also sinusoidal, but is out of phase with the strain

$$\sigma = \sigma_0 \sin(\omega t + \delta). \tag{4.16}$$

The strain lags behind the stress by a phase angle δ. A vector method of representing the dependence of γ and σ on ωt is shown in Figure 4.8. If we expand σ (from eqn 4.16),

$$\sigma = (\sigma_0 \cos \delta)\sin \omega t + (\sigma_0 \sin \delta)\cos \omega t, \tag{4.17}$$

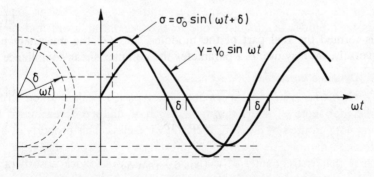

4.8 Vector representation of an alternating stress leading an alternating strain by phase angle δ.

we see that the stress consists of two components: one in phase with the strain ($\sigma_0 \cos \delta$); the other 90° out of phase ($\sigma_0 \sin \delta$).

The relationship between stress and strain in this dynamic case can be defined by writing

$$\sigma = \gamma_0 [G' \sin \omega t + G'' \cos \omega t],$$

in which (from eqn 4.17)

$$G' = \frac{\sigma_0}{\gamma_0} \cos \delta, \tag{4.18}$$

and

$$G'' = \frac{\sigma_0}{\gamma_0} \sin \delta. \tag{4.19}$$

Thus the component of the stress $G' \gamma_0$ is in phase with the oscillatory strain; the component $G'' \gamma_0$ is 90° out of phase.

This formulation is analogous to the relationship between current and voltage in an electrical circuit, which suggests the use of a complex representation. If we write

$$\gamma^* = \gamma_0 \exp i \omega t \tag{4.20}$$

and

$$\sigma^* - \sigma_0 \exp i(\omega t + \delta), \tag{4.21}$$

then the complex shear modulus is

$$G^* = \frac{\sigma^*}{\gamma^*} \tag{4.22}$$

$$= \frac{\sigma_0}{\gamma_0} \exp i \delta$$

$$= \frac{\sigma_0}{\gamma_0} (\cos \delta + i \sin \delta) \tag{4.23}$$

$$= G' + iG''. \tag{4.24}$$

G' is termed the real part of the modulus, and G'' the imaginary part. Conversely, we may use as a parameter the complex shear compliance

$$J^* = \frac{\gamma^*}{\sigma^*} \tag{4.25}$$

$$= \frac{\gamma_0}{\sigma_0} \exp -i \delta$$

$$= \frac{\gamma_0}{\sigma_0} (\cos \delta - i \sin \delta) \tag{4.26}$$

$$= J' - iJ'', \tag{4.27}$$

in which $J' = \gamma_0 \cos \delta / \sigma_0$ and $J'' = \gamma_0 \sin \delta / \sigma_0$. Note that the tangent of the phase angle is

$$\tan \delta = \frac{J''}{J'} = \frac{G''}{G'}. \tag{4.28}$$

For solids which are purely elastic $\tan \delta$ equals zero. Metals, for example, conform fairly closely to this ideal: another good example of a low-damping solid is quartz. Polymers on the other hand have values of δ of the order of several degrees: in certain temperature ranges (for instance the glass to rubber transition) δ may approach $30°$. This very high damping is frequently of use technically (4.N.3).

Example 4.3

A bar of polymer is of length 60 mm and has a rectangular cross-section of dimensions $10 \text{ mm} \times 4 \text{ mm}$. It is subjected to a force acting along the length of the bar and oscillating sinusoidally between -150 and 150 N at a frequency of 50 Hz. The imaginary part of the complex tensile compliance of the polymer at 50 Hz is $D'' = 3.7 \times 10^{-9} \text{ Pa}^{-1}$. Calculate the power input required to sustain the oscillations.

Solution

Let the angular frequency (in rad s^{-1}) corresponding to 50 Hz be ω. The power input P is given by the work done per cycle (integral of force $\times$ displacement) multiplied by the number of cycles per second. If F is the oscillating force and x the relative displacement of the ends of the bar which results,

$$P = 50 \int_{t=0}^{t=2\pi/\omega} F \, dx = 50 \int_{0}^{2\pi/\omega} F \frac{dx}{dt} \, dt.$$

Since the cross-sectional area of the bar is $4 \times 10 = 40 \text{ mm}^2$, F is related to the oscillating stress σ thus:

$$F = 40 \times 10^{-6} \sigma$$

and, since the length of the bar is 60 mm, x is related to the oscillating strain ε thus:

$$x = 60 \times 10^{-3} \varepsilon.$$

Hence

$$P = 50 \int_{0}^{2\pi/\omega} (40 \times 10^{-6} \sigma)(60 \times 10^{-3} \, d\varepsilon / dt) \, dt$$

$$= 1.2 \times 10^{-4} \int_{0}^{2\pi/\omega} \sigma \left(\frac{d\varepsilon}{dt} \right) dt.$$

Let the stress amplitude be σ_0, then the sinusoidal oscillations of σ can be written

$$\sigma = \sigma_0 \sin \omega t.$$

By employing the complex tensile compliance D^* in an analogous manner to J^* (see text), D' and D'' can be used to obtain the in-phase and (90°) out-of-phase components of ε, thus

$$\varepsilon = \sigma_0 D' \sin \omega t - \sigma_0 D'' \cos \omega t$$

and, differentiating,

$$\frac{d\varepsilon}{dt} = \omega \sigma_0 (D' \cos \omega t + D'' \sin \omega t).$$

By substitution in the integral, we obtain

$$P = 1.2 \times 10^{-4} \int_0^{2\pi/\omega} \omega \sigma_0^2 (D' \sin \omega t \cos \omega t + D'' \sin^2 \omega t) \, dt.$$

Finally, by carrying out the integration, we arrive at

$$P = 1.2 \times 10^{-4} \times \pi \sigma_0^2 D''.$$

From the cross-sectional area of 40 mm^2 and the amplitude (150 N) of the oscillating force

$$\sigma_0 = 150/40 \times 10^{-6} = 3.75 \times 10^6 \text{ Pa}.$$

Substitution for σ_0 and D'' yields

$$P = 1.2 \times 10^{-4} \times \pi \times (3.75 \times 10^6)^2 \times 3.7 \times 10^{-9} = 19.6 \text{ W}.$$

Comment

This power appears as heat; the temperature of the specimen rises.

Note that a determination of G^* at a particular temperature and angular frequency gives J^* at the same temperature and frequency, since they are related by

$$G^* = (1/J^*). \tag{4.29}$$

This is not true of $G(t)$ and $J(t)$, which are **not** related by the equation $G(t) = J(t)^{-1}$, **although this latter equation can be of practical use as a reasonable approximation in design** (see Chapter 8).

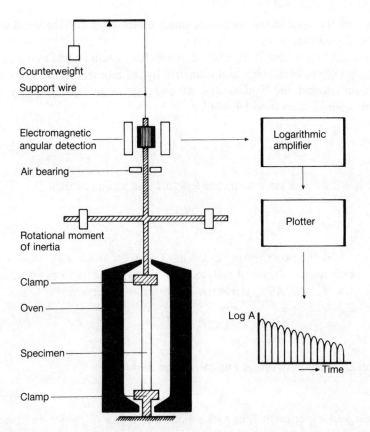

Counterweight

Support wire

Electromagnetic angular detection

Logarithmic amplifier

Air bearing

Plotter

Rotational moment of inertia

Clamp

Oven

Specimen

Clamp

Log A

Time

4.9 Torsion pendulum for the determination of shear modulus and damping as functions of temperature at frequencies around 1 Hz; the support wire has negligible torsional rigidity (after Struik).

The phase angle is related to the logarithmic decrement, which determines the rate at which a stimulated oscillation dies away when the stimulus is removed. As an example, consider the torsion pendulum shown in Figure 4.9. The lower end of the specimen is clamped rigidly and the upper clamp is attached to the inertia arm. By moving the masses of the inertia arm the rotational moment of inertia can be adjusted so as to obtain the required resonant frequency of rotational oscillation. Rotation is detected by an electromagnetic transducer. The system is counterbalanced so that the specimen is not subject to axial stresses.

A measurement is obtained as follows. The system is rotated slightly from its null position by a deflecting torque. The specimen and support wire are thus twisted. The support wire is long and of small diameter so that a negligible torque is induced in it: the only significant torque is induced by the twist in the specimen, which tends to restore the pendulum to its null position.

The deflecting torque is then released and the system begins to oscillate. The frequency of oscillation ω is governed by the moment of inertia of the pendulum around the central axis M and the torque produced by the specimen for unit rotation ($\theta = 1$) Γ_1:

$$\omega = \sqrt{\frac{\Gamma_1}{M}}. \tag{4.30}$$

If the specimen is a circular rod of length l and radius a, then

$$\Gamma_1 = \frac{NG'}{l}, \tag{4.31}$$

in which N is the polar second moment of area ($N = \pi a^4 / 2$).

The resonant oscillation at frequency ω decays. The ratio of successive amplitudes A_n and A_{n+1} yields the logarithmic decrement

$$\Lambda = \ln \frac{A_n}{A_{n+1}}. \tag{4.32}$$

The relationship between Λ and $\tan \delta$ (for $\Lambda \ll 1$) is

$$\Lambda = \pi \tan \delta. \tag{4.33}$$

This elegant experiment thus yields both $\tan \delta$ and G', since from eqns 4.30 and 4.31,

$$G' = \frac{lM\omega^2}{N}.$$

If required, G'' is obtained from

$$G'' = G' \tan \delta.$$

The method is precise even for low values of $\tan \delta$ since the amplitude can be observed over many oscillations.

A direct determination of phase angle can be made—for example in the tubular specimen of Figure 4.3—by applying an oscillatory torque Γ. Let Γ be

$$\Gamma^* = \Gamma_0 \exp i(\omega t + \delta). \tag{4.34}$$

This is most commonly achieved with an electrical torque transducer in which an oscillatory current

$$I^* = I_0 \exp i(\omega t + \delta) \tag{4.35}$$

produces the oscillatory torque

$$\Gamma^* = k_i I^*, \tag{4.36}$$

where k_i is a constant. Note that torque and current are in phase. This is true also of torque and stress, since from eqn 4.5, for a specimen in the form of a thin walled tube,

$$\sigma^* = \left(\frac{1}{2\pi r^2 s} \right) \Gamma^*.$$

Thus the current is exactly in phase with the stress. What of the strain in the tube? From eqn 4.6,

$$\gamma^* = \left(\frac{r}{l} \right) \theta^*. \tag{4.37}$$

It follows that the strain and the rotational angle θ are in phase. Thus the phase angle between the two observables θ^* and I^* is identical to the phase angle between γ^* and σ^*. Writing

$$\theta^* = \theta_0 \exp i\omega t, \tag{4.38}$$

it follows that

$$G' + iG'' = \frac{\sigma^*}{\gamma^*}$$

$$= \left(\frac{l}{2\pi r^3 s} \right) \frac{\Gamma^*}{\theta^*}. \tag{4.39}$$

Hence

$$G' = \left(\frac{lk_i}{2\pi r^3 s} \right) \frac{I_0 \cos \delta}{\theta_0} \tag{4.40}$$

and

$$G'' = \left(\frac{lk_i}{2\pi r^3 s} \right) \frac{I_0 \sin \delta}{\theta_0}. \tag{4.41}$$

Thus G' and G'' are obtained by measuring I_0, θ_0, and δ. The difficult quantity to determine is δ. The best method is to convert θ into a voltage using a rotational transducer. The phase difference δ between this voltage and the torque current can then be obtained by one of a number of precise electrical methods. This forced-oscillation technique is used most efficiently at frequencies somewhat remote from the resonant frequency of the system.

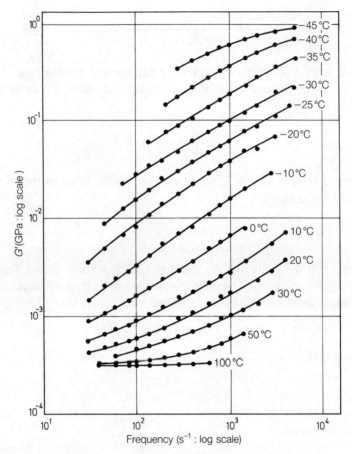

4.10 Frequency dependence of the dynamic shear modulus $G'(\omega)$ of polyisobutylene at different temperatures in the range −45°C to +100°C. This relaxation is the glass-to-rubber relaxation: it is observed here centred in the region of −10°C, well above the glass transition (−80°C) because of the high frequency of observation. The measurements were by forced oscillation (after Fitzgerald, Grandine, and Ferry).

Forced oscillation has two advantages over the torsion pendulum. First, it is a reliable technique for high values of δ: the torsion pendulum is not useful as a precise instrument for high damping. Second, it is very easy to change frequency: the frequency control knob on the sine wave generator is simply rotated. By means of appropriate design the obtainable frequency spectrum can lie in the range 10^{-3} to 10^4 Hz.

The dependence of the dynamic shear modulus G' of polyisobutylene at frequencies between 10 and 10^4 Hz is shown in Figure 4.10 at temperatures between −45°C and 100°C. Note that the centre of the relaxation is in the region of −10°C. Here G' changes by a factor of ×10 as the frequency increases from 30 to 3000 Hz. Note from Figure 4.7 that the

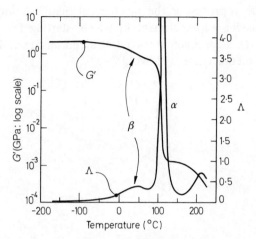

4.11 Temperature dependence of shear modulus G' and logarithmic decrement Λ as a function of temperature for poly(methyl methacrylate) at constant frequency near 1 Hz. The α-relaxation is due to the onset of movement of the main backbone of the molecule; the β-relaxation is due to the onset of hindered rotation of the side group. The polymer glass, at temperatures where the side group is mobile, is more ductile than at the lower temperatures when its movement is frozen in. Measurements taken by torsion pendulum (after Schmieder and Wolf).

relaxation seen in $E(t)$ in the same polymer is centred 60°C lower, at −70°C. This large difference we discuss later, but we can state at once that the responsible factor is the difference in the effective experimental timescales; in stress relaxation $t \approx 10^2$ s but in the dynamic experiment the frequency is centred at $\sim 10^2$ Hz.

The determination of G' and Λ (or any equivalent pair of dynamic mechanical parameters) is used very commonly both as a research tool and also as a quality control tool. The measurements are made at various frequencies over an extremely wide range of temperature, usually by the forced oscillation technique (known as Dynamic Mechanical Analysis). An indication of the scope of such measurements is shown in Figure 4.11 for the amorphous polymer poly(methyl methacrylate). It will be seen that G' decreases with increasing temperature. The decrease in G' occurs particularly rapidly in two temperature regions: one centred at 50°C and the other at 110°C. In these same temperature regions Λ exhibits a peak. It is clear from Figure 4.11 that Λ is in some way a measure of the slope of the $\log G'$ versus temperature curve; this point will be taken up in the next section. It is normal custom to term the viscoelastic process at the highest temperature the α-process (or α-relaxation), that at the next lowest temperature the β-process (or β-relaxation) and so on to lower temperatures (γ, δ, etc.).

It will assist the reader if we state at once the origin of these two relaxations. The α-process is the mechanical manifestation of the glass transition. Consider an experiment in which, starting from the rubbery state (150°C, say), a specimen is cooled at a constant rate. The molecules of poly(methyl methacrylate) (PMMA)

$$
\begin{array}{c}
\text{CH}_3 \\
\left[\begin{array}{c} \text{H} \quad | \\ -\text{C}-\text{C}- \\ \text{H} \quad | \end{array}\right]_n \\
\text{C} \\
\diagup \quad \diagdown \\
\text{O} \qquad \text{O} \\
\diagdown \\
\text{CH}_3
\end{array}
\qquad\qquad \text{(I)}
$$

on cooling from the rubbery state through the glass transition region (see Section 2.6), become increasingly sluggish. When the time constant for large scale molecular movement approaches 1 second (the periodicity of the pendulum) the peak in Λ is observed and G' increases from the rubber value to the glassy value. In the glass the molecules are frozen in particular conformations. The side groups, however, are not affected by the freezing of the molecular backbone: they continue to rotate as indicated by the curved arrow in Structure I. At T_g the time constant τ for the rotation of the side group is extremely short (in the microsecond region). On further cooling, the side-group τ increases; when it has risen to ~ 1 second (the periodicity of the pendulum) another peak in Λ is observed and again a sudden increase in G'. Thus the α and β relaxations are observed when **a time constant for a specific molecular motion passes through the time constant of the mechanical experiment**.

Crystalline polymers exhibit more mechanical relaxations than amorphous polymers. It is not an overstatement to remark that the greater number of mechanical relaxations in crystalline polymers is the cause of the substantial difference in properties between crystalline and amorphous polymers (4.N.4). For example in linear (LPE) and branched (BPE) polyethylene at temperatures above −200°C, there is a sequence of relaxations (see Figure 4.12). In branched PE the processes are γ-relaxation at −120°C, β-relaxation at −10°C, and α-relaxation at 70°C. The presence of a relaxation is detected most easily by the peak in Λ; this is one reason why this parameter is of value. The relaxation observed in creep in linear PE at room temperature and above (shown in Figure 4.4) is the α-process. The torsion pendulum is a useful tool for yielding quickly a description of temperature regions where creep or stress-relaxation processes are to be expected. In addition, the relaxation temperatures often mark transitions in ductility: the polymer becomes increasingly brittle as it is cooled.

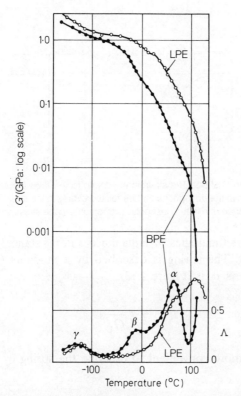

4.12 Temperature dependence of the shear modulus G' and logarithmic decrement Λ of linear (LPE) and branched (BPE) polyethylene. Measurements by torsion pendulum at frequency ~1 Hz. The relaxation processes are labelled α for that at the highest temperature, β for that at the next highest temperature, and so on. The observation of the α-process in creep for LPE is shown in Figure 4.4 (after Flocke).

We have now completed our description of what viscoelasticity is, that is to say, what the parameters are and how they are measured, together with a brief description of typical results. In the following section we present the theory.

4.3 Theory of linear viscoelasticity

The theory of linear viscoelasticity is phenomenological: there is no attempt to discover the time and frequency response of the solid in an altogether a priori fashion. The aim is to predict behaviour under certain circumstances, having observed it under others; for example, to correlate creep, stress relaxation, and dynamic properties so that if one of these has been determined then all the others are known. This is closely related to electrical network theory, both in aim and, as will soon be apparent, in method.

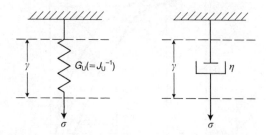

4.13 The spring and dashpot. The strain in the spring is $\gamma = J_U \sigma$; the strain rate $(d\gamma/dt) = J_U(d\sigma/dt)$. The strain in the dashpot cannot be related simply to the stress (it depends on the stress history). The strain rate in the dashpot is proportional to the stress and is $(d\gamma/dt) = \sigma/\eta$.

The parameters (analogies of inductance and resistance) are the spring and the dashpot. The spring is described by a modulus G_U. The strain induced by a stress σ is (Figure 4.13),

$$\gamma = \frac{\sigma}{G_U}. \tag{4.42}$$

It is frequently more convenient to describe the spring by its compliance J_U, so that

$$\gamma = \sigma J_U. \tag{4.43}$$

No matter how rapidly σ varies with time, the strain in a spring is always given by eqn 4.42 or 4.43. We here, and in what follows, use a subsymbol (J_U, J_d, G_U, G_d) for an elastic quantity with no time dependence. These elastic quantities can always be inverted, for example $J_U = G_U^{-1}$.

The instantaneous **strain rate** in the dashpot, when multiplied by the dashpot coefficient η, equals the instantaneous stress (see Figure 4.13)

$$\sigma = \eta \frac{d\gamma}{dt}. \tag{4.44}$$

No matter how rapidly σ varies with time, the strain rate in the dashpot is always given by eqn 4.44. Now these models by themselves give fairly trivial results: it is only when they are combined together that they prove fruitful.

4.3.1 The Zener model

This model can be posed in two ways ((a) or (b) of Figure 4.14) which are —as will become apparent—identical. The horizontal parallel lines in Figure 4.14 indicate elements which have constant strain: e.g. J_d and η of

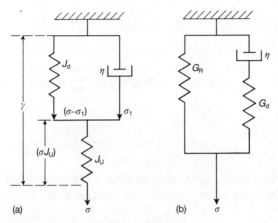

4.14 The Zener model (or standard linear solid). The model may be represented as a spring in series with a Kelvin model, as in (a), or as a spring in parallel with a Maxwell model, as in (b). The significant properties inherent in the Zener model include: (i) two time constants, one for constant stress τ_σ and one for constant strain τ_γ; (ii) an instantaneous strain at $t = 0$ when subject to a step-function stress; and (iii) full recovery following removal of the stress. For the Kelvin and Maxwell models, see Problem 4.8.

(a) have the same strain. We now derive the differential equation describing the relationship between σ, γ, $\dot\sigma$, and $\dot\gamma$ for (a). Let the stress in the dashpot be σ_1. The stress in J_d is then $\sigma - \sigma_1$, so that

$$\gamma - \sigma J_U = J_d(\sigma - \sigma_1)$$

and

$$\sigma_1 = \eta \frac{d}{dt}(\gamma - \sigma J_U).$$

Therefore

$$\gamma - \sigma J_U = J_d \sigma - J_d \eta \left[\frac{d\gamma}{dt} - J_U \frac{d\sigma}{dt} \right].$$

Write[1]

$$\tau_\sigma = J_d \eta.$$

Therefore

$$\gamma = \sigma(J_U + J_d) - \tau_\sigma \frac{d\gamma}{dt} + \tau_\sigma J_U \frac{d\sigma}{dt}.$$

---

1   This step (the introduction of a relaxation time in place of the product of a compliance and dashpot coefficient) is best made in all problems at this point.

Write

$$J_R = J_U + J_d.$$

Therefore

$$\gamma + \tau_\sigma \frac{d\gamma}{dt} = \sigma J_R + \tau_\sigma J_U \frac{d\sigma}{dt},$$

therefore

$$\frac{1}{J_R}\left[\gamma + \tau_\sigma \frac{d\gamma}{dt}\right] = \sigma + \tau_\sigma \left(\frac{J_U}{J_R}\right)\frac{d\sigma}{dt}. \tag{4.45}$$

Now the quantity $\tau_\sigma(J_U/J_R)$ is another time constant, which it is useful to write as τ_γ. The ratio of the two time constants is thus

$$\frac{\tau_\sigma}{\tau_\gamma} = \frac{J_R}{J_U} = \frac{G_U}{G_R}. \tag{4.46}$$

Hence from eqn 4.45

$$\frac{1}{J_R}\left[\gamma + \tau_\sigma \frac{d\gamma}{dt}\right] = \sigma + \tau_\gamma \frac{d\sigma}{dt}. \tag{4.47}$$

It is left as a problem for the reader to show that the alternative variant of the Zener model (Figure 4.14(b)) also leads to eqn 4.47. A solid which conforms to the predictions of this model is termed a Zener solid.[1] The model shows all the significant characteristics of polymer relaxations: it has to be generalized slightly (see Section 4.3.2) to be a precise fit. The solutions which we now state for creep, stress relaxation, and dynamic response do however have considerable illustrative significance.

4.3.1.1 Creep   A constant stress σ_0 is applied at $t = 0$. $d\sigma/dt$ thereafter equals zero, so that from eqn 4.47

$$\gamma + \tau_\sigma \frac{d\gamma}{dt} = J_R \sigma_0. \tag{4.48}$$

The strain $\gamma(t)$ then relaxes to its equilibrium value of $J_R \sigma_0$ with time constant τ_σ. The solution to eqn 4.48 is

$$\gamma(t) = J_R \sigma_0 - \sigma_0[J_R - J_U]\exp[-(t/\tau_\sigma)]$$

---

1   The Zener model leads to a linear equation in stress and strain and their first time derivatives

$$a_1\gamma + a_2\dot\gamma = b_1\sigma + b_2\dot\sigma.$$

This equation, rather than the model with spring and dashpot elements, may be regarded as the foundation of the theory.

or, from eqn 4.3, dividing through by σ_0

$$J(t) = J_R - [J_R - J_U]\exp[-(t/\tau_\sigma)]. \tag{4.49}$$

This equation may be written

$$J(t) = J_U + [J_R - J_U]\{1 - \exp[-(t/\tau_\sigma)]\}. \tag{4.50}$$

By inspection of Figure 4.14(a) we see the first term of eqn 4.50 to be the time-independent deflection of the spring;[1] the second term is the time-dependent deflection of the parallel-spring dashpot element.

4.3.1.2 Stress relaxation A constant strain γ_0 is applied at $t = 0$. $d\gamma/dt$ therefore equals zero, so that from eqn 4.47

$$G_R \gamma_0 = \sigma + \tau_\gamma \frac{d\sigma}{dt}. \tag{4.51}$$

The stress relaxes with time constant τ_γ to its equilibrium value $G_R \gamma_0$. The solution of eqn 4.51 is

$$\sigma(t) = G_R \gamma_0 + [\sigma(0) - G_R \gamma_0]\exp[-(t/\tau_\gamma)],$$

where σ_0 is the initial value ($t = 0$) of the stress. But this is

$$\sigma(0) = G_U \gamma_0,$$

since G_U is the modulus when no relaxation has been permitted to take place. Hence

$$\sigma(t) = G_R \gamma_0 + \gamma_0[G_U - G_R]\exp[-(t/\tau_\gamma)]. \tag{4.52}$$

Physically, it can be seen from Figure 4.14(b) that $G_R \gamma_0$ is the time-independent stress in the element G_R. The time-dependent stress in the other arm starts off at $t = 0$ with value $G_d \gamma_0$, because the total deflection is instantaneously in the spring G_d. As t increases, the dashpot relaxes this stress to zero at $t = \infty$. Using eqn 4.11 in eqn 4.52,

$$G(t) = G_R + (G_U - G_R)\exp[-(t/\tau_\gamma)]. \tag{4.53}$$

---

1 It is very helpful when dealing with viscoelastic models to think physically. For example, examining Figure 4.14, we see that the instantaneous response to an applied stress σ_0 is, in (a), the elongation of the spring J_U (**the dashpot cannot move instantaneously**); in (b) it is the elongation of both springs G_R and G_d. So, clearly, in the latter case the instantaneous strain (at $t \approx 0$) is

$$\frac{\sigma_0}{G_R + G_d} = \frac{\sigma_0}{G_U} = J_U \sigma_0.$$

The same thought process can be used profitably for both models for large values of t.

This equation may be written

$$G(t) = G_U - (G_U - G_R)\{1 - \exp[-(t/\tau_\gamma)]\}. \qquad (4.54)$$

A plot of $J(t)$ and $G(t)$ (eqns 4.50 and 4.54) against $\log t$ is given in Figure 4.15 for a Zener model with $J_U = 1$ GPa^{-1}, $J_R/J_U = 10$, and $\tau_\sigma = 100$ s (so that, from eqn 4.46, $\tau_\gamma = \tau_\sigma/10 = 10$ s). Notice that in both creep and stress relaxation, the solutions are identical at:

- very short times, $t \ll \tau_\sigma$—in both cases the ratio of stress to strain is G_U $(=J_U^{-1})$; and

- very long times, $t \gg \tau_\sigma$—in both cases the ratio of stress to strain is G_R $(=J_R^{-1})$.

At intermediate times it will be seen that, in creep, the compliance passes from J_U to J_R with time constant τ_σ. In stress relaxation the modulus passes from G_U to G_R with time constant τ_γ. Thus, at very short and very long times the stress and strain are Hookean, but at intermediate times when the time t is of the order of the relaxation times[1] this is not true and it in this region that we see viscoelastic effects. The relationship between theory (Figure 4.15) and experiment (Figures 4.4 and 4.7, for example) will be explored later; the reader may well however compare these figures now and see in outline how theory is in broad agreement with experiment.

4.3.1.3 Frequency response   The frequency dependence of the dynamic mechanical properties of the Zener model can be obtained by substituting eqns 4.20 and 4.21 into eqn 4.47. It then follows from eqn 4.25 after reduction that

$$J^* = J_U + \frac{(J_R - J_U)}{1 + i\omega\tau_\sigma}. \qquad (4.55)$$

Since $J^* = J' - iJ''$,

$$J' = J_U + \frac{(J_R - J_U)}{1 + \omega^2\tau_\sigma^2}, \qquad (4.56)$$

and

$$J'' = \frac{(J_R - J_U)\omega\tau_\sigma}{1 + \omega^2\tau_\sigma^2}.$$

Analogous equations were derived by Debye in 1913 for dielectric loss (4.N.5).

---

1   In rheological circles it is customary to refer to τ_σ as the **retardation time** and τ_γ as the **relaxation time**. We will not make that distinction here and will use the term relaxation time for both (relaxation time for creep or relaxation time for stress relaxation).

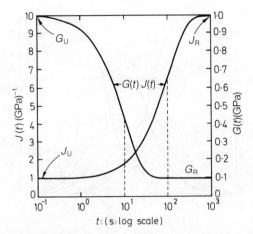

4.15 Solution to the Zener model for a creep experiment ($J(t)$) or for a stress-relaxation experiment ($G(t)$). Model with $J_U = 1$ GPa^{-1}, $J_R/J_U = 10$, and $\tau_\sigma = 100$ s. Note that from eqn 4.46 $\tau_\gamma = 10$ s.

For G^*, inserting eqns 4.20 and 4.21 into eqn 4.47 it follows from eqns 4.25 and 4.29 that

$$G^* = G_U - \frac{(G_U - G_R)}{1 + i\omega\tau_\gamma}. \tag{4.57}$$

Since $\qquad\qquad G^* = G' + iG''$,

$$G' = G_R + \frac{(G_U - G_R)\omega^2\tau_\gamma^2}{1 + \omega^2\tau_\gamma^2}, \tag{4.58}$$

and $\qquad\qquad G'' = \frac{(G_U - G_R)\omega\tau_\gamma}{1 + \omega^2\tau_\gamma^2}$

In Figure 4.16 the parameters J', J'', G', and G'' are plotted against log ω for a Zener model with $J_U = 1$ GPa^{-1}, $J_R/J_U = 10$, and $\tau_\sigma = 100$ s. Note that the inflection points (in J' and G') and the maxima (in J'' and G'') occur at

$$\omega = \frac{1}{\tau_\sigma} \qquad (J' \text{ and } J''), \tag{4.59}$$

and

$$\omega = \frac{1}{\tau_\gamma} \qquad (G' \text{ and } G''). \tag{4.60}$$

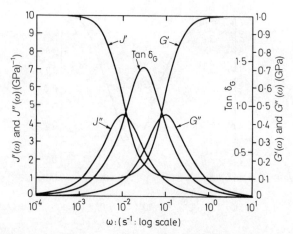

4.16 Solution to the Zener model for a dynamic experiment. Model with $J_U = 1$ GPa^{-1}, $J_R/J_U = 10$ and $\tau_\sigma = 100$ s. Note that maxima occur: in J'' at $\omega = \tau_\sigma^{-1} = 0.01$ rad s^{-1}; in G'' at $\omega = \tau_\gamma^{-1} = 0.1$ rad s^{-1}; and in tan δ at $\omega = (\tau_\sigma \tau_\gamma)^{-\frac{1}{2}} = 0.0316$ rad s^{-1}.

For this model tan δ is also plotted in Figure 4.16 against log ω: the maximum in tan δ occurs at

$$\omega = \frac{1}{(\tau_\sigma \tau_\gamma)^{1/2}} \qquad (\tan \delta). \qquad (4.61)$$

The advantage of using log ω, as opposed to ω, as parameter in a plot is apparent from Figure 4.16. This plot gives an immediate impression of the relaxation time.

The reader will note that the definition of a phenomenological theory given at the start of this section has been satisfied by the Zener model. Measurements in creep, stress relaxation, or dynamic response are inter-related by three of the model parameters, say J_U, J_R, and τ_σ.

4.3.2 Distribution of relaxation times

In order to fit isothermal data (for example the linear polyethylene data for 15°C, see Figure 4.4) it is necessary to fit to the data the three adjustable parameters of the Zener model, which are J_U, J_R, and τ_σ. Since the Zener model is supposedly equally valid for creep, stress relaxation, and dynamic response, these same parameters should then fit (for linear polyethylene at 15°C) $G(t)$, $G^*(\omega)$ and $J^*(\omega)$.

The Zener model almost succeeds, but not quite: it fails only in that the relaxations observed are broader than the predictions. For example, note that in Figure 4.15 the predicted rate effects effectively run their complete course in 3 decades of time: for $J(t)$ the active relaxation region extends from $t \approx 1$ s to $t \approx 10^3$ s; and for $G(t)$ the relaxation extends from $t \approx 0.1$ s to $t \approx 10^2$ s. It will be seen that for linear polyethylene the relaxation illustrated in Figure 4.4 is much broader than 3 decades. Thus at 15°C the $J(t)$ curve appears to be entering the relaxation, being of slowly increasing slope; at 45°C the $J(t)$ curve has the steepest slope; and at 75°C the curve is of steadily decreasing slope, signifying that the relaxation is almost complete. At each of these temperatures the measurements extend over ~4 decades, which is comfortably greater than the ~3 decades necessary to observe a single relaxation-time process (such as the Zener model) in its entirety.

The failure of the Zener model is easily rectified by assuming that the mechanism is **a set of relaxation processes with a band of relaxation times which are closely spaced.** The α-process shown in Figure 4.4 is the interlamellar (intercrystal) slip process, which is analogous to the grain boundary creep process in metals. The heterogeneity of the polymeric solid is the origin of the fact that the relaxation times occur in a distribution: all

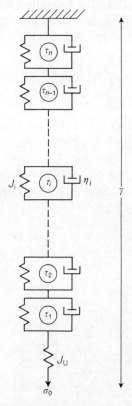

4.17 Distribution of n Kelvin relaxation elements $\tau_1, \tau_2, \ldots, \tau_i, \ldots, \tau_{n-1}, \tau_n$. Each element is stressed by the stress σ_0 which also acts on the instantaneous compliance J_U. The strains in the elements all add to give the total strain.

relaxations in polymers are found to be described by distributed relaxation times.

The generalization of the Zener model to a distribution is a routine extension; see for example the generalization of Figure 4.14(a) shown in Figure 4.17. There is a spring J_U, as in Figure 4.14(a), but it is now in series with n parallel spring–dashpot elements rather than one. The ith of these has compliance J_i, dashpot coefficient η_i, and relaxation time $\tau_i = \eta_i J_i$. When a constant stress σ_0 is imposed at $t = 0$ the spring extends instantaneously, thus giving a strain $\sigma_0 J_U$. The ith element is acted on by the same stress σ_0, as are all the other elements. It is left as an exercise for the reader to argue (Problem 4.11) that at time t

$$J(t) = J_U + \sum_{i=1}^{n} J_i \{1 - \exp[-(t/\tau_i)]\}. \tag{4.62}$$

It is found by experiment that the relaxation times are so closely spaced that it is physically sensible, as well as mathematically convenient, to replace the summation by an integral, so that eqn 4.62 becomes

$$J(t) = J_U + \int_0^\infty \{1 - \exp[-(t/\tau)]\} j(\tau) \, d\tau. \tag{4.63}$$

By the definition of an integral, it will be seen that eqns 4.62 and 4.63 are identical when J_i is identified with $j(\tau) \, d\tau$. The distribution of creep relaxation times $j(\tau)$ is shown in Figure 4.18. The shaded area at τ, $j(\tau) \, d\tau$, is identified with the relaxation strength of the relaxation at τ. When $t = \infty$, $J(\infty) \equiv J_R$ by definition, so from eqn 4.63

$$J_R = J_U + \int_0^\infty j(\tau) \, d\tau,$$

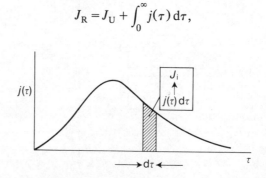

4.18 The distribution of relaxation times plotted against τ from the model of Figure 4.17. The element of area at τ, $j(\tau) \, d\tau$, represents the strength of the relaxation at τ. The integrated area (from eqn 4.64) equals $J_R - J_U$.

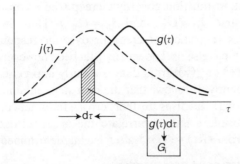

4.19 The distribution of relaxation times $g(\tau)$ for the model of Problem 4.12; a continuous distribution of Maxwell elements. The element of area at τ, $g(\tau)\,d\tau$, represents the strength of the relaxation at τ. The integrated area (from eqn 4.67) equals $G_U - G_R$.

so that the area under the $j(\tau)$ curve is

$$\int_0^\infty j(\tau)\,d\tau = (J_R - J_U). \qquad (4.64)$$

The way in which experimental data, such as Figure 4.4, are fitted is to adjust J_R, J_U, and the shape of the $j(\tau)$ curve so that the data are reproduced. This is quite a formidable task and it is not one which is done routinely.

For stress relaxation the equations are as for the model of Problem 4.12:

$$G(t) = G_R + \sum_1^m G_i \exp[-(t/\tau_i)]. \qquad (4.65)$$

For the continuous distribution of relaxation times

$$G(t) = G_R + \int_0^\infty \exp[-(t/\tau)]g(\tau)\,d\tau, \qquad (4.66)$$

the area under the $g(\tau)$ curve (Figure 4.19) is

$$\int_0^\infty g(\tau)\,d\tau = (G_U - G_R). \qquad (4.67)$$

Stress relaxation data such as those in Figure 4.7 are fitted by fitting G_U, G_R, and the shape of the $g(\tau)$ function. As for creep, the precision fitting of data to the model is not a simple task and it is not often required to be done. The dynamic parameters $J'(\omega)$, $J''(\omega)$, $G'(\omega)$, and $G''(\omega)$ are fitted by the same procedure (see Problem 4.13).

Suppose stress relaxation experimental data are fitted to G_U, G_R, and $g(\tau)$ (the variables of the model), then methods are available to obtain

$j(\tau)$, the distribution function governing creep (the creep limiting compliances are of course $J_U = G_U^{-1}$ and $J_R = G_R^{-1}$). Thus, in principle, one experiment (stress relaxation or creep or dynamic response) is required and then everything else is known. In practice, however, obtaining the spectra (either $j(\tau)$ or $g(\tau)$) from data is a fairly imprecise procedure and it is done infrequently. It follows that the next step, obtaining say $j(\tau)$ from $g(\tau)$ (or vice versa), involves further error. In practice, the best (but approximate) procedure is to interrelate the observed data, for example obtaining $G(t)$ from $G(t) = J(t)^{-1}$ after having determined $J(t)$.

4.3.3 Origin of temperature dependence

The Zener model gives a lucid description of the origin of the large temperature dependence of the viscoelastic properties of polymers. Of the three parameters J_U, J_R, and τ_σ the dominant parameter is τ_σ. Both J_U and J_R do vary with temperature, but the effect is small. For convenience, let the relaxation time τ_σ at temperatures T and T_0 be written τ_T and τ_{T_0}. Then the temperature dependence can be described by a parameter a_T:

$$\tau_T = a_T \tau_{T_0}. \tag{4.68}$$

Most polymer relaxations have relaxation times governed by the Arrhenius equation (4.N.6), for which (from eqns 4.68 and 4.N.6.4)

$$a_T = \exp \frac{\Delta H}{R} \left[\frac{1}{T} - \frac{1}{T_0} \right], \tag{4.69}$$

where ΔH is the activation enthalpy of the relaxation and R is the gas constant. The essential fact is that as T increases, the value of τ_T decreases, so that all viscoelastic effects occur more rapidly. This can be quantified for creep as follows.

From eqn 4.50 the compliance at time t and temperature T_0 is $J^{T_0}(t)$, where

$$J^{T_0}(t) = J_U + (J_R - J_U) \left\{ 1 - \exp \left[-\left(\frac{t}{\tau_{T_0}} \right) \right] \right\} \tag{4.70}$$

and, similarly, at time $a_T t$ and temperature T the compliance is

$$J^T(a_T t) = J_U + (J_R - J_U) \left\{ 1 - \exp \left[-\left(\frac{a_T t}{a_T \tau_{T_0}} \right) \right] \right\}, \tag{4.71}$$

in which we have used eqn 4.68. Comparing eqns 4.70 and 4.71, it will be seen that

$$J^{T_0}(t) \equiv J^T(a_T t). \tag{4.72}$$

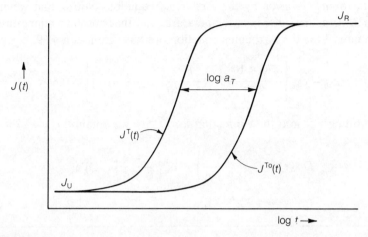

4.20   Illustration of the time–temperature shift (eqn 4.72). The shear compliance curves at T and T_0 when plotted against $\log t$ are simply displaced horizontally by $\log a_T$: a_T is the shift factor (eqn 4.68). The small temperature dependence of J_U and J_R is neglected here.

This result holds for all values of t. If the compliances at T and T_0 are plotted against $\log t$ it follows that the two curves are separated by a constant displacement along the $\log t$ axis; this displacement is $\log a_T$ (see Figure 4.20). One curve may then be shifted by $\log a_T$ to superimpose on the other: a_T is, for this reason, frequently known as **the shift factor**.

Example 4.4

A grade of polypropylene is found to have the following tensile creep compliance when measured at 35°C:

$$D(t) = 1.2t^{0.1}\ \mathrm{GPa}^{-1},$$

where t is expressed in seconds. It exhibits time–temperature superposition and obeys the Arrhenius equation with an activation enthalpy $\Delta H = 170\ \mathrm{kJ\ mol}^{-1}$. Predict the tensile creep compliance of this polypropylene at 40°C.

Solution

Let 35°C be taken as reference temperature T_0 and 40°C as temperature T. Then, applying eqn 4.72 to the tensile creep compliance yields

$$D_{40}(a_{40}t) = D_{35}(t).$$

Rearranging, the required tensile creep compliance at 40°C is given by

$$D_{40}(t) = D_{35}\left(\frac{t}{a_{40}}\right).$$

The factor a_{40} is given by the Arrhenius equation. Noting that we have taken 35°C as the reference temperature, and that absolute temperatures are required for the Arrhenius equation, we have from eqn 4.69

$$a_{40} = \exp\left[\frac{170 \times 10^3}{8.31}\left(\frac{1}{273 + 40} - \frac{1}{273 + 35}\right)\right] = 0.346.$$

Substituting a_{40} into the expression for $D_{40}(t)$, we obtain

$$D_{40}(t) = D_{35}\left(\frac{t}{0.346}\right) = 1.2\left(\frac{t}{0.346}\right)^{0.1} \text{ GPa}^{-1}$$

$$= 1.33t^{0.1} \text{ GPa}^{-1}.$$

The foregoing theory of shifting also holds for a distribution of relaxation times, if ΔH (eqn 4.69) is the same for all elements of the distribution. The discerning reader may already have noted this possibility from Figures 4.4 and 4.7; the curves give the appearance of being shiftable horizontally to form a smooth curve. It was this simple and rudimentary observation that stimulated Leaderman fifty years ago to formulate this theory of superposition.

In principle, superposition permits short-time experiments at a high temperature T to yield information about properties which, at a lower temperature T_0, could take much longer to obtain.[1] For example, suppose we require to know for 300 K the compliance at 10 years for a new polymer—one which has been synthesized in the past year, so that the 10-year data could not possibly be known—then we can obtain this by tests at 350 K (say) over a much shorter period. Suppose that $\Delta H = 120$ kJ mol^{-1}, a typical value for relaxations in a crystalline polymer or in a glass below the T_g region, then from eqn 4.69 (taking $T_0 = 300$ K and $T = 350$ K)

$$a_T = \exp\left(\frac{0.12 \times 10^6}{8.31}\right)\left[\frac{1}{350} - \frac{1}{300}\right]$$

$$= 1.03 \times 10^{-3}.$$

Thus, by performing the experiment at 350 K, the time required is not 10 years but $1.03 \times 10^{-3} \times 10$ years, which is 90 hours. The test (at 350 K)

---

1 Comparable techniques developed for metal creep are, due to Dorn, the temperature-compensated time parameter, and the related Larson–Miller parameter.

would therefore be complete within one week. This result holds, of course, for both creep and stress relaxation.

Despite the attractiveness of time–temperature superposition and the potential saving in time, the method has not in fact been widely used to obtain creep data for design. One good reason for doubt about the precision of the method is the existence of **physical ageing** (see Section 4.4.1). Nevertheless, general points well worth retaining in the mind are: (i) creep deformation processes are speeded up at higher temperatures; (ii) the **effective time** at a temperature T_0 is t/a_T, where t is the time for the same mechanical effect at another temperature T.

4.4 Polymer selection: stiffness

4.4.1 Temperature dependence

Knowledge of the temperature dependence of the modulus[1] gives the first and most important clue to the engineering application of a polymer. The temperature scan of shear modulus G' for three polymers is shown in Figure 4.21. **The temperature dependence is characterized by a shallow decline of $\log G'$ with T over most of the scan with here and there an abrupt drop**.

- The shallow decline is due to thermal expansion: the molecules in the solid move further apart as the temperature increases, and this lowers the modulus. This effect occurs in all solids (other than entropy elastic rubbers).

- Each abrupt drop is generated by a viscoelastic relaxation process due to some specific type of molecular movement (4.N.7).

Note that the shear modulus for all three polymers at −200°C is 1 to 5 GPa.

- **Natural rubber.** On heating from −50°C to −25°C the modulus falls abruptly (glass-to-rubber relaxation) and levels off at the rubber elastic modulus specific to this rather high degree of cross-linking ($G' \approx 1.2$ MPa). In the rubbery state the modulus increases with temperature (eqn 3.30).

- **Poly(vinyl chloride) (PVC).** In the region of −40°C a small relaxation

---

1 We here, and in what follows, do not specify the effective time (or frequency) of measurement. Strictly this should be done, but since the modulus does not depend dramatically on effective time—other than in the centre of a major relaxation—and since it would become tedious to keep mentioning the point, we will leave it out. As a rule, the modulus will normally refer to a determination with effective time ~1 s.

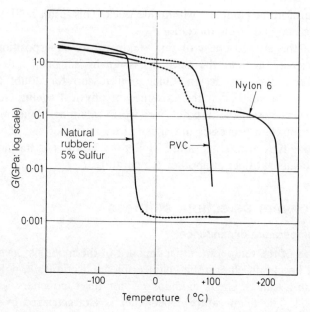

4.21 Dependence of the shear modulus on temperature for three representative engineering polymers: natural rubber (cross-linked); PVC (essentially amorphous and not cross-linked); and nylon 6 (crystalline). The temperatures at which these polymers are used in technology are indicated (•—•—•—•—•) (after Wolf).

process (β) causes a dip in G'. The major drop in G' is at $\sim +70°C$ (the glass-to-rubber relaxation (α)).

- **Polycaprolactam (nylon 6).** This polymer has a crystallinity $\sim 50\%$ and shows a series of relaxations as the specimen is heated. The major relaxations are at 50°C (glass-to-rubber relaxation of the amorphous fraction), and crystal melting point at 220°C. Other less intense relaxations of the amorphous fraction occur below 0°C. The difference between the temperature characteristics of amorphous and crystalline polymers illustrated in Figure 4.21 is most marked and entirely characteristic (compare also Figures 4.12 and 4.11). Glassy polymers (4.N.4) have one dominant relaxation (glass-to-rubber) and, sometimes, a smaller secondary relaxation (as in PMMA, PVC, and polycarbonate[1]). Crystalline polymers usually have several relaxations.

The temperature range in which each of these three engineering polymers is used is indicated in Figure 4.21. Natural rubber has wide application as an elastomer. Its upper use temperature is determined by the onset

---

1 The ductile polymeric glasses **always** exhibit a secondary viscoelastic relaxation below room temperature, and in some instances more than one.

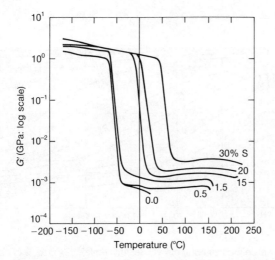

4.22 Shear modulus *G* of natural rubber for varying degrees of vulcanization at temperatures between −200°C and +250°C (after Wolf).

of slow chemical degradation ∼+100°C. Its lower use temperature as a rubber is determined by the onset of stiffening as the glass transition of natural rubber approaches.

The glass transition of natural rubber is modified by the degree of crosslinking, as indicated by the sulfur content in Figure 4.22. At 0°C, for example, note that the modulus of cross-linked rubber increases by ×10³ as the sulfur content is increased from zero to 30%. The cross-linking raises the glass transition (increases the relaxation time of molecular movement) and this in turn raises the temperature of the abrupt modulus drop (4.N.8). The change induced by the added sulfur (a change at 20°C from a typical rubber to a typical glassy polymer) is a kinetic effect in which the rate of molecular movement is systematically decreased by the addition of sulfur cross-links (see Figure 0.2).

Below −40°C PVC (Figure 4.21) is a brittle glass of no value. The small β-relaxation at ∼−40°C transforms the polymer from a brittle to a reasonably ductile glass. The upper use temperature is determined by the onset of the **glass-to-rubber relaxation** (α). For example, in Figure 4.23 the stress relaxation modulus *E*(*t*) for PVC is plotted against log *t*. Note that time effects are not easily detected at 24°C but that as the temperature is increased, time effects become more and more pronounced. The presence of these time effects in the region of 70°C could be easily deduced from the simple modulus–temperature plot in Figure 4.21,[1] but

---

1 Wherever the modulus–temperature plot dips, time effects will be observed.

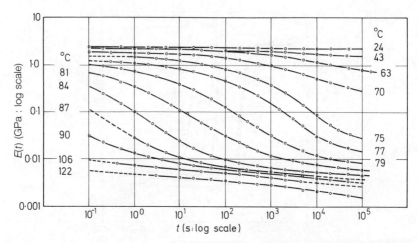

4.23 Stress relaxation modulus $E(t)$ of PVC determined from 24°C to 122°C (after Sommer).

for quantitative design at temperatures up to 70°C data such as those in Figure 4.23 are essential.

Below −30°C nylon 6 is a brittle solid: its use temperature spans the range −30°C to ~+150°C (see Figure 4.21). The designer must be aware of the dramatic drop in modulus in this temperature range: a change of the order of 1 decade. The upper use temperature can be extended by reinforcement with glass or carbon fibre or with mineral particles: this is true of all thermoplastics, particularly the crystalline thermoplastics.

The temperature dependence of the modulus recorded in Figure 4.21, for example, is reproducible **if** the specimen has not been quenched from the melt during its moulding cycle. If the specimen were quenched from the melt[1] then it can be stabilized by annealing at a temperature above T_g but below the melting point. For example, the temperature dependence of G' and Λ for quenched nylon 6.6 is shown in Figure 4.24. Quenched specimens were produced by quenching sheets from the melt into iced water. On reheating through the glass transition (~60°C) during the measurements, the quenched solid commences to crystallize further. **This is marked by the rise in G'.** By the time the specimen has been heated to 100°C the crystallization has reached a quasi-equilibrium.

Heating a crystalline polymer above the T_g of the amorphous fraction will always permit additional crystallization to occur if the specimen is not initially at equilibrium. Probably the most extreme example of this normal

---

1   This point is significant in relation to the moduli of injection mouldings which can be quenched, to a greater or lesser extent, in a moulding cycle (Chapter 7), and also to polymer fibres which when spun cool very rapidly because they are of small diameter (~10 μm).

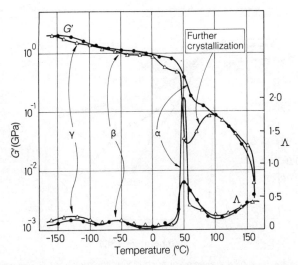

4.24 Temperature scan of G' and Λ for nylon 6.6: one specimen is quenched ($\triangle$); the other ($\bullet$) is quenched but heated at 90°C prior to the temperature scan (so as to complete the crystallization). Note that the quenched specimen starts to crystallize when it is heated to 50°C (positive slope of log G–T curve). The dominant relaxation (α) is the glass relaxation of the amorphous fraction which occurs at $T \approx 50°C$ (after Wolf and Schmieder).

behaviour is PET. If PET is quenched rapidly from the melt to below T_g (67°C) the specimen can be obtained with no detectable crystallinity. Such a specimen yields a log $G' - T$ curve as shown in Figure 4.25 (labelled as 0%). When quenched specimens were heat-treated at the three temperatures marked T_c (124°C, 174°C, and 230°C), specimens of crystallinities 26%, 33%, and 40% resulted, for which the $G' - T$ curves are indicated. The latter crystallization ($T_c = 230°C$ for 1 day) produces a specimen which is essentially in pseudoequilibrium: that is, its crystallinity will not be increased above 40% by further annealing.

The presence of small absorbed molecules in the polymer can have a large effect on the temperature dependence of the modulus.[1] We consider here the effect of water, which is the most significant of the molecules that may be absorbed. The effect depends to a large extent on the amount of water absorbed, which depends on:

- the relative humidity—absorption increases with relative humidity (RH);
- the polymer—polymers with hydrogen bonds (e.g. nylons, polyurethanes) or with polar groups (e.g. plyoxymethylene, PMMA) absorb up to several weight per cent at 100% RH (see 1.N.9, Table 1.1); and

---

1 Absorbed molecules can have a large effect on strength also (see Chapter 5).

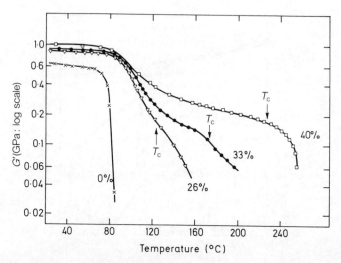

4.25 Poly(ethylene terephthalate) when quenched from the melt has 0% crystallinity; when it is heated through T_g the modulus drops dramatically as indicated. Specimens of crystallinity 26%, 33%, and 40% have modulus–temperature plots as shown. These three values of crystallinity were produced by heating quenched specimens at the temperatures T_c prior to the temperature scan (after Illers and Breuer).

- the crystallinity—the water molecules absorb into the amorphous fraction and so the **higher** the crystallinity the **less** the absorption.

The modulus–temperature plot for a dry polyurethane is shown in Figure 4.26. The pattern is comparable with that of the nylons—with two major relaxations above room temperature (glass–rubber (α) and melting), and two smaller ones below. The dominant effect of water (see Figure 4.27) is to move the α-relaxation from 30°C in the dry state down to ~0°C. Note from Figure 4.27 that at 20°C adding water to the dry polyurethane lowers the modulus from 0.9 to 0.3 GPa (4.N.9). This effect is typical of all polyurethane polymers and of the nylons.

When an amorphous polymer is cooled from the liquid into the glassy state it is in a metastable condition. The process by which the glass contracts towards the equilibrium line (see Figure 4.28) is known as **physical ageing**. Physical ageing results in a stiffening of the solid and is distinct from chemical ageing at high temperatures, which is due to molecular degradation. Physical ageing, as observed mechanically, is shown for PVC in Figure 4.29. A specimen was quenched from 90°C (above T_g) to 20°C (below T_g) and held at 20°C for a period of 4 years. Tensile creep experiment were run from time to time during the 4 years; after a creep experiment the specimen was allowed to recover before the next creep

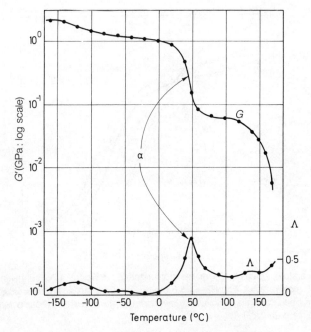

4.26 Temperature scan of G' and Λ for a dry polyurethane with structure

$$-\!\!\left[(HN)\!-\!(CH_2)_u\!-\!(NH)\!-\!(CO)\!-\!O(CH_2)_v\!-\!O\!-\!(CO)\right]_n-$$

with $u = 6$ and $v = 4$ (after Wolf and Schmieder).

experiment. The first such experiment was performed 0.03 days after quenching and the last 1000 days after quenching. The dramatic change in compliance with ageing time is shown in Figure 4.29. The specimen stiffness increases with ageing.

Physical ageing occurs also in crystalline polymers. It is sometimes, in crystalline polymers, difficult to distinguish physical ageing from effects due to recrystallization. However, the two are quite different: physical ageing in crystalline polymers occurs even when the temperature history precludes recrystallization. In crystalline polymers, as in glassy polymers, it is due to a small isothermal contraction (decrease in specific volume) of the polymer after cooling.

The consequences of physical ageing are twofold. First, in order to obtain reproducible data, the specimen must be stored at the measuring temperature for some time, and it should not have been quenched. Second, in design, the consequences of a slow stiffening—allied to a slow embrittlement—must be taken into account.

4.4.2 Stress analysis

Metallic components can usually be designed for load-bearing duties on

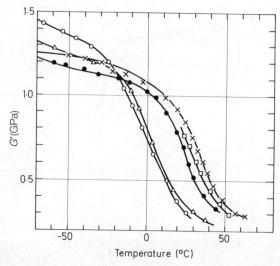

4.27 Temperature scan of G' for the polyurethane of Figure 4.26 containing water as follows: x = 0 mol%; □ = 3.7 mol%; • = 5.6 mol%; △ = 24.5 mol%; and ○ = 28.8 mol%. The α-(glass-to-rubber) relaxation moves to lower temperatures with increasing water content (after Jacobs and Jenckel).

the basis of a small group of properties—yield stress, tangent modulus, and proportional limit determined in standard stress–strain tests. The temperature and time (or rate) sensitivity of metals is not a matter of prime importance, apart from special applications at temperatures above approximately half the melting point (i.e. in applications such as jet turbine blades). With polymer design, things are quite different. Polymers are used fairly close to their melting points. For example, the crystalline engineering polymers in Figures 4.4 and 4.21 (nylon 6 and linear polyethylene) are both

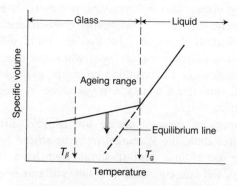

4.28 The origin of physical ageing in an amorphous polymer. After quenching to a temperature below T_g (but above T_β, the temperature of the highest secondary relaxation), the volume slowly contracts: the movement is towards the equilibrium, which is the extrapolated $v - T$ line for the liquid (after Struik).

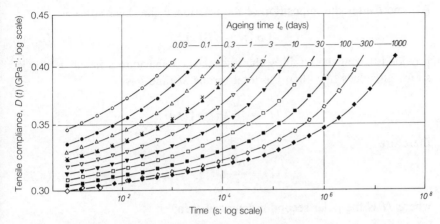

4.29 Tensile compliance of PVC at 20°C. The specimen was quenched from 90°C (20°C above T_g) to 20°C, and kept at 20°C for 4 years. At specific times (t_e) after the quench: 0.03, 0.1, 0.3 days, etc., the compliance was determined. Note the systematic shift of the creep curves. This effect is termed **physical ageing** (see also Figure 4.28) (after Struik).

used at temperatures up to 50°C below T_m. The temperature range of use can be extended closer to T_m with fibre or ceramic reinforcement. The glassy polymer in Figure 4.11 (PMMA) can also be used up to 40°C below T_g in stressed applications. In both cases the major limitation will be viscoelastic creep (or stress relaxation) of the component. Fortunately, this can be allowed for in design using straightforward methods.

In Chapter 8 the topic of design is described systematically, taking into account many factors such as chemical resistance, fracture, fatigue, and forming constraints, in addition to the viscoelastic limitations. We here examine briefly two matters: stress analysis with a viscoelastic material; and the method of design when the stress applied to the viscoelastic material is neither a step-function nor a sinusoidal function of time. The following treatments rest on the assumption of linear viscoelastic behaviour.

4.4.2.1 Torsion of a circular shaft   As a simple illustration of the quantitative application of viscoelastic data we examine the torsion of a solid circular shaft. In the linear range the analysis is straightforward. A solid circular shaft is twisted through an angle θ at $t = 0$. What is the torque at a later time t required to maintain θ constant? Let the shaft be of length l and radius a. Consider the elemental tube of radius r and thickness dr. We assume, as in the elastic case, that the cross-sections remain plane during twisting and that the radii of the shaft remain straight. It follows that the shear strain, which remains constant, in the elemental tube at r is

$$\gamma = \left(\frac{r}{l}\right)\theta. \qquad (4.73)$$

At time t after the imposition of the rotation θ the shear stress is

$$\sigma(t) = \gamma G(t), \tag{4.74}$$

provided that at the strain γ the viscoelastic behaviour is linear. If this is so for all values of r then the total torque at time t is

$$\Gamma(t) = \int_0^a 2\pi r^2 \sigma(t)\,dr,$$

therefore

$$\Gamma(t) = \frac{N\theta}{l}\,G(t), \tag{4.75}$$

where N is the polar second moment of area:

$$N = \frac{\pi a^4}{2}. \tag{4.76}$$

It is thus possible to ·letermine $\Gamma(t)$ for all values of t for which $G(t)$ is known. The solution is identical to the elastic solution.

This will be true only if the material is linear up to the maximum strain. The maximum strain occurs at $r = a$ and is

$$\gamma = \frac{a\theta}{l}.$$

It can be appreciated that the problem is much more complex if part of the cross-section (the outer part) deforms in the non-linear range. If this occurs, the strain in the outer part is still given by eqn 4.73. The complexity arises because the stress generated at each value of r is less than that given by eqn 4.74 (see Figure 4.6). The problem is in some respects analogous to the twisted elastic–plastic rod.

Non-linear stress analysis is described in Chapter 8. It will be seen that, providing the required viscoelastic data are to hand, the methods are empirical, fairly elementary, and sufficiently accurate.

Example 4.5

A polymeric cantilever spring of length l, whose cross-section has a second moment of area I, is held to a constant deflection δ_0 by a force F applied at its free end. Show that the time-dependent decay of F is given as follows in terms of the stress relaxation modulus $E(t)$:

$$F(t) = \frac{3I\delta_0}{l^3}\,E(t).$$

Solution

If a section of the beam is maintained at constant radius of curvature R_0, the bending moment M at that section, at a time t after the curvature was applied, will be given by

$$M(t) = \frac{I}{R_0} E(t).$$

The derivation of this result is analogous to that for the torsion of a circular shaft (eqn 4.75). At a distance x from the built-in end of the cantilever,

$$M(x,t) = [l-x]F(t).$$

Substituting gives an expression for the radius of curvature at this position

$$\frac{1}{R_0(x)} = \frac{[l-x]}{I} \frac{F(t)}{E(t)}.$$

For small deflections y of the beam we may use the usual approximation employed in beam theory:

$$\left(\frac{d^2y}{dx^2}\right)_0 = \frac{1}{R_0(x)} = \frac{[l-x]}{I} \frac{F(t)}{E(t)}.$$

Integrating once gives the gradient

$$\left(\frac{dy}{dx}\right)_0 = \left[lx - \frac{x^2}{2}\right] \frac{F(t)}{IE(t)};$$

since

$$\frac{dy}{dx} = 0 \text{ at } x = 0;$$

and integrating again yields the deflection at the end of the cantilever

$$\delta_0 = \int_0^l \left(\frac{dy}{dx}\right)_0 dx = \frac{l^3}{3I} \frac{F(t)}{E(t)}.$$

Finally, rearranging yields

$$F(t) = \frac{3I\delta_0}{l^3} E(t).$$

Comment

This analysis turns out to be a simple extension of that for the deflection of a linear **elastic** cantilever. Young's modulus is replaced by the appropriate viscoelastic counterpart—in this case the tensile stress relaxation modulus. That this is the appropriate viscoelastic property to be employed can be thought of as arising from the fact that, when the cantilever is subjected to constant deflection, every element of it is subjected to a constant tensile strain (i.e. to tensile stress relaxation conditions).

4.4.2.2 Boltzmann superposition principle   The BSP dates from 1876 and is of the greatest significance, both theoretically and practically. Suppose $J(t)$ is known from 1 s to 1 Ms; we can then find the shear strain $\gamma_0(t)$ for any value of t (1 s $< t <$ 1 Ms) resulting from a constant stress σ_0 applied at $t = 0$:

$$\gamma_0(t) = \sigma_0 J(t). \tag{4.77}$$

Now suppose that an **additional** stress σ_1 is applied at time t_1: what effect does it have? According to the BSP we ask first what is the effect of σ_1 acting by itself and this is, of course,

$$\gamma_1(t) = \sigma_1 J(t - t_1). \tag{4.78}$$

Note that at time t, since σ_1 was applied at t_1, the elapsed time for σ_1 is $t - t_1$ and it is the shear compliance at $t - t_1$ that dictates $\gamma_1(t)$. The BSP then states that the total strain at t is

$$\begin{aligned} \gamma(t) &= \gamma_0(t) + \gamma_1(t) \\ &= \sigma_0 J(t) + \sigma_1 J(t - t_1). \end{aligned} \tag{4.79}$$

In words, the BSP states that if σ_0 (acting alone) yields $\gamma_0(t)$, and if σ_1 (acting alone) yields $\gamma_1(t)$, then σ_0 and σ_1 (acting together) yield $[\gamma_0(t) + \gamma_1(t)]$.

If, instead of merely two stress pulses, we have a sequence (see Figure 4.30), then

$$\gamma(t) = \Delta\sigma_1 J(t - t_1) + \Delta\sigma_2 J(t - t_2) + \Delta\sigma_3 J(t - t_3) + \dots. \tag{4.80}$$

If the stress changes continuously with time then eqn 4.80 can be generalized to give the integral form of the BSP. We require the strain at time t for a specimen which at time u is acted upon by a stress $\sigma(u)$ (u takes values $0 < u < t$). The specimen is virgin at $u = 0$, it having forgotten its previous stress history. Being left for a long period under zero stress would cause the previous stress history to be lost from the memory.[1] The strain at t is then

$$\gamma(t) = \int_0^t J(t - u)\, \frac{d\sigma}{du}\, du. \tag{4.81}$$

---

1 We here give the reader a taste of the interesting anthropomorphic language used by early viscoelasticians and rheologists.

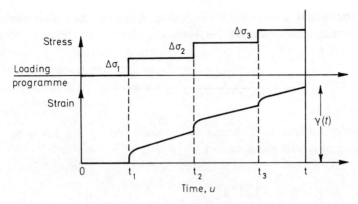

4.30 Three stress increments at a sequence of times (t_1, t_2, t_3) leading to a total strain $\gamma(t)$.

For tensile loading the strain is

$$\varepsilon(t) = \int_0^t D(t-u) \frac{d\sigma}{du} \, du, \tag{4.82}$$

in which $D(t)$ is the tensile compliance and $\sigma(u)$ is the tensile stress acting on the specimen at time u. For tensile stress relaxation due to a continually varying strain history the stress at t is

$$\sigma(t) = \int_0^t E(t-u) \frac{d\varepsilon}{du} \, du. \tag{4.83}$$

$E(t-u)$ is the stress relaxation modulus at time $t-u$, and $\varepsilon(u)$ is the imposed strain at time u $(0 < u < t)$.

We now give two examples of the use of the BSP.

Relationships between creep and recovery. A stress σ_0 is applied at $u = 0$ and is removed at $u = t_1$ (see Figure 4.31). From eqn 4.80,

$$\varepsilon(t) = \sigma_0 D(t) - \sigma_0 D(t - t_1). \tag{4.84}$$

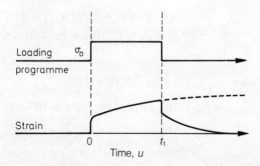

4.31 Response of a viscoelastic solid to loading at 0 and unloading at t_1.

Thus the actual recovery curve is known precisely. As t becomes much greater than t_1, $D(t)$ and $D(t - t_1)$ converge so that $\varepsilon(t)$ tends to zero (see Figure 4.31).

Example 4.6

The polypropylene of Example 4.4 is subjected to the following time-sequence of tensile stress at 35°C:

$$\sigma = 0 \qquad\qquad\qquad t < 0$$
$$\sigma = 1 \text{ MPa} \qquad\qquad 0 \leq t < 1000 \text{ s}$$
$$\sigma = 1.5 \text{ MPa} \qquad 1000 \text{ s} \leq t < 2000 \text{ s}$$
$$\sigma = 0 \qquad\qquad\qquad 2000 \text{ s} \leq t$$

Find the tensile strain at the following times t: (a) 1500 s; (b) 2500 s. Assume that under these conditions polyproylene is linearly viscoelastic and therefore obeys the Boltzmann superposition principle.

Solution

Since the time-sequence of stress consists of a series of discrete steps, the discrete form of the Boltzmann superposition principle is the most straightforward to apply. For a series of n steps in tensile stress of magnitude $\Delta \sigma_i$ applied at times t_i, the BSP states (eqn 4.80) that

$$\varepsilon(t) = \sum_{i=1}^{n} \Delta \sigma_i D(t - t_i) \qquad (t_n < t).$$

(a) When $t = 1.5 \times 10^3$ s: $n = 2$, $\Delta\sigma_1 = 1$ MPa, $t_1 = 0$, $\Delta\sigma_2 = 0.5$ MPa, and $t_2 = 10^3$ s. Substituting into the BSP yields

$$\varepsilon(1.5 \times 10^3 \text{ s}) = 10^6 \times 1.2 \times 10^{-9} \times (1.5 \times 10^3)^{0.1}$$
$$+ 0.5 \times 10^6 \times 1.2 \times 10^{-9}(1.5 \times 10^3 - 10^3)^{0.1}$$
$$= 3.61 \times 10^{-3}.$$

(b) When $t = 2.5 \times 10^3$ s: $n = 3$, $\Delta\sigma_1 = 1$ MPa, $t_1 = 0$, $\Delta\sigma_2 = 0.5$ MPa, $t_2 = 10^3$ s, $\Delta\sigma_3 = -1.5$ MPa, and $t_3 = 2 \times 10^3$ s. Substituting into the BSP now yields

$$\varepsilon(2.5 \times 10^3 \text{ s}) = 10^6 \times 1.2 \times 10^{-9} \times (2.5 \times 10^3)^{0.1}$$
$$+ 0.5 \times 10^6 \times 1.2 \times 10^{-9}(2.5 \times 10^3 - 10^3)^{0.1}$$
$$- 1.5 \times 10^6 \times 1.2 \times 10^{-9} \times (2.5 \times 10^3 - 2 \times 10^3)^{0.1}$$
$$= 0.52 \times 10^{-3}.$$

Comment

In each case, the time argument of D used in finding the response to a given increment in stress $\Delta\sigma_i$ is the time $t - t_i$ elapsed since this stress was applied.

Example 4.7

Show, without recourse to the model itself, that the Zener model whose stress relaxation modulus is given by eqn 4.53,

$$G(t) = G_R + (G_U - G_R)\exp\left(\frac{-t}{\tau_\gamma}\right),$$

has a dynamic modulus G^* with real and imaginary parts given by eqn 4.58.

Solution

Since the Zener model is a linear viscoelastic model, it obeys the Boltzmann superposition principle. In this problem we are concerned with a strain history which is a smoothly varying function of time, with γ undergoing sinusoidal oscillations. Therefore the integral form of the BSP is the most straightforward one to apply:

$$\sigma(t) = \int_0^t G(t-u)\,\frac{\mathrm{d}\gamma}{\mathrm{d}u}\,\mathrm{d}u.$$

The lower limit of the integral in the BSP is always the instant at which deformation commences. Thus, if we choose the sinusoidal oscillations of strain to commence at $u = 0$, the strain history is as follows:

$$\gamma = 0 \qquad\qquad u \leq 0$$
$$\gamma = \gamma_0\sin\omega u \qquad\qquad 0 \leq u,$$

therefore

$$\frac{\mathrm{d}\gamma}{\mathrm{d}u} = 0 \qquad\qquad u \leq 0$$

$$\frac{\mathrm{d}\gamma}{\mathrm{d}u} = \gamma_0\omega\cos\omega u \qquad\qquad 0 \leq u.$$

It will be convenient to express $\cos\omega u$ in complex notation

$$\frac{\mathrm{d}\gamma}{\mathrm{d}u} = \gamma_0\,\frac{\omega}{2}\,[\exp(\mathrm{i}\omega u) + \exp(-\mathrm{i}\omega u)],$$

while $G(t - u)$ is given by

$$G(t - u) = G_R + (G_U - G_R)\exp\left(\frac{-t}{\tau_\gamma} + \frac{u}{\tau_\gamma}\right).$$

Substitution for G and $d\gamma/du$ in the BSP yields

$$\sigma(t) = G_R\gamma_0\sin \omega t + \gamma_0 \frac{\omega}{2}(G_U - G_R)\exp\left(\frac{-t}{\tau_\gamma}\right)I,$$

where I is the integral

$$I = \int_0^t\left\{\exp\left[u\left(\frac{1 + i\omega\tau_\gamma}{\tau_\gamma}\right)\right] + \exp\left[u\left(\frac{1 - i\omega\tau_\gamma}{\tau_\gamma}\right)\right]\right\}du$$

$$= \frac{2\tau_\gamma}{\left(1 + \omega^2\tau_\gamma^2\right)}\left[\exp\left(\frac{t}{\tau_\gamma}\right)(\omega\tau_\gamma\sin \omega t + \cos \omega t) - 1\right].$$

Finally,

$$\sigma(t) = \gamma_0\left[G_R + (G_U - G_R)\frac{\omega^2\tau_\gamma^2}{\left(1 + \omega^2\tau_\gamma^2\right)}\right]\sin \omega t + \gamma_0(G_U - G_R)$$

$$\times \frac{\omega\tau_\gamma}{\left(1 + \omega^2\tau_\gamma^2\right)}\cos \omega t - \gamma_0(G_U - G_R)\frac{\omega\tau_\gamma}{\left(1 + \omega^2\tau_\gamma^2\right)}\exp\left(\frac{-t}{\tau_\gamma}\right).$$

The third term in the expression for $\sigma(t)$ can be seen to be a transient which decays within a time of about τ_γ from the commencement of the oscillations. The first two terms constitute the steady-state response. The terms in $\sin \omega t$ and $\cos \omega t$ represent the stresses in phase and 90° out of phase with the strain, respectively. Consequently, we identify

$$G' = G_R + (G_U - G_R)\frac{\omega^2\tau_\gamma^2}{1 + \omega^2\tau_\gamma^2},$$

and

$$G'' = (G_U - G_R)\frac{\omega\tau_\gamma}{1 + \omega^2\tau_\gamma^2},$$

in agreement with eqn 4.58, which was deduced directly from the Zener model.

Comment

This solution illustrates an important point: for a linear viscoelastic material, it is possible to convert between the viscoelastic response functions, without knowledge of the stress/strain/time differential equations to which they correspond.

Conventional stress–strain curves.   In the standard tensile test the machine is programmed to move the clamps apart at a constant rate and the load

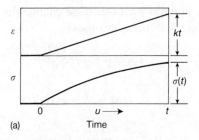

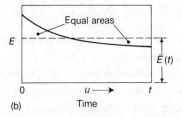

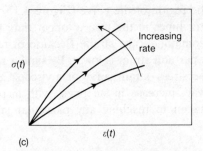

4.32 (a) A constant strain rate leads to a non-linear stress; (b) the definition of $\bar{E}(t)$, the mean modulus; and (c) increasing the strain rate (k) increases the slope of the stress–strain curve.

generated is recorded. This test can be analysed using the BSP by assuming that at $u = 0$ a strain programme is initiated such that (see Figure 4.32(a))

$$\varepsilon(u) = ku,$$

and

$$\frac{d\varepsilon}{du} = k.$$

Substitution of $d\varepsilon/du$ into eqn 4.83 gives the stress at time t in the test

$$\sigma(t) = k \int_0^t E(t - u) \, du. \qquad (4.85)$$

The meaning of the integral is clearer if we change the integration variable to $v = t - u$. Equation 4.85 then becomes

$$\sigma(t) = k \int_0^t E(v)\,dv, \tag{4.86}$$

and because $\varepsilon(t)$ is simply kt, that is, $k = \varepsilon(t)/t$,

$$\sigma(t) = \frac{\varepsilon(t)\int_0^t E(v)\,dv}{t} \tag{4.87}$$

$$= \varepsilon(t)\bar{E}(t). \tag{4.88}$$

$\bar{E}(t)$ is the mean value of E averaged over the interval of time from 0 to t in a stress relaxation test (see Figure 4.32(b)). Equation 4.88 expresses a fact which must always be borne in mind in the constant-strain-rate testing of viscoelastic materials: since $\bar{E}(t)$ decreases as t increases (Figure 4.32(b)), the stress–strain curve measured in a constant strain rate test must always appear **non-linear**—bending towards the strain axis—even in the **linear** viscoelastic region of the curve (Figure 4.32(c)). The reason for this is simple: at later times in the test, a longer time has been available for the stress to accommodate the strain. Because of relaxation of stress, therefore, the stress per unit strain is lower. By similar reasoning, eqn 4.88 also predicts that the stress–strain curve of a viscoelastic material in this type of test will always increase in steepness with increasing strain rate, since less time is taken in reaching any particular strain level (Figure 4.32(c)).

Notes for Chapter 4

4.N.1

Most commonly, in design, it is the stress which is held constant. For example, consider a high-density polyethylene pipeline supported above ground on brackets at regular intervals along its length (Figure 4.33). At the moment during construction when the pipe is raised on to its supports, any element of the pipe is suddenly subjected to a bending stress, say σ_0 (due to the self-weight of the pipe), which subsequently remains constant. The resulting strains in the pipe cause it to sag between supports. These strains, and hence the deflection, would be constant for an elastic pipe, but since polyethylene is viscoelastic the strains increase with time, causing the pipe gradually to sag. The designer of the pipeline would need to allow for creep of the pipe when deciding the distance between support brackets (see Problem 4.19).

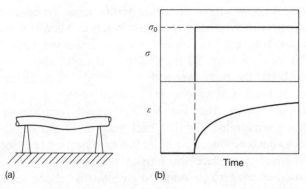

(a)   (b)

4.33 The increasing sag of a viscoelastic pipe. Essentially constant stress due to the weight of the pipe leads to a time-dependent strain.

4.N.2

In design the constraint is sometimes a constant strain, the stress being the dependent variable. Consider a nylon bolt used to join two rigid plates (Figure 4.34). When the bolt is tightened suddenly it is stretched, the tensile strain in the bolt increasing to a value ε_0 and then remaining constant. If the bolt were elastic, the resulting tensile stress would follow suit, also increasing and then remaining constant. In practice, because it is viscoelastic, nylon exhibits stress relaxation. This has obvious practical consequences (see Problem 4.2). The stress relaxation of polymer gaskets and washers is also of considerable importance.

4.N.3

Human tissues (which are polymeric) have a remarkable damping capacity which is required to combat the stresses involved in locomotion. Every time you run, or even walk, shock waves are generated when your heel meets the ground. This is known as heel-strike. The shock bows the legs,

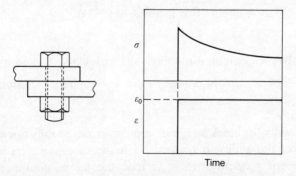

4.34 Stress decay in a bolt as a function of time after tightening: the strain is constant.

compresses discs between the vertebrae and bounces the base of the brain half a millimetre. In a running sport, for instance, when you take a stride you subject the front leg to a force of $\sim 17g$, which is the equivalent of the force of hitting a wall at 30 mph. The heel-strike shock waves travel through the body: the body tissues damp them down, but damage can be caused to soft tissue surrounding the skeletal system, particularly in the spine and lower limbs. The maladies that result include osteoarthritis, stress fracture, tendonitis, migraine, back pain, and inner ear disturbance. Polymers have been produced by molecular modification (e.g. polyurethane elastomers) which reproduce the physical characteristics of body tissue. They are used in modern footwear to cushion the shock waves and are extremely successful.

4.N.4

Block and graft copolymers have mechanical spectra quite unlike **random copolymers**; in some respects the spectra ($\log G'$ and Λ versus temperature at ~ 1 Hz) are like those of crystalline polymers. The effect is illustrated in Figure 4.35. Poly A and poly B are two amorphous homopolymers; their mechanical spectra have the form indicated. A random copolymer, poly (A–co–B) has a comparable spectrum: all three show a ~ 3 decade drop in G' and a Λ peak at the glass–rubber relaxation: the only difference is in the relaxation temperatures, high for poly B, low for poly A, and **intermediate** for poly (A–co–B). But poly (A–b–B) and poly (A–g–B) show two relaxations. The reason is that both the block and the graft copolymers form **two-phase solids**, the poly A and the poly B segments aggregating to form domains (as illustrated in Figure 3.11). An important example of this is the styrene–butadiene system: the dependence of T_g on copolymer content is shown in Figure 3.9 for the random copolymers, **which form single-phase solids**. But the block and graft copolymers form two-phase solids and, as indicated in Figure 4.35, these show a pair of relaxations one typical of polystyrene and the other of polybutadiene. By varying the proportions of comonomer it is possible to obtain:

- at low styrene content, a rubbery solid with glassy inclusions;
- at high styrene content, a glassy solid with rubbery inclusions.

The opportunity for block and graft copolymerization thus opens up to the manufacturer the ability to synthesize families of copolymers with significantly different properties; these include the thermoplastic rubbers and the rubber toughened alloys, both amorphous and crystalline.

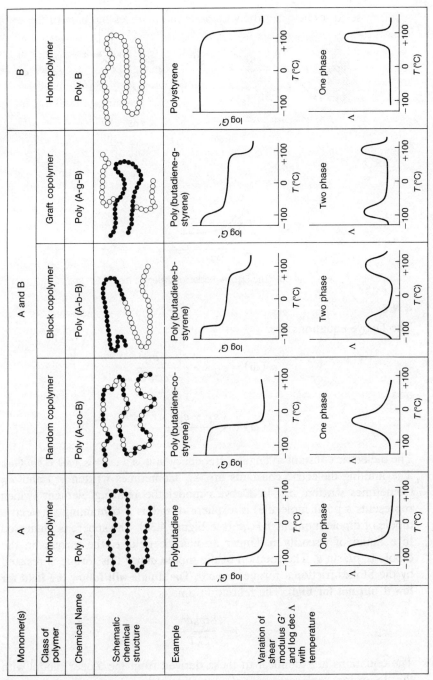

4.35 The mechanical spectra ($\log G'$ and Λ at ~1 Hz versus temperature) for copolymers (random, block, and graft).

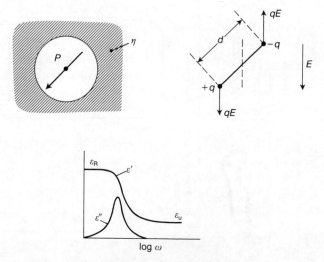

4.36   The Debye model for dielectric loss.

4.N.5

The Debye equations are:

$$\varepsilon'(\omega) = \varepsilon_U + \frac{(\varepsilon_R - \varepsilon_U)}{1 + \omega^2\tau^2},$$

and

$$\varepsilon''(\omega) = \frac{(\varepsilon_R - \varepsilon_U)\omega\tau}{1 + \omega^2\tau^2}.$$

The dielectric constant at angular frequency ω is $\varepsilon^*(\omega) = \varepsilon'(\omega) - i\varepsilon''(\omega)$. The limiting dielectric constants are ε_U (sometimes written ε_∞) and ε_R (sometimes written ε_s). In Debye's model the relaxing element (which represents a polar molecule) is a sphere of radius a containing an electric dipole of dipole moment $p = qd$ (see Figure 4.36). The sphere is immersed in a liquid of viscosity η. Under an electric field E the torque on the dipole is $pE \sin \theta$. The rotation of the dipole under this torque is resisted by the Stokes frictional torque $8\pi\eta a^3\dot{\theta}$. **The dipole will follow the field for low $\dot{\theta}$ but not for high.** The relaxation time is

$$\tau = \frac{8\pi\eta a^3}{2kT}.$$

The equations are identical to those derived from the Zener model with the following replacements: $\gamma(\omega) \rightarrow D(\omega)$ (the electric displacement); $\sigma(\omega) \rightarrow E(\omega)$ (the electric field); and $J^*(\omega) \rightarrow \varepsilon^*(\omega)$ (the complex dielectric constant at frequency ω).

4.N.6

We outline briefly the physical basis of the Arrhenius equation as applied to atomic or molecular movements in a solid. All viscoelastic effects, including large strain effects (see Chapter 5), are due to thermally activated movements of segments of macromolecules under imposed mechanical stress. **There is no question of the stress generating molecular movements which, in the absence of the stress, would not take place.** What in fact occurs is that molecular movements or jumps occur spontaneously, and it is the function of the stress to bias them so that they no longer occur in random directions. This results in a molecular flux which leads to time-dependent mechanical strain.

A simplified description is given here for a monatomic system. The fundamental question to be answered is, in a solid, why should (and how can) an atom in an equilibrium position at α in Figure 4.37, apparently completely 'caged-in' by its neighbours, move to an equivalent equilibrium position at β?

The atom at α in (a) oscillates in the potential well defined by its nearest neighbours; along the line kk', over which it will in time move to position β, it approaches the barrier imposed by the atoms X and Y, ν times per second. In order to pass through, it must surmount an energy barrier; the intense elastic distortion required is indicated at (b). This distorted 'activated complex', has a very small probability of occurrence. Normally the atom approaches the barrier and is reflected back. But, very infrequently, it approaches and a thermal fluctuation permits it to move to

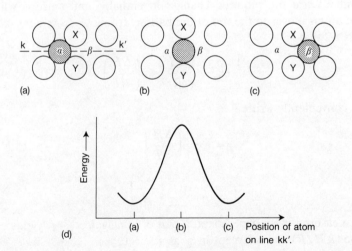

4.37 A schematic illustration of the way in which an atom (shaded) moves from one equilibrium position at α to another at β. The Gibbs free energy of the system changes in the manner indicated as the shaded atom moves along the line kk'.

the top of the barrier, to form the activated complex. Once the atom reaches the top of the barrier (b), it moves on into the neighbouring potential well at β (as indicated at (c) in Figure 4.37). Suppose, for example, the probability of occurrence of the activated complex is 10^{-16}. The frequency with which the atom surmounts the barrier is then equal to the product of the number of times the atom approaches the barrier ($\nu \approx 10^{13}$ s^{-1}) and the probability of occurrence of the complex (10^{-16}) so that,

$$\text{mean jump frequency} = 10^{13} \times 10^{-16} \text{ s}^{-1}$$

$$= 10^{-3} \text{ s}^{-1}.$$

The **mean time** an atom stays in a potential well before jumping, τ, is the reciprocal of the mean jump frequency, which in this case is

$$\tau = 10^3 \text{ s}.$$

The precise equation for τ is

$$\tau = \frac{1}{\nu} \exp\left(\frac{\Delta G}{RT} \right). \qquad (4.N.6.1)$$

ΔG is the change in Gibbs free energy required to take a mole of atoms from the equilibrium position to the top of the barrier. The Gibbs free-energy change is

$$\Delta G = \Delta H - T \Delta S \qquad (4.N.6.2)$$

ΔH and ΔS are the required changes in enthalpy and entropy when a mole of atoms moves to the top of the barrier. It follows from eqns 4.N.6.1 and 4.N.6.2 that

$$\tau = \frac{1}{\nu} \exp - \left(\frac{\Delta S}{R} \right) \exp\left(\frac{\Delta H}{RT} \right). \qquad (4.N.6.3)$$

This is conveniently written

$$\tau = \tau_\infty \exp\left(\frac{\Delta H}{RT} \right) \qquad (4.N.6.4)$$

$$\tau_\infty = \frac{1}{\nu} \exp - \left(\frac{\Delta S}{R} \right). \qquad (4.N.6.5)$$

If τ is measured in an experiment, a plot of $\ln \tau$ against T^{-1} yields a line of slope $(\Delta H/R)$ and intercept $\ln \tau_\infty$ at $T^{-1} = 0$.

The Arrhenius equation is identical to eqn 4.N.6.4 except that the parameters ΔH and τ_∞, as outlined here, take on specific atomic meaning; in the Arrhenius equation they are empirical parameters.

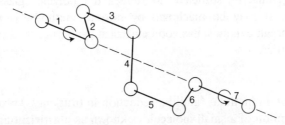

4.38 Limited mobility in the amorphous region of polyethylene. The γ-mechanism (see Figure 4.12) is attributed to movements of this type ('crankshaft rotation').

The application to movements within a **solid polymer** must take into account the elastic distortion of the surrounding molecules, as outlined above for a monatomic system, plus the energy needed to facilitate molecular rotation. One way in which this is thought to happen in polyethylene is indicated in Figure 4.38. Within the amorphous region of solid polyethylene, molecular movement can occur by short sections of the molecule flipping from one equilibrium position to another like a crankshaft. Bonds 1 and 7 are colinear; it will be seen that four atoms can then flip round by rotations around bonds 1 and 7. The onset of this mechanism is one explanation of the γ-viscoelastic process in polyethylene.

4.N.7

The intense temperature dependence of the viscoelastic properties of polymers is due to the temperature dependence of the relaxation times. (The limiting compliances J_U and J_R do change with temperature but the effect is small—less than 1% per °C). Consider, for example, PMMA (see Figure 4.11): at −200°C both the possible modes of molecular motion, the side group rotation (β) and the backbone 'snake-like' motion (α) are frozen-in. An alternative, but identical, statement is that both β and α relaxation times are extremely long. As the specimen is heated from −200°C both processes move to shorter times: as each approaches the time constant of the torsion pendulum (the β-process first, the α-process next), G' drops suddenly and Λ passes through a peak. The temperature dependence of G' and Λ is controlled completely by the temperature dependence of the α and β relaxation times: and these are controlled by the rate of molecular movement of the side-group (β-process) and of the backbone (α-process). The onset of these movements is marked also by a change in failure mode from brittle to ductile.

4.N.8

The raising of the glass transition of rubbers (polyisoprene, polybutadiene, styrene–butadiene, and acrylonitrile–butadiene) by cross-linking ultimately produces a material with the generic name **ebonite**: the degree of cross-linking must be sufficient to render the polymer glassy at room temperature. It may be machined by normal methods, is resistive to swelling by organic liquids, has good electrical properties, and resists many corrosive liquids.

4.N.9

The reduction in modulus (and the reduction in brittleness) brought about by the absorption of a small molecule is known as **plasticization**: the small molecule is known as a **plasticizer**. When the humidity increases over a nylon or polyurethane specimen the absorbed water content increases, which leads to a small increase in volume and a drop in modulus: if the humidity is decreased all changes reverse. These effects are reduced by fibre (or particle) reinforcement. Small molecules other than water (ethylene glycol, dioctyl phthalate, etc.) are frequently used as plasticizers. The plasticizer should not exude from the polymer once it has been mixed. All plasticizers act by lowering T_g.

Problems for Chapter 4

4.1   A straight rod of solid polymer is of length 1 m and diameter 10 mm. The polymer is linearly viscoelastic with a tensile creep compliance

$$D = 2 - \exp(-0.1t)\ \text{GPa}^{-1},$$

where t is in hours. The rod is suspended vertically and a mass of 10 kg is hung from it for 10 hours: find the change of length of the rod.

4.2   A nylon bolt of diameter 8 mm is used to join two rigid plates (4.N.2 Figure 4.34). The nylon can be assumed to be linear viscoelastic with a tensile stress relaxation modulus approximated by

$$E = 5\exp[-(t)^{\frac{1}{3}}]\ \text{GPa},$$

where t is in hours. The bolt is tightened quickly so that the initial force in the bolt (at $t = 0$) is 1 kN.

(1) Find the strain in the bolt.
(2) Find the force remaining after 24 hours.

4.3   A pipe of external diameter $d = 90$ mm and wall thickness $W = 8$ mm is produced from the polymer of Problem 4.1. It is internally pressurized with closed ends to a pressure of 0.4 MPa. Find the hoop strain and axial strain in the wall of the pipe 20 hours after the pressure is applied. Take Poisson's ratio to be constant at 0.4 for this material and hoop stress $= P(d - 2W)/2W$.

4.4 Repeat the solution to Problem 4.3, but instead of assuming Poisson's ratio to be constant use the alternative (and usually more accurate) assumption that the compressibility, B, is constant, at 0.9 GPa^{-1}. (Hint: using $K = E/3(1 - 2\nu)$ show that $\nu = 1/2 - B/6D$, where K is the bulk modulus.)

4.5 Derive, from first principles, the equations giving G' and $\tan \delta$ from the frequency and attenuation of oscillations of a torsion pendulum, when the specimen is a linearly viscoelastic rod of circular cross-section and $\Lambda \ll 1$ (see Section 4.2.3):

$$G' = \frac{Ml\omega^2}{N},$$

and

$$\tan \delta = \frac{\Lambda}{\pi}.$$

(Hint: write down the usual equation of motion of a torsion pendulum and solve it using complex notation, noting that the complex shear modulus has both real and imaginary parts.)

4.6 Show that to maintain a linear viscoelastic body in steady sinusoidal oscillations of shear strain with amplitude γ_0, the exciting torque must input a net energy ΔW per unit volume per cycle which is given by

$$\Delta W = \pi \gamma_0^2 G''.$$

4.7 A tube of poly(methyl methacrylate) (PMMA) with length 200 mm, diameter 20 mm, and wall thickness 1 mm, is to be subjected to sinusoidal torsional oscillations at 1 Hz, in which the relative rotation of its ends will be by $\pm 10°$. Before the oscillations commence the tube is in equilibrium with air at 20°C. Predict the initial rate of temperature rise and the final equilibrium temperature. Use the torsion pendulum data for PMMA given in Figure 4.11, and take the density and specific heat of PMMA to be 1200 kg m^{-3} and 1450 J kg^{-1}K^{-1}, respectively. The heat transfer coefficient for convective heat transfer from the external surface of the tube to air at about 20°C may be assumed to be 10 W m^{-2}K^{-1}. The heat loss from the inside surface can be neglected. Hint: you may refer to the equation given in Problem 4.6.

4.8 Derive the differential equations linking shear strain γ, stress σ, and time t for each of the following viscoelastic models. J and G represent spring compliance and stiffness, respectively.

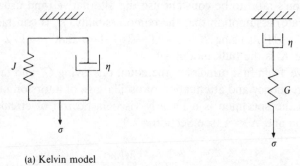

(a) Kelvin model

(b) Maxwell model

4.9 Derive the creep compliances for each of the following viscoelastic models (see Problem 4.8).

(1) A Kelvin model with $J = 1$ GPa^{-1}, $\eta = 10^{12}$ Pa s.
(2) A Maxwell model with $G = 1$ GPa, $\eta = 10^{12}$ Pa s.

4.10 By solving the differential eqn 4.47 show that the stress–strain relation $\sigma(\gamma)$ for constant-strain-rate deformation of a Zener solid has the following form

$$\sigma(\gamma) = G_R \gamma + (G_U - G_R)\tau_\gamma k \left[1 - \exp\left(\frac{-\gamma}{k\tau_\gamma} \right) \right],$$

where k is the strain rate. Sketch the form of this curve.

4.11 Write down the set of differential equations relating strain γ, stress σ, and time t for the generalized Zener model shown in Figure 4.17. Solve them to show that the creep compliance function $J(t)$ is given by

$$J(t) = J_U + \sum_{i=1}^{n} J_i \left[1 - \exp\left(\frac{-t}{\tau_i} \right) \right],$$

where $\tau_i = \eta_i J_i$.

4.12 For the alternative generalization of the Zener solid shown below, derive the differential equations relating strain, stress, and time.

Solve them to show that the stress relaxation modulus function $G(t)$ is given by

$$G(t) = G_R + \sum_{i=1}^{m} G_i \exp\left(\frac{-t}{\tau_i}\right),$$

where, in this case, $\tau_i = \eta_i/G_i$.

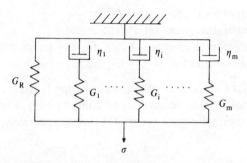

4.13 For the generalized viscoelastic model given in Problem 4.12, show that real and imaginary parts of the complex dynamic modulus $G^*(\omega)$ are given by

$$G' = G_R + \sum_{i=1}^{m} \frac{G_i \omega^2 \tau_i^2}{1 + \omega^2 \tau_i^2},$$

and

$$G'' = \sum_{i=1}^{m} \frac{G_i \omega \tau_i}{1 + \omega^2 \tau_i^2}.$$

4.14 For a certain polymer, viscoelastic behaviour in shear can be represented by the model shown in Figure 4.17, with

$$n = 2, J_U = J_0, J_1 = 2J_0, \tau_1 = \tau_0, J_2 = 3J_0, \tau_2 = 2\tau_0.$$

Determine (a) the compliance after a constant stress has been maintained for a time equal to $3\tau_0$, and (b) the complex compliance at an angular frequency $\omega = 1/3\tau_0$.

4.15 Use the shear creep data in Figure 4.4, together with the method of time–temperature superposition, to **estimate** the shear creep compliance for linear polyethylene at 20°C and a creep time 10^6 s. List the assumptions that you make in this long extrapolation of the creep data.

4.16 A certain pipe-grade PVC deforms in shear with a creep compliance function at 20°C of the form

$$J(t) = 0.75 + 0.15 \log_{10} t + 0.018 (\log_{10} t)^2 \text{ GPa}^{-1},$$

where t is expressed in seconds. Time–temperature superposition is obeyed. The shift factor for 60°C, relative to 20°C, is $a_{60} = 10^{-3}$. A section of PVC pipe is of length 2 m, outside diameter 40 mm, and wall thickness 2 mm. Calculate the relative rotation of the ends of the pipe when an axial torque of 8 N m is applied for 20 hours under each of the following conditions:

(1) temperature constant at 20°C;
(2) temperature constant at 60°C; and,
(3) temperature constant at 20°C for $19\frac{3}{4}$ hours, followed by temperature constant at 60°C for $\frac{1}{4}$ hour.

4.17   A straight section of polypropylene pipe is fixed rigidly at its ends. Its tensile stress relaxation modulus at time t and coefficient of linear thermal expansion at 20°C are, respectively,

$$E(t) = 2t^{-0.09} \text{ GPa,}$$

and

$$\alpha = 10^{-4} \text{ K}^{-1},$$

where t is expressed in seconds. The pipe is initially stress-free and at 20°C. Suddenly, there is a rapid rise in temperature to 50°C as a result of passage of some hot water, which continues to flow for one hour. The pipe then returns rapidly to 20°C. Sketch a graph of the thermally induced stress versus time, and calculate the stress in the pipe at 20°C 100 s after it returns to 20°C. (Assume that the pipe does not buckle, and that polypropylene obeys the BSP and exhibits time–temperature superposition with a_T obeying the Arrhenius equation, with an activation energy $\Delta H = 145 \text{ kJ mol}^{-1}$.)

4.18   Show that, for a linear viscoelastic beam under a constant bending moment M_0, the curvature of the beam $1/R$ (R is the radius of curvature) increases with time t according to an equation similar to that for a linear **elastic** beam, except that the reciprocal tensile creep compliance (or 'creep modulus') takes the place of Young's modulus:

$$\frac{1}{R(t)} = \frac{M_0}{I} D(t),$$

where I is the second moment of area of the beam section.

4.19   A continuous linear polyethylene (HDPE) pipeline is supported above ground on brackets, which are equally spaced. The pipe has an external diameter of 90 mm and wall thickness of 8 mm. Find the maximum allowable spacing between the brackets, if the pipe must nowhere sag by more than 10 mm under its own weight after a period of 10^6 s at 20°C. (Take the density of HDPE to be 970 kg m^{-3}; take Poisson's ratio to be constant at 0.41, and use the shear creep data of Problem 4.15. The central deflection δ of a linear elastic beam of length l, built-in at each end, and under a uniformly distributed load per unit length p, is given by

$$\delta = \frac{pl^4}{384EI},$$

where E and I have their usual meanings.)

4.20 Continue the solution of Problem 4.1. After 10 hours, the 10 kg mass is removed. Calculate the strain remaining in the rod after a further 10 hours.

4.21 Continue the solution of Problem 4.2. After 24 hours the bolt is rapidly retightened in order to return the clamping force to 1 kN.

(1) Find the new strain in the bolt.

(2) Find the clamping force remaining after a further 24 hours.

4.22 The tensile-stress relaxation modulus of a certain polymer can be approximated by an expression of the form given in eqn 4.53. Relaxed and unrelaxed tensile moduli are $E_R = 0.5$ GPa and $E_U = 1.5$ GPa, respectively, with relaxation time $\tau = 5$ s. The polymer is subjected to a constant rate of tensile strain $\varepsilon = 10^{-3}$ s^{-1}. Derive the stress–strain relation $\sigma(\varepsilon)$ for these conditions, using the integral form of the Boltzmann superposition principle.

4.23 Use the integral form of the Boltzmann superposition principle to show that the creep compliance and stress relaxation modulus of any linear viscoelastic material are related through

$$\int_0^t J(t-u)\,\frac{dG}{du}\,du + J(t)G_U = 1.$$

Check that this is satisfied by J and G as given by eqns 4.49 and 4.53 respectively for the Zener solid.

5   *Yield and fracture*

5.1   Introduction

The isochronous stress–strain curves for plastics begin to deviate from linearity at strains of between 0.001 and 0.01. At higher strains, typically between 0.01 and 0.1, the deviation becomes so large that $d\sigma/d\varepsilon$ falls to zero. This condition defines the yield point. Some materials fracture immediately after yielding (or even before), but the more ductile plastics such as polyethylene can reach strains as high as 25 before final failure. By contrast with metals, it is difficult to make a clear distinction between recoverable (elastic) and non-recoverable (plastic) strain, because the extent to which a polymer recovers its original dimensions depends upon the temperature and upon the time allowed for recovery: thermoplastics of high RMM can recover almost completely from very high strains if they are warmed.

The driving force for recovery is entropic, as in a cross-linked rubber. There are no chemical cross-links in a thermoplastic, but chain entanglements ('knots') and other physical interactions between molecules have a similar effect in causing the material to behave like a network.

A few plastics appear to fracture in a brittle manner, with no sign of ductility (polystyrene is a familiar example). Close examination, however, reveals that brittle crack propagation in plastics is invariably accompanied by a certain amount of **localized yielding over a restricted region near the crack tip**. In polystyrene and in other glassy thermoplastics, this takes the form of craze formation, which is illustrated in Figure 5.1, and is discussed more fully later in this chapter.

To the naked eye, a craze looks like an extension of the crack, but electron microscopy reveals that **load-bearing fibrils about 10 nm in diameter span the gap between the surfaces of the polymer**. A network of open holes of similar diameter runs through the craze. Molecular entanglements are essential, since without them there would be little to stabilize the loaded fibrils. If the polymer chains are too short to form effective entanglements, the material is extremely fragile.

Although it is brittle in tension, polystyrene is ductile in compression, and the same is true of other apparently brittle thermoplastics, and also of lightly cross-linked thermosetting resins. Very tightly cross-linked resins show little yielding under any conditions because the molecular network is

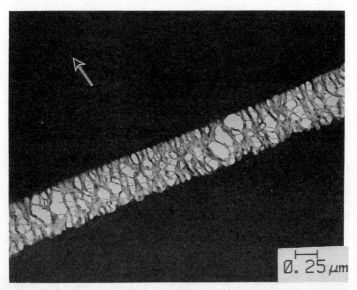

5.1 A craze in polystyrene; the arrow indicates direction of tensile stress (after R. P. Kambour).

unable to deform sufficiently. Whether a polymer is ductile or brittle in any given circumstance depends upon its resistance

- to yield and
- to crazing and subsequent crack propagation.

The balance between these competing mechanisms is affected by temperature, strain rate, type of loading, component geometry, and the presence of aggressive liquids.

5.2 Yielding

The standard method for measuring the modulus and yield stress of polymers is a tensile test, in which grips are attached to the ends of a bar, and pulled apart at constant speed. In this type of experiment, some non-linearity is to be expected in the stress–strain curve, even within the region of linear viscoelastic behaviour, because the measurements are made over a range of times (see Chapter 4). Also, as the strain increases, non-linear viscoelasticity becomes more pronounced, until eventually the specimen reaches a load maximum. Most ductile polymers show a load drop immediately after reaching the maximum load, owing to a combination of strain softening and localized necking, as shown for polyethylene in Figure 5.2.

Polystyrene behaves quite differently, forming crazes at strains of about 0.005, and fracturing without showing any significant deviation from a

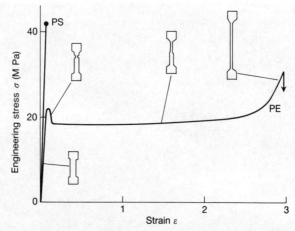

5.2 Stress–strain curves for polystyrene (PS) and polyethylene (PE).

linear stress–strain curve (see Figure 5.2). Despite the obvious differences between these two thermoplastics, there are interesting parallels between **macroscopic drawing** in PE and the **microscopic internal drawing** of craze fibrils in PS. An important macroscopic distinction between the two mechanisms is that ordinary yielding occurs by slip at angles of about 40° to the tensile axis (approximately on planes of maximum shear stress), whereas crazes form on planes normal to the tensile stress.

The stress that induces yielding is called the yield stress σ_y. It is generally defined in a tensile test—in which the clamps are moved apart at a set rate—as the stress where the rate of flow equals the imposed strain rate. Thus, in Figure 5.2, the yield stress for polyethylene is the stress at the stress maximum; at this point flow takes place instantaneously at the imposed strain rate, with the stress neither increasing nor decreasing. For some polymers there is no maximum, the stress–strain curve merely changing from a higher to a lower slope; in this, less well-defined, case the stress at the 'knee' is taken, somewhat arbitrarily, to be the yield stress. In our treatment we will discuss only the first case.

5.2.1 Considère's construction

The phenomenon of yield and flow in a tensile test can be interpreted using the construction introduced by Considère (see Figure 5.3). In this elementary classical treatment it is assumed that the material is rate-insensitive and there is no intention of developing a mechanistic interpretation of yield. Consider a tensile test in which the specimen has been deformed from initial length L_i and cross-section A_i to values L and A at which point the force is F. The strain (engineering strain) is

$$\varepsilon = \frac{L - L_i}{L_i} = \lambda - 1, \qquad (5.1)$$

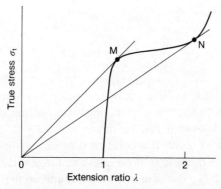

5.3 Considère's construction for a material that yields and cold-draws.

λ being the extension ratio (eq. 3.14). For a deformation at constant volume (this is a coarse approximation for stiff, glassy polymers but is better for those that are more ductile),

$$A_i L_i = AL$$
$$A_i = A\lambda.$$

The true stress is

$$\sigma_t = \frac{F}{A} \tag{5.2}$$

and the nominal stress is

$$\sigma = \frac{F}{A_i} \tag{5.3}$$

It follows that

$$\sigma = \frac{\sigma_t}{\lambda}, \tag{5.4}$$

so that the slope of the σ against λ plot at any point is

$$\frac{d\sigma}{d\lambda} = \frac{1}{\lambda}\frac{d\sigma_t}{d\lambda} - \frac{\sigma_t}{\lambda^2}. \tag{5.5}$$

Because $d\varepsilon = d\lambda$ (differentiating eqn 5.1), it follows that yield, defined by

$$\frac{d\sigma}{d\varepsilon} = 0 \text{ (at yield)},$$

is defined also by

$$\frac{d\sigma}{d\lambda} = 0 \text{ (at yield)}. \tag{5.6}$$

It follows from eqns 5.5 and 5.6 that

$$\frac{\mathrm{d}\sigma_t}{\mathrm{d}\lambda} = \frac{\sigma_t}{\lambda} \text{ (at yield).} \tag{5.7}$$

In a plot of σ_t against λ (Figure 5.3) yield will occur according to eqn 5.7 at point M: that is to say the engineering stress–strain curve will show a maximum only if a tangent can be drawn from $\lambda = 0$ to touch the true stress–extension ratio curve at a point such as M.

In some cases it is possible to draw a second tangent through 0, touching the curve at point N. This defines a minimum in the nominal stress–extension ratio curve, where the molecular orientation stiffens the drawn polymer (in the neck) so as to resist further extension. At this point, the neck stabilizes, and begins to extend by **drawing fresh material from the tapering regions on either side until the whole of the parallel section of the specimen has yielded**. If the polymer does not strain-harden sufficiently to permit the construction of a second tangent, the neck will continue to thin down until it breaks (as in hot silica glass or in a ductile metal).

Considère's construction is applicable equally to metals and polymers, but the necking of polymers is affected by two physical factors which are not normally so significant in metals:

- dissipation of mechanical energy as heat can raise the temperature in the neck, causing significant softening (the magnitude of this effect increases with strain rate); and

- the deformation resistance of the neck, which has a higher strain rate than the surrounding polymer, can rise as a result of the **strain-rate dependence of the yield stress**.

The relative importance of these two (opposing) effects depends upon the material, the length and thickness of the specimen, and the test conditions, especially the strain rate. The drawing of fibres, film, and sheet in forming operations occurs at very high strain rates.

Load–deformation curves for polyethylene and polystyrene are shown in Figure 5.2. Polyethylene necks and cold draws in the manner described above (5.N.1). As it is a semi-crystalline polymer, yielding involves considerable disruption of the crystal structure. Slip occurs

- between the crystalline lamellae, which slide by each other like a pack of cards, and

- within the individual lamellae by a process comparable to glide in monatomic crystals.

The second—and dominant—process leads to molecular orientation, since the slip direction within the crystal is along the axis of the molecule. As plastic flow continues, the slip direction rotates systematically towards the tensile axis. Ultimately, the slip direction—that is, the molecular axis—coincides with the tensile axis, and the polymer is then oriented and resists further extension. Whilst these two processes continue during plastic flow, the lamellae and spherulites increasingly lose their identity and a new fibrillar structure is formed (see Chapter 2, Figure 2.21).

5.2.2 Eyring's model of the flow of solids

The flow model of Eyring provides a basis for analysis, and contains the necessary elements for further development and refinement. Its aim is to correlate the effects of temperature and strain rate on flow stress, and it seeks to do this from a molecular model of the flow mechanism. We present here a simple statement of the model and its success in describing experiments. It is related to the Arrhenius equation as described in 4.N.6.

The fundamental idea is that an atom or a molecule—and, for a polymer, a segment of a macromolecule—must pass over an energy barrier in moving from one position to another in the solid. In the absence of stress the segments of the polymer jump over the barrier very infrequently and they do so in random directions; for example (see 4.N.6) the mean time-of-stay τ between segment jumps in the solid may be measured in years. The essential fact is that the solid is not a solid in the sense that all such movement has ceased absolutely, even if the existing movement occurs very infrequently.

The rate of jump is proportional to τ^{-1} and from eqn 4.N.6.4 this is

$$\text{jump rate} = \alpha \exp\left[-\left(\frac{\Delta H}{RT}\right)\right],$$

where α is a constant and ΔH is the enthalpy required to take a mole of segments from the potential well (see Figure 5.4) to the top of the barrier (see 4.N.6). The segment oscillates in a potential well for long periods of time, but occasionally, by a random fluctuation, it draws from the thermal bath (comprising the internal energy of the system) an excess of energy, which causes it to move rapidly to the top of the barrier and then on into the neighbouring well. It then starts the same process all over again, oscillating in the well before making another jump; note that the final state for each jump becomes the initial state for a subsequent jump.

The significant point introduced by Eyring is that the application of a shear stress σ_s modifies the barrier height (see Figure 5.4) so that, in the direction of the stress, the rate of segment jumping, formerly so slow, now becomes fast enough to give rise to a measurable strain change. The

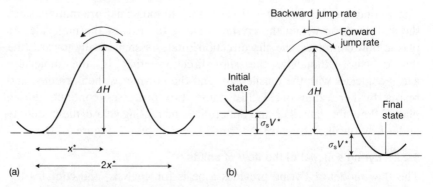

5.4 Eyring model of solid flow. (a) Before stress is applied, the polymer segments reside in potential wells separated from equivalent wells by: (i) an enthalpy barrier ΔH; and (ii) a distance $2x^*$. (b) On application of shear stress σ_s the potential wells are changed so that the forward barrier is $(\Delta H - \sigma_s V^*)$ and the backward barrier $(\Delta H + \sigma_s V^*)$. The polymer segments then jump preferentially over the barrier in the direction of σ_s; with no stress the segments jump at equal rate forwards and backwards.

polymer segment has an effective area A^* and is therefore under a traction from the shear stress of $\sigma_s A^*$: so long as the segment remains in the well, the traction does no work. When the segment jumps it moves to a new site which is separated from the initial site by the enthalpy barrier: the distance from the initial site to the barrier is x^*, so that the traction does work $(\sigma A^*)x^*$ as the segment moves to the top of the barrier. This work is absorbed by the moving segment, so that the energy required to facilitate the jump is now not ΔH, but $\Delta H - \sigma A^* x^*$. The quantity $A^* x^*$ has the dimensions of volume:

$$V^* = A^* x^*, \tag{5.8}$$

and is known as the activation volume (5.N.2). The jump rate along the direction of the shear stress is thus increased to

$$\text{forward jump rate} = \alpha \exp\left[-\left(\frac{\Delta H - \sigma_s V^*}{RT}\right)\right].$$

The energy barrier for molecular segments jumping in the direction opposing the shear stress is raised to $\Delta H + \sigma_s V^*$, so that

$$\text{reverse jump rate} = \alpha \exp\left[-\left(\frac{\Delta H + \sigma_s V^*}{RT}\right)\right]. \tag{5.9}$$

The net rate of flow in the stress direction is the difference between these two opposing rates, so that

$$\text{net jump rate} = \alpha\left\{\exp\left[-\left(\frac{\Delta H - \sigma_s V^*}{RT}\right)\right] - \exp\left[-\left(\frac{\Delta H + \sigma_s V^*}{RT}\right)\right]\right\}. \tag{5.10}$$

In a solid, the jump rate in the reverse direction is slower even than the jump rate before the application of the stress. The reverse jump rate can therefore be neglected in comparison with the forward rate. Equation 5.10 then becomes

$$\text{net jump rate} = \alpha \exp\left[-\left(\frac{\Delta H}{RT}\right)\right]\exp\left(\frac{\sigma_s V^*}{RT}\right). \qquad (5.11)$$

The final form of the Eyring equation rests on the reasonable assumptions that:

- the imposed strain rate is proportional to the net rate at which segments jump preferentially in the direction of the shear stress; and
- the dominant shear stress in a tensile test is the maximum shear stress, and at yield this is $\sigma_s = \sigma_y/2$.

It then follows that at tensile stress σ_y, the imposed strain rate at yield $\dot{\varepsilon}_y$, is

$$\dot{\varepsilon}_y = \dot{\varepsilon}_0 \exp\left[-\left(\frac{\Delta H}{RT}\right)\right]\exp\left(\frac{\sigma_y V^*}{2RT}\right), \qquad (5.12)$$

where $\dot{\varepsilon}_0$ is a constant.

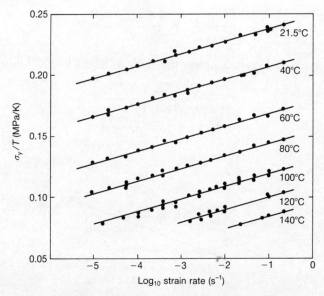

5.5 Eyring plot of σ_y/T against $\log \dot{\varepsilon}$ for polycarbonate (after C. Bauwens-Crowet, J. C. Bauwens, and G. Homes).

In the analysis of measurements of σ_y at varying strain rate it is preferable to rearrange eqn 5.12:

$$\left(\frac{\sigma_y}{T}\right) = \left(\frac{2}{V^*}\right)\left[\left(\frac{\Delta H}{T}\right) + 2.303R\log\left(\frac{\dot{\varepsilon}_y}{\dot{\varepsilon}_0}\right)\right]. \tag{5.13}$$

Figure 5.5 shows plots of σ_y/T against $\log\dot{\varepsilon}_y$ for polycarbonate for a series of temperatures between 21.5°C and 140°C (the T_g of PC is 160°C). Note that the predictions are, in outline, obeyed:

- σ_y/T increases linearly with $\log\dot{\varepsilon}_y$ (the yield stress is rate-dependent).
- At constant $\log\dot{\varepsilon}_y$, σ_y/T increases with decreasing temperature (the yield stress is temperature-dependent).

Example 5.1

Calculate the activation volume V^* and activation enthalpy ΔH for the yielding of polycarbonate, using the data shown in Figure 5.5.

Procedure

Using eqn 5.13, the slope of the lines gives V^*, and the separation of the lines, either pq (constant $\log\dot{\varepsilon}_y$) or qr (constant σ_y/T), gives ΔH (see Figure 5.6).

Solution

$$\frac{\mathrm{d}(\sigma_y/T)}{\mathrm{d}(\log\dot{\varepsilon}_y)} = \frac{2\times2.303\times R}{V^*}.$$

Using a ruler on Figure 5.5 gives the slope to be $9.8\,\mathrm{kPa\,K^{-1}}$ per decade.

$$V^* = \frac{2\times2.303\times8.314}{9.8\times10^3}$$

$$= 3.9\times10^{-3}\,\mathrm{m^3\,mol^{-1}}.$$

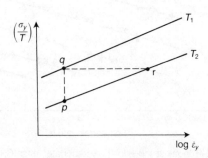

5.6 Procedure for determining V^* and ΔH from Eyring plot of yield stress data.

Taking the separation qr (constant σ_y/T), from eqn 5.13 (see Figure 5.6)

$$\left(\frac{\Delta H}{T_1}\right) + 2.303R \log \dot{\varepsilon}_y^{T_1} = \left(\frac{\Delta H}{T_2}\right) + 2.303R \log \dot{\varepsilon}_y^{T_2}.$$

Therefore

$$\Delta H = \frac{2.303R\left(\log \dot{\varepsilon}_y^{T_2} - \log \dot{\varepsilon}_y^{T_1}\right)}{\left(\dfrac{1}{T_1} - \dfrac{1}{T_2}\right)}.$$

Using a ruler on the lines at $T_1 = 60°C$ and $T_2 = 100°C$ gives a separation of 5.2 decades.

Therefore

$$\Delta H = \frac{2.303 \times 8.314 \times 5.2}{(333^{-1} - 373^{-1})}$$

$$= 309 \text{ kJ mol}^{-1}.$$

Comment

The activation enthalpy obtained in this calculation is typical of deformation processes in plastics. By comparison with ΔH values for chemical reactions, it is high. Note, however, that the 'mole' to which it relates is a mole of polymer segments, each segment containing many carbon atoms. This point is brought out by considering the size of V^*. The size of V^* per jumping segment is

$$V^*/\text{Avogadro's constant} = 3.9 \times 10^{-3}/6.02 \times 10^{23}$$

$$= 6.5 \text{ nm}^3.$$

In order to obtain an idea of the scale of this volume, note (see Figure 2.1) that the volume of the polyethylene unit cell (which contains 4 CH_2 units) is $0.254 \times 0.736 \times 0.492 = 0.1 \text{ nm}^3$; thus V^* is a volume equivalent to approximately 260 polymer carbon atoms.

Example 5.2

Estimate the yield stress of polycarbonate at 21°C in an impact test in which instrumentation shows fracture occurring in approximately 1 ms.

Solution

Because of the complex geometries involved, especially in notched specimens, it is virtually impossible in most cases to measure strain rates directly during impact. It is however known that yield strains in ductile polymers are usually between 0.05 and 0.1. Data sheets on PC give a figure of 0.07. If the polymer is to yield in impact, it must therefore reach this strain in 1 ms. This gives a value of $0.07/0.001 = 70 \text{ s}^{-1}$ for the required strain rate. Extrapolating the top curve in Figure 5.5 we find that for $\log \dot{\varepsilon} = +1.8$, $\sigma/T = 0.265 \text{ MPa K}^{-1}$, from which $\sigma_y = 0.265 \times 294.5 = 78 \text{ MPa}$.

Not all 'Eyring plots' of applied stress against log strain rate give straight lines. Linearity is to be expected only when there is a unique flow mechanism controlling the strain rate. In some materials, one mechanism is dominant up to a certain temperature, above which a second mechanism becomes active.

Note from Figure 5.5 that for polycarbonate the change in σ_y generated by an increase in strain rate of one decade is equivalent to a reduction in temperature in the region of 10 K. The 'one decade equals 10 K' rule is approximate, since the effect of temperature change decreases as T approaches T_g (see Figure 5.5).

5.2.3 Yielding under multiaxial stresses

The discussion has so far concentrated upon yielding under applied shear or tensile stresses. The more general case of yielding under multiaxial loading is of interest, especially in connection with fracture, because the fracture resistance of a material is determined by its **ability to develop a yield zone in the region of a crack tip, where it is usually in a state of triaxial tension**. The approach adopted in developing a general yield criterion for plastics is to take one of the established criteria for metals and to modify it by introducing a term in the hydrostatic pressure p. Some use has been made of the Tresca and Coulomb yield criteria, but the main starting point has been the von Mises criterion:

$$(\sigma_{11} - \sigma_{22})^2 + (\sigma_{22} - \sigma_{33})^2 + (\sigma_{33} - \sigma_{11})^2$$
$$+ 6(\sigma_{12}^2 + \sigma_{13}^2 + \sigma_{23}^2) \geq 6C^2, \quad (5.14)$$

where the σ_{ij} are components of the stress matrix. If the left-hand side exceeds $6C^2$, then yield has occurred. **In metals, C is a constant; in plastics, it varies with p.** Several versions of eqn 5.14 have been proposed, the simplest of which shows C increasing linearly with p. As already discussed, C **is also a function of temperature and strain rate**. One interesting effect of hydrostatic pressure is that is suppresses crazing, a topic that will be discussed later in this chapter.

Example 5.3

A sample of linear polyethylene tested at 23°C and 10^{-3} s^{-1} yielded at

30.0 MPa in uniaxial tension, and at 31.5 MPa in uniaxial compression. Assuming that the yield stress is a linear function of hydrostatic pressure, calculate σ_y under superimposed hydrostatic pressure of 500 MPa.

Solution

The difference between tensile and compressive yield stress arises because of the hydrostatic component of the applied stress. For any state of stress, p is an **invariant** given by

$$p = -\tfrac{1}{3}(\sigma_{11} + \sigma_{22} + \sigma_{33}).$$

In tension

$$p = -\tfrac{1}{3}\sigma_{11} = -\tfrac{1}{3}30 = -10 \text{ MPa}.$$

In compression

$$p = -\tfrac{1}{3}\sigma_{11} = -\tfrac{1}{3}(-31.5) = +10.5 \text{ MPa}.$$

Writing $C_0 + C_1 p$ in place of C in eqn 5.14, we have for both cases

$$|\sigma_{11}| = 3^{\frac{1}{2}}(C_0 + C_1 p).$$

Tension

$$30.0 = 3^{\frac{1}{2}}(C_0 - 10C_1).$$

Compression

$$31.5 = 3^{\frac{1}{2}}(C_0 + 10.5C_1).$$

From this we obtain $C_0 = 17.743$ MPa and $C_1 = 0.04225$ (for p in MPa). When tests are conducted in a pressure chamber, the hydrostatic stress term p is the sum of the superimposed pressure and the hydrostatic component of the stress applied by the testing machine. Let σ_y be the nominal yield stress, considering the case of extensional loading. Then

$$p = 500 - \tfrac{1}{3}\sigma_y \text{ MPa}.$$

The condition for yielding then becomes

$$\sigma_y = \sqrt{3}\left(C_0 + 500C_1 - \tfrac{1}{3}\sigma_y C_1\right).$$

Substituting and rearranging, we have

$$\sigma_y = \frac{17.743\sqrt{3} + 0.04225 \times 500\sqrt{3}}{1 + \left(\dfrac{0.04225\sqrt{3}}{3}\right)}$$

$$\sigma_y = 65.7 \text{ MPa}$$

Comment

This reveals a large pressure sensitivity of σ_y, quite unlike the case of metals, in which the yield stress is essentially insensitive to hydrostatic pressure. The Eyring equation takes no account of change in σ_y with hydrostatic pressure. It succeeds only because the experimental range of applied tensile stress (and the consequent range of hydrostatic stress) is relatively small.

The modified von Mises criterion applies to isotropic materials, in which the yield stress is independent of direction. Most moulded or extruded components contain regions of adventitious molecular orientation, where yield and fracture properties depend upon the direction of testing. Insertion of controlled orientation is often a critical feature of a manufacturing operation; biaxial stretching of thermoformed sheet is an example. As in the case of drawn polyethylene, which was mentioned earlier, molecular orientation raises the yield stress in the draw direction. To take an extreme example, if all of the molecules in a sample were aligned parallel, it would be impossible to produce yielding by applying a tensile stress in the direction of the chain axes. On the other hand, application of the stress at an angle of 45° to the chain axes could cause yielding, because the molecules would then be able to slide past each other. Anisotropy is an important consideration in any discussion of the strength properties of polymers.

The yield stresses of thermoplastics vary widely, depending upon the molecular structure. The methylene group—CH_2—provides flexibility to

the chain, whereas the the paraphenylene group —〈◯〉— is a very rigid

unit. Electric dipoles and hydrogen bonds increase intermolecular attractions, and tend to raise the yield stress. For example, the yield stress of nylon 6.6 at room temperature is 75 MPa when the polymer is dry, but falls to 35 MPa when it is saturated with water, because water breaks the hydrogen bonds between amide groups, and thereby greatly increases chain mobility. Water-saturated nylon 6.6, like polyethylene and polypropylene, is a semicrystalline polymer which is above its glass transition at room temperature. Crystalline regions stiffen these materials, and raise the yield stress. In commercial grades of polyethylene, crystallinity varies from 50 to 80%, and the range of σ_y is from about 7 to 40 MPa.

One very important method for modifying the yield behaviour of a polymer is to blend it with a small amount of a second polymer. This approach avoids the enormous costs of launching a completely new polymer onto the market, and can offer considerable improvements in the balance of properties. By far the most successful class of blends are the rubber-modified polymers, which contain up to 20 vol.% of well-dispersed rubber particles. Addition of rubber reduces both yield stress and stiffness, but greatly increases fracture resistance, a theme which is developed in Section 5.7.

5.3 Crazing

Crazes have been observed in most glassy thermoplastics and in some semicrystalline polymers, notably polypropylene. Figure 5.1 is an electron micrograph of a section through a typical craze, showing the intricate network of fibrils connecting the two bulk surfaces of the polymer. The fibrils are drawn out of the solid polymer to an extent that is controlled by the concentration of molecular entanglements. The maximum extension ratios within the craze fibrils vary from 2 to 5, depending upon the polymer and on the RMM. Once craze fibrils have formed, they must either draw fresh material from the walls of the craze so as to increase the fibril length, or undergo fracture. At fracture, the stresses in the fibrils are typically about 200 MPa. Microscopy shows that a craze is a long, thin wedge of deformed polymer, with a tip that is sharp on a scale of tens of nanometres, broadening out over a distance of about 0.1 mm to its full thickness by the fibril drawing process described above.

Crazes nucleate at points of high stress concentration on free surfaces: if a crack is present, the craze will extend from its tip; if not, crazes initiate at the surface or at voids within the polymer. Microscopy shows that subsequent growth occurs not by repeated nucleation of new holes but by a process in which the existing voids advance finger-like extensions into the bulk polymer, eventually linking up and leaving stretched fibrils in their wake. Rates of craze initiation and growth are strongly dependent upon applied stress and temperature. The Eyring activated flow model has been used successfully to correlate rates of crazing in a number of glassy polymers, and in this respect there are similarities with shear yielding. Another feature common to all glassy polymers is the dependence of deformation rate upon molecular orientation: the stress necessary to initiate crazes **within a given time** in a hot-drawn specimen tested parallel to the draw direction can be more than double that required to cause crazing in an undrawn specimen. The kinetics of crazing and of yielding in shear play an important part in determining whether a polymer is brittle or ductile.

One major difference between the two mechanisms of deformation is illustrated in Figure 5.7, which shows a failure envelope for PMMA under biaxial loading. The pure-shear line, defined by $\sigma_{11} = -\sigma_{22}$, marks the

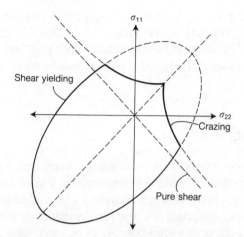

5.7 Failure envelope (heavy continuous line) for PMMA under biaxial stress ($\sigma_{33}=0$) at room temperature, showing intersection of crazing and shear yielding envelopes (after S. Sternstein and L. Ongchin).

boundary between hydrostatic compression and hydrostatic tension. Below this line, **crazing and other hole-forming processes do not take place**, because the pressure component of the stress matrix tends to reduce rather than to increase volume; above the line, crazing is the principal mechanism of failure.

Several equations have been proposed to define the criterion for crazing under multiaxial loading, but as yet none has gained general acceptance. The main problem is in obtaining reliable data over a range of triaxial tensile stresses. It is, however, agreed that the crazing envelope is asymptotic to the pure shear line. Within the tensile quadrant of the diagram, and over part of the two neighbouring quadrants, the crazing envelope lies below the shear yielding envelope, and the material fractures without yielding. Below the pure-shear line, failure of PMMA is represented in the diagram by a pressure-dependent von Mises criterion in which the quantity C in eqn 5.14 increases linearly with p.

Although crazing is a response to stress, an externally applied stress is not always required to produce it. Internal stresses arising from differential contraction during cooling in the mould can be of sufficient magnitude to initiate crazes in injection mouldings, and similar effects occur in other types of formed items. Internal stresses can also be generated by relaxation of oriented molecules: entropic recovery forces (produced on heating, or by absorption of a liquid) facilitate the elimination of orientation, the molecules returning spontaneously to the random state.

Crazing and fracture caused by absorbed liquids and vapours is one of the most serious limitations on the use of plastics in engineering applications. The terms 'solvent crazing' and 'environmental stress cracking' are applied to the phenomena, although non-solvents are effective crazing agents provided that they are able to plasticize the polymer. In the presence of an aggressive liquid, the surface becomes very susceptible to crazing under either externally applied, or internal, tensile stress. Once the process has initiated, liquid is drawn into the voids by capillary action, and travels rapidly to the tip, where it promotes further craze growth.

A standard test for measuring craze resistance is to clamp a strip of the polymer between a curved former and a matching frame, and to apply the liquid to the convex surface, which conforms to the shape of a quarter ellipse, so that the strain varies continuously along the strip. Results are expressed as minimum strains to cause crazing in a given time: typical figures are 0.01 in air and 0.003 in a moderately aggressive liquid, after 24 hours under test. Because of the simplicity of the elliptical strain jig, it is possible to carry out a large number of long-term tests relatively cheaply.

Most practical problems are concerned not with short-term failure in active environments, but with **long-term crazing in relatively mild environments**. Modification of the chemical structure is often sufficient to overcome this problem. For example, polystyrene crazes at low strains in air and at even lower strains in a range of liquid environments, whereas styrene–acrylonitrile copolymers (SAN) are more resistant. The nitrile group ($-C\equiv N$) carries an electric dipole which provides additional intermolecular attractions, thereby both increasing the stresses necessary to cause cavitation, and at the same time reducing the absorption of non-polar liquids (5.N.3).

Polystyrene and SAN form only a few crazes before fracturing at strains of about 0.02; the stress–strain curve is almost linear up to to the point of failure. On the other hand, rubber-modified grades of these polymers form large numbers of crazes, allowing them to yield and extend to strains of up to 0.5 before fracturing. The effects of rubber particles upon the tensile properties of polystyrene are illustrated in Figure 5.8. Millions of small reflecting planes give the yielded material a whitened appearance. Because necking takes place internally through fibrillation, the specimen shows little change in cross-sectional area throughout the deformation, unless shear yielding occurs simultaneously. **The rubber particles not only initiate multiple crazing at low applied stresses, but also extend and deform with the crazed matrix, providing stability against premature fracture**. Rubbers are unique in their ability to perform both functions, and therefore to toughen brittle plastics. Other types of particle, including glass beads, can accelerate crazing sufficiently to cause yielding, but only well-bonded rubber particles enable essentially brittle polymers to reach large strains.

From the engineering standpoint, crazing itself is of minor importance. There are a few applications in which a certain level of craze formation renders the component unserviceable, e.g. PMMA helicopter cabins, where visibility is reduced, and ABS pipes, in which porosity can be a problem.

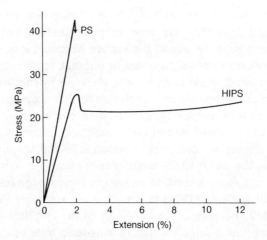

5.8 Stress–strain curves for polystyrene (PS) and high-impact polystyrene (HIPS) which is a mixture of polystyrene with extremely small rubber particles.

However, the main reason for the continuing interest in crazing is that it is the **precursor to fracture in a large number of polymers.**

5.4 Linear elastic fracture mechanics

Brittle solids fracture because the applied stress is amplified by minute cracks—of order 1μm in size—which occur naturally, as a result of fabrication, solidification, fatigue damage, etc. These cracks are frequently termed Griffith cracks, after the originator of the theory we are about to describe.

Consider a stress $\bar{\sigma}$ applied to a wide sheet which contains a through-thickness elliptic crack, oriented as shown in Figure 5.9. The axes of the ellipse are of length $2a$ and $2b$. Let the sheet be of width W and thickness B. The description 'wide' (or 'infinite') is to be taken to mean that $W \gg 2a$. A force F applied to the end surfaces of the sheet develops the stress $\bar{\sigma}$:

$$\bar{\sigma} = \frac{F}{WB}. \tag{5.15}$$

The presence of the crack modifies the elastic stress distribution in its vicinity.

From the fracture viewpoint, the stress distribution along the indicated line Ox_1 is particularly significant. In the x_2-direction, the stress σ_{22} reaches a maximum value σ_m at the ellipse surface; σ_{22} falls as x_1 increases, and ultimately obtains the value $\bar{\sigma}$ at a distance from the crack (as expected). In the x_1-direction, σ_{11} is zero at the ellipse surface and rises to a value of the order of $\bar{\sigma}$ before falling again to zero with

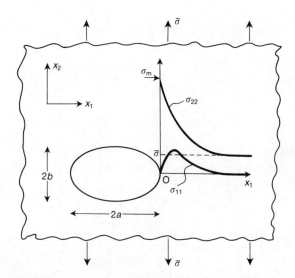

5.9 An elliptical, through-thickness crack in an elastic sheet subject to a stress $\bar{\sigma}$ in the x_2-direction causes a stress distribution along Ox_1 as shown; σ_{22} is amplified from $\bar{\sigma}$ to σ_m at both tips of the crack.

increasing x_1 (as expected). The amplification of the applied stress $\bar{\sigma}$ is greatest at the crack surface ($x_1 = 0$) and is

$$\text{maximum stress amplification} = \frac{\sigma_m}{\bar{\sigma}} = \left(1 + \frac{2a}{b}\right). \tag{5.16}$$

For a central crack, such as that in Figure 5.9, the same amplification occurs at both tips of the crack.

For a circle ($a = b$), the maximum stress amplification is 3; for a thin crack, for which $a/b = 500$, for example, the maximum amplification is $\sim 10^3$. Real cracks, of course, do not conform to precise elliptic shape, but nevertheless this enormous level of stress amplification can occur at the tips of the sharp cracks which are found in all solids. It follows that even at low applied stress, the stress at the crack tip may approach the theoretical strength of the solid, where interatomic bonds are brought to their breaking point. However, these conditions occur only when (as in diamond and in some ceramics) the material is unable to relieve the stress concentration by plastic flow or other mechanisms of crack blunting. In the great majority of engineering materials, including polymers, failure is inhibited by energy-absorbing processes around the crack tip. **A crack will spread only if the total energy of the system is lowered thereby.**

The way in which fracture at the crack tip is inhibited can be understood by examining the change in the total internal energy of the system as the crack begins to spread. Consider a thin sheet of thickness B and infinite width (see Figure 5.10) containing a through-thickness sharp crack of

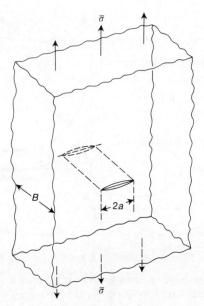

5.10 A narrow, through-thickness crack in a thin, wide sheet subject to a stress $\bar{\sigma}$.

length $2a$ transverse to a tensile stress $\bar{\sigma}$ applied by fixed grips (which means that as the crack spreads, the grips remain stationary). The total energy of the system then comprises two terms, the **elastic strain energy and the work of crack formation**.

- **Elastic strain energy**. At a distance remote from the crack the elastic strain energy per unit volume is $+(\bar{\sigma}^2/2E)$, in which E is Young's modulus. If the specimen is of volume V, then prior to the introduction of the crack the total strain energy is $+(V\bar{\sigma}^2/2E)$. The introduction of the crack modifies the stresses in its vicinity; for example, the modification to σ_{22} and σ_{11} along one line Ox_1 is shown in Figure 5.9. If the stresses are calculated at all points around the crack, and from this the associated strain energy is calculated, it is found that the net effect of the insertion of the crack is a lowering of the total strain energy of the sheet by $\bar{\sigma}^2\pi a^2 B/E$ (see 5.N.4).

- **Work of crack formation**. As the crack spreads, the two new surfaces are prised apart by mechanical forces at the crack tip. The work done per unit area of crack surface is G_c. Part of this work is used to form the structure of the new crack surfaces and the remainder is dissipated as heat.

In the Griffith formulation it was assumed that no heat is dissipated and that the crack surfaces are 'normal'—that is, identical to a surface not formed by crack propagation. In this case G_c is equal to twice the normal surface energy, i.e. 2γ (the factor of 2 occurs because 1 m^2 of crack is composed of 2 m^2 of surface). For polymers, 2γ is of order 1 J m^{-2}, but G_c is far greater, of order 500 J m^{-2}; in polymers, heat is generated and dissipated at the crack tip as the crack spreads, and the crack surface is quite different from a normal surface, being grossly deformed. The total work of crack formation (see Figure 5.10) is then

$$(\text{crack area}) \times G_c = 2aBG_c.$$

Thus the change in energy brought about by the introduction of the crack is

$$\Delta U = -\frac{\bar{\sigma}^2 \pi a^2 B}{E} + 2aBG_c. \tag{5.17}$$

Note that as the crack spreads, the negative sign implies that the elastic strain energy decreases. This decrease continues until (when the crack has spread right across the specimen) the strain energy is zero.

The dependence of ΔU on a is shown in Figure 5.11:

- at small a the term linear in a dominates: it is positive and represents the increase in the work of crack formation as the crack spreads
- at large a the term in a^2 dominates: it is negative and represents the diminution in total strain energy as the crack spreads.

When the crack is on the point of growth under stress $\bar{\sigma}$, the work of

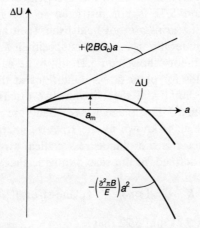

5.11 Dependence of ΔU (the change in energy of a wide sheet) on a (the length of the crack is $2a$) at constant $\bar{\sigma}$. For $a < a_m$ the crack is stable; for $a > a_m$ the crack propagates catastrophically. (Fixed-grip conditions.)

crack propagation just balances the decrease in elastic strain energy, that is

$$\frac{d}{da}\left(\frac{\bar{\sigma}^2\pi a^2 B}{E}\right) = \frac{d}{da}(2aBG_c).$$ (5.18)

Therefore

$$\bar{\sigma}^2\pi a = EG_c.$$ (5.19)

This point is represented by the maximum in the ΔU versus a plot; under a given stress $\bar{\sigma}$ the maximum value of a which is stable is a_m:

$$a_m = \frac{EG_c}{\pi\bar{\sigma}^2}$$ (5.20)

For values of a above a_m, dU/da is negative, so that as a increases U decreases—that is, the total energy of the system decreases and the crack will spread catastrophically.

For values of a below a_m, dU/da is positive—cracks of this size will not spread, because U increases with crack size. This is the region in which fracture is inhibited.

If the stress $\bar{\sigma}$ is increased from zero it follows from eqn 5.19 that the fracture stress $\bar{\sigma}_F$ is

$$\bar{\sigma}_F = \left(\frac{EG_c}{\pi a}\right)^{\frac{1}{2}}   \text{(plane-stress)}.$$ (5.21)

With G_c replaced by 2γ, this is known as the Griffith equation, see 5.N.5.

The validity of eqn 5.21 is not restricted to the fixed-grip situation considered here. If the grips are not held fixed, then at constant load, an increase in $2a$ will generate an increase in specimen length: nevertheless, an extension of the theory shows eqn 5.21 still to be valid.

Equation 5.21 holds for plane-stress conditions at the crack tip, which are most likely to occur if the sheet is thin (we discuss later what 'thin' implies in this context). For plane-stress, then, the operative fracture parameter is G_c, which is known as the **fracture energy**. A related (and more useful) parameter is the plane-stress **critical stress intensity factor** K_c (5.N.6), which is defined, in the case of a wide sheet, by

$$K_c = \bar{\sigma}_F(\pi a)^{\frac{1}{2}}   \text{(plane-stress)}.$$ (5.22)

It follows from eqns 5.21 and 5.22 that

$$K_c = (EG_c)^{\frac{1}{2}}   \text{(plane-stress)}.$$ (5.23)

The preferred fracture parameter can be K_c or G_c, since they are related by eqn 5.23.

In principle, the determination of K_c is straightforward; it is merely required to measure the value of $\bar{\sigma}_F$ at which a crack of length $2a$ in a thin, wide plate begins to propagate, and to use eqn 5.22 to determine K_c. The use of K_c to determine whether or not a given thin sheet will fracture under a stress $\bar{\sigma}$ implies that the size of the largest crack in the sheet is known to the designer. If it is known (and is $2a$), then the stress intensity factor

$$K = \bar{\sigma}(\pi a)^{\frac{1}{2}}$$

can be computed and compared with K_c: the crack will not spread for $K < K_c$.

Example 5.4

A sharp, central crack of length 60 mm in a wide, thin sheet of a glassy plastic commences to propagate at $\bar{\sigma}_F = 3.26$ MPa. (i) Find K_c; (ii) find G_c given that $E = 3$ GPa; and (iii) will a crack of length 2 mm in a similar sheet fracture under $\bar{\sigma} = 10$ MPa?

(i) $\qquad K_c = \bar{\sigma}_F(\pi a)^{\frac{1}{2}} = 3.26\left(\pi \times \tfrac{1}{2} \times 60 \times 10^{-3}\right)^{\frac{1}{2}}$

$\qquad\qquad = 1.00$ MPa m$^{0.5}$.

(ii) $\qquad G_c = K_c^2/E = (10^6)^2/3 \times 10^9$

$\qquad\qquad = 333$ J m^{-2}.

(iii) $\qquad K = \bar{\sigma}(\pi a)^{\frac{1}{2}} = 10(\pi \times 1 \times 10^{-3})^{\frac{1}{2}}$

$\qquad\qquad = 0.56$ MPa m$^{0.5}$.

K is 56% of K_c: the sheet will not fracture.

The foregoing theory for a crack in a thin sheet needs to be modified when considering a crack in a thick plate. The differences between the two are as follows.

- In the thin sheet, at the tip of the stressed crack, the thickness of the specimen decreases because of the Poisson contraction; **plane-stress conditions occur at the crack tip**.

- In the thick plate, at the tip of the stressed crack, the thickness of the specimen does not decrease by Poisson contraction: through-thickness

stresses are generated which offset Poisson contraction. This phenomenon is known as elastic constraint; it is generated by the material
surrounding the crack (outside the region of high stress at the crack
tip). **Plane-strain conditions occur at the crack tip.** Note that, in this
case, at the outer surface of the plate a thin skin of material deforms
in plane stress because there cannot be, at the surface, through-
thickness stresses. The thick plate therefore deforms in the manner of
a sandwich structure, with plane-stress in the two thin outer layers and
plane-strain in the inner section. Plane-strain conditions reduce the
extent of yielding at the crack tip, because when all three principal
stresses are tensile the shear stresses generated are lower. Yielding is
driven by shear components of stress, and cannot occur in pure triaxial
tension (see eqn 5.14).

The parameters which describe plane-strain fracture are G_{IC} and K_{IC}
(the subscript I refers to the method of opening of the crack assumed here
to operate, that is, the prising open of the crack by tensile forces (see
5.N.7)). The method of analysis is analogous to that given for plane-stress
and the result is that, for a thick plate of infinite width containing a crack
of length $2a$, the fracture stress is

$$\bar{\sigma}_F = \left(\frac{EG_{IC}}{\pi(1 - \nu^2)a} \right)^{\frac{1}{2}} \quad \text{(plane-strain)}, \quad (5.24)$$

in which ν is Poisson's ratio. The critical value of the stress intensity factor
in plane-strain is

$$K_{IC} = \bar{\sigma}_F (\pi a)^{\frac{1}{2}} \quad \text{(plane-strain)}. \quad (5.25)$$

From Eqns 5.24 and 5.25 this is

$$K_{IC} = \left(\frac{EG_{IC}}{(1 - \nu^2)} \right)^{\frac{1}{2}} \quad \text{(plane strain)}. \quad (5.26)$$

G and K with a subscript c refer to the critical plane-stress parameters;
with a subscript IC, they refer to the critical plane-strain parameters. For
the calculation of in-use fracture stresses, plane-strain is normally the best
assumption, because materials show minimum toughness in plane-strain,
so that $G_{IC} < G_c$ and $K_{IC} < K_c$. If a component will not fracture in
plane-strain, it will certainly not fracture in plane-stress. The stress intensity factor in plane-strain under stress $\bar{\sigma}$ for a wide plate containing a
crack of length $2a$ is

$$K_I = \bar{\sigma}(\pi a)^{\frac{1}{2}}. \quad (5.27)$$

For $K_I < K_{IC}$ the plate will not fracture. Alternatively, we can define an energy release rate $G_I = -dU/dA$, where A is crack area, and state that for $G_I < G_{IC}$ the plate will not fracture. Note that G_I and K_I are test variables, whereas G_{IC} and K_{IC} are properties of the material. The relationship between G_I and K_I is identical to that between G_{IC} and K_{IC} in eqn 5.26.

Example 5.5

A thick, wide plate of polystyrene contains a central, sharp crack of length $2a = 40$ mm. The crack is found to propagate at $\bar{\sigma}_F = 4.20$ MPa. (i) Find K_{IC}; (ii) find G_{IC} given that $E = 3.0$ GPa and $\nu = 0.40$; and (iii) will a crack of length 2 mm in a similar sheet fracture if $\bar{\sigma} = 10$ MPa?

Solution

(i)
$$K_{IC} = \bar{\sigma}_F(\pi a)^{\frac{1}{2}} = 4.20\left(\pi \times \tfrac{1}{2} \times 40 \times 10^{-3}\right)^{\frac{1}{2}}$$
$$= 1.05 \text{ MPa m}^{\nu.5}.$$

(ii)
$$G_{IC} = \frac{(1-\nu^2)K_{IC}^2}{E} = \frac{(1-0.4^2)(1.053 \times 10^6)^2}{3.0 \times 10^9}$$
$$= 310 \text{ J m}^{-2}.$$

(iii)
$$K_I = 10 \times 10^6(\pi \times 1 \times 10^{-3})^{\frac{1}{2}}$$
$$= 0.56 \text{ MPa m}^{0.5}$$

K_I is 53% of K_{IC}: the plate will not fracture.

The crucial practical consequence of constraint at the crack tip (under the imposed plane-strain conditions) is that it inhibits crack tip plastic deformation, and consequently reduces the critical strain energy release rate to a value below G_c. Similarly, the critical stress intensity factor falls below K_c.

5.4.1 Measurement and application of K_{IC}

K_{IC} may be determined for a thick plane of infinite width (i.e. for width $\gg 2a$) by measuring the stress $\bar{\sigma}_F$ at which a crack of length $2a$ begins to grow; K_{IC} is then determined from eqn 5.25. Conversely, if K_{IC} is known, then the likelihood of fracture for given $\bar{\sigma}$ and a can be computed for a plate of infinite width. It is usually not convenient to determine K_{IC} on an extremely wide plate and, in any case, in design the computation of fracture characteristics must be possible for specimens of arbitrary shape.

The presence of external surfaces (the edge surfaces; see for example

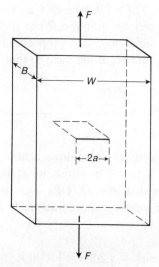

5.12 A crack of length $2a$ in a sheet of width W; $2a$ is of the same order of magnitude as W.

Figure 5.12 which shows a centre crack of length $2a$ in a specimen of width W) modifies both the magnitude of the stresses at the crack tip and the lowering of elastic strain energy as the crack grows. For this specimen geometry, the applied force F develops $\bar{\sigma}$:

$$\bar{\sigma} = \left(\frac{F}{BW} \right).$$

The value of $\bar{\sigma}$ at which this crack begins to grow is $\bar{\sigma}_F$, and the equation for K_{IC} is

$$K_{IC} = \bar{\sigma}_F \left[W \tan\left(\frac{\pi a}{W} \right) \right]^{\frac{1}{2}}. \tag{5.28}$$

The value of K_I for the specimen under stress $\bar{\sigma}$ is

$$K_I = \bar{\sigma} \left[W \tan\left(\frac{\pi a}{W} \right) \right]^{\frac{1}{2}}. \tag{5.29}$$

Note that for $(W/a) = \infty$ these equations must, and do, reduce to eqns 5.25 and 5.27.

In summary, K_{IC} and K_I refer through eqns 5.25 and 5.27 to infinitely wide plates; they may be calculated from data on test plates which are not infinitely wide using eqns 5.28 and 5.29 (or other appropriate equations). The derived values may be used for any plate using the appropriate equations.

Example 5.6

A plate of polystyrene of width $W = 100$ mm contains a central sharp crack of length $2a = 40$ mm. The crack is found to propagate at $\bar{\sigma}_F = 3.91$ MPa. (1) Find K_{IC}. (2) Will a central crack of length 14 mm in an identical plate propagate under $\bar{\sigma} = 9$ MPa? (3) Will a crack of length 3 mm in an infinitely wide polystyrene plate propagate under a stress of 10 MPa?

Solution

(1)
$$K_{IC} = \bar{\sigma}_F \left[W \tan\left(\frac{\pi a}{W} \right) \right]^{\frac{1}{2}}$$

$$= 3.91 \left[100 \times 10^{-3} \times \tan\left(180° \times \frac{20}{100} \right) \right]^{\frac{1}{2}}$$

$$= 1.05 \text{ MPa m}^{0.5}.$$

(2)
$$K_I = \bar{\sigma} \left[W \tan\left(\frac{\pi a}{W} \right) \right]^{\frac{1}{2}}$$

$$= 9 \left[100 \times 10^{-3} \times \tan\left(180° \times \frac{7}{100} \right) \right]^{\frac{1}{2}}$$

$$= 1.35 \text{ MPa m}^{0.5}.$$

$K_I > K_{IC}$, hence plate will fracture.

(3) For a plane of infinite width use eqn 5.27 for K_I,

$$K_I = \bar{\sigma}(\pi a)^{\frac{1}{2}} = 10\left(\pi \times \tfrac{3}{2} \times 10^{-3} \right)^{\frac{1}{2}}$$

$$= 0.686 \text{ MPa m}^{0.5}.$$

$K_I < K_{IC}$, hence plate will not fracture.

For fracture mechanics tests on plastics the most widely used test pieces are:

- the compact tension (CT) specimen (see Figure 5.13(a)); and
- the single-edge-notched bend (SENB) specimen (see Figure 5.13(b)).

For either test piece it is required that the crack tip be sharp. The crack must be grown in a stable manner. Standards recommend machining a sharp notch and then sharpening it using a new razor blade. Where

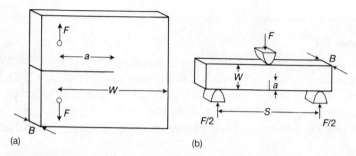

5.13 Specimen geometries used to determine K_{IC}: (a) compact tension (CT) specimen; and (b) single-edge-notched bend (SENB) specimen.

possible, the blade should be tapped into the notch to produce a small amount of natural crack growth. If the polymer is too tough to propagate a natural crack in this way, the blade should be drawn across the notch tip.

It is required to know from theoretical calculation the equivalent equations to eqn 5.28, to take account of the finite width W of the specimen. For the CT specimen,

$$K_{IC} = Y\bar{\sigma}_F(\pi a)^{\frac{1}{2}}, \tag{5.30}$$

$$Y = 16.70 - 104.7(a/W) + 369.9(a/W)^2$$
$$- 573.8(a/W)^3 + 360.5(a/W)^4, \tag{5.31}$$

and $\bar{\sigma} = F/BW.$

Example 5.7

A CT specimen machined from PMMA plate has thickness $B = 6$ mm, effective width $W = 50$ mm and effective crack length $a = 25$ mm. The force F at the loading pins increases linearly with deflection until fracture occurs at $F = 225$ N. (1) Find K_{IC}; (2) find G_{IC} given $E = 3.2$ GPa and $\nu = 0.42$; and (3) estimate the critical length of crack for a wide plate stressed at $\bar{\sigma} = 15$ MPa (see Figure 5.10).

Solution

(1)   For $(a/W) = 25/50 = 0.5$, $Y = 7.63$.

$$K_{IC} = Y\bar{\sigma}_F(\pi a)^{\frac{1}{2}} = 7.63\left(\frac{225}{0.006 \times 0.050}\right)(\pi \times 0.025)^{\frac{1}{2}}$$

$$K_{IC} = 1.60 \text{ MPa m}^{0.5}.$$

(2) $$G_{IC} = \frac{K_{IC}^2(1 - \nu^2)}{E} = \frac{(1.60 \times 10^6)^2(1 - 0.42^2)}{3.2 \times 10^9}$$

$$G_{IC} = 659 \text{ J m}^{-2}.$$

(3) For the wide (infinite) plate the crack length is $2a$ and

$$K_{IC} = \bar{\sigma}_F(\pi a)^{\frac{1}{2}}$$

$$a = \frac{K_{IC}^2}{\pi \bar{\sigma}_F^2} = \frac{(1.60 \times 10^6)^2}{\pi (15 \times 10^6)^2}$$

$$= 3.62 \text{ mm}$$

and
$$2a = 7.24 \text{ mm}.$$

Example 5.8

A rectangular bar of PMMA in the form of an SENB specimen, of thickness $B = 6$ mm and width $W = 10$ mm, contains a central edge crack of length $a = 1$ mm (see Figure 5.13(b)). Calculate the force F required to fracture the bar in single-edge-notched bending with span $S = 80$ mm. For this geometry, with $S/W = 80/10 = 8.0$, Y is given by

$$Y = 1.11 - 1.55(a/W) + 7.71(a/W)^2 - 13.5(a/W)^3 + 14.2(a/W)^4.$$

For $(a/W) = 1/10$ we obtain $Y = 1.02$.

$$K_{IC} = Y\sigma_{max}(\pi a)^{\frac{1}{2}}$$

$$1.60 \times 10^6 = 1.02 \times \sigma_{max} \times (\pi \times 0.001)^{\frac{1}{2}}$$

$$\sigma_{max} = 28.0 \text{ MPa}.$$

From elementary beam theory, the outer fibre stress at the site of the crack is

$$\sigma = \frac{3S}{2BW^2} F$$

$$28.0 \times 10^6 = \frac{3 \times 0.08}{2 \times 0.006 \times (0.01)^2} \times F$$

$$F = 140 \text{ N}.$$

In the discussion so far we have considered only linear elastic fracture mechanics (LEFM): the term linear elastic means that the cracked specimen obeys Hooke's law to a good approximation. In the context of fracture mechanics, the requirement is that the extent of yielding in the neighbourhood of the crack tip is negligible, so that the force–deflection curves for test specimens are linear. In addition, the value of B must be sufficiently high that the deformation at the crack tip occurs under plane-strain conditions. It is found experimentally that these conditions are met if

$$a, B, (W-a) > 2.5 \left(\frac{K_{IC}}{\sigma_y} \right)^2 \qquad (5.32)$$

Provided these conditions are met, the spatial extent of the plastically deformed zone at the crack tip is less than 2% of the above dimensions, the specimen fractures in plane-strain and the measured K_{IC} is a true material property.

Example 5.9

Calculate the minimum specimen dimensions for plane-strain fracture of PVC using the following data: $K_{IC} = 3.2$ MPa m$^{0.5}$, $\sigma_y = 68$ MPa.

Solution

$$2.5 \left(\frac{K_{IC}}{\sigma_y} \right)^2 = 2.5 \left(\frac{3.2 \times 10^6}{68 \times 10^6} \right)^2 = 5.5 \text{ mm}.$$

Hence a, B, and $(W-a)$, must exceed 5.5 mm.

Because it is difficult to make and test very large polymer specimens in order to measure G_{IC} and K_{IC}, standard test methods make some allowance for minor deviations from linearity. These may be due not only to plastic deformation at the crack tip, but also to non-linear elasticity, general viscoelasticity, or stable crack growth after initiation but before instability. The procedure recommended by ESIS (European Structural Integrity Society) is shown in Figure 5.14. A best straight line is drawn to determine the initial compliance C_0 of the specimen, and a second straight line is then drawn to indicate a 5% increase in compliance. If fracture occurs at a load maximum F_{max} falling between these two lines, F_{max} is used to calculate an effective fracture toughness K_Q. If the load–deflection curve crosses the second line at load $F_{5\%}$, then that is taken as the load at crack initiation, and is used to calculate K_Q, provided that $F_{max} < 1.1F_{5\%}$. If $F_{max} > 1.1F_{5\%}$, the test is invalid. The value of K_Q obtained using either of these two procedures is then checked against the size criteria given in eqn 5.32, taking σ_y as 0.7 times the compressive yield stress if no uniaxial data are available. If all criteria are satisfied, then $K_Q = K_{IC}$.

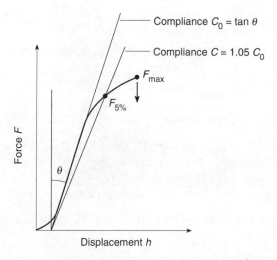

5.14   Procedure for determination of K_{IC} from a non-linear force-displacement curve. Specimen compliance $C = h/F$.

5.5   Elastic–plastic fracture mechanics

When yielding causes large departures from linearity in the force–displacement curve, such that valid K_{IC} data cannot be obtained, it is still possible to make geometry-independent measurements of the fracture resistance of the material, using the methods of elastic–plastic fracture mechanics. These usually require additional information to determine whether non-linearity is due to crack tip plasticity alone, or to a combination of plasticity and crack growth. Several different approaches have been developed, of which we will discuss only two: the crack tip opening displacement (CTOD) and *J*-integral methods.

Plasticity causes blunting of the crack tip, as illustrated in Figure 5.15. The resulting crack tip separation defines both δ, the CTOD, and δ_c, the critical CTOD at the onset of crack growth, which may be used as a measure of toughness. Where fracture is within the linear elastic range, δ_c can be correlated with G_{IC} through the equation:

$$G_{IC} = \delta_c \sigma_y \tag{5.33}$$

The CTOD is defined by constructing two 45° lines from the centre of the crack tip, as shown. During the early stages of the test, where the tip is simply becoming blunter, the crack appears to extend by an amount Δa, which (assuming the crack tip to have a semicircular profile) is equal to $\delta/2$, as shown in Figure 5.15. In some cases, the crack tip becomes sharper when it begins to propagate, and there is a corresponding change in the force–deflection curve; in others, there is no obvious transition between blunting and crack propagation.

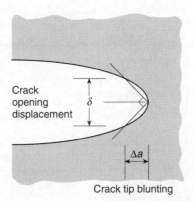

Crack tip blunting

5.15 Apparent increase in crack length Δa due to blunting, with definition of crack opening displacement.

Another standard approach to elastic–plastic fracture mechanics is to extend the energy release rate principle to specimens that exhibit marked non-linear elasticity, viscoelasticity, or plasticity. The basic approach is illustrated in Figure 5.16, which compares linear and non-linear force–displacement curves for specimens containing cracks of length a_0 and $(a_0 + \mathrm{d}a)$. Whether the curves are linear or non-linear, if the material behaves elastically the energy release rate $-1/B(\mathrm{d}U/\mathrm{d}a)$ can be obtained directly by subtracting U_1, the energy absorbed initially at crack length a_0, from U_2, the energy recovered later, on unloading at crack length $(a_0 + \mathrm{d}a)$. The energy difference, which is available to extend the crack area by an

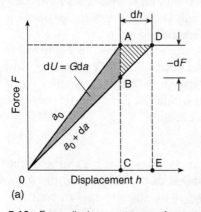
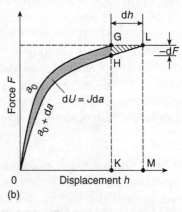

5.16 Force-displacement curves for specimens with slightly differing crack lengths a_0 and $(a_0 + \mathrm{d}a)$: (a) linear elastic behaviour; (b) non-linear elastic-plastic behaviour. At fixed displacement h, the energy $\mathrm{d}U$ absorbed in crack propagation $\mathrm{d}a$ is given by areas OABO (linear case) and OGHO (non-linear case). At fixed force F, the corresponding areas are OADBO and OGLHO. For very small $\mathrm{d}h$, the areas ABD and GHL are negligible, and $\mathrm{d}U$ is independent of loading conditions.

increment $B\,\mathrm{d}a$, is defined by the shaded area $\mathrm{d}U$. This is a valid method for measuring $\mathrm{d}U$ because all of the energy stored elastically in the specimen is recovered on unloading. Thus for the case of a stationary crack tip, unloading and reloading curves are identical.

This approach is clearly applicable to rubbers with low mechanical hysteresis, which exhibit non-linear elastic behaviour. However, because the energy release rate G_I is defined specifically for the case of linear elastic fracture, we define a new parameter J for the non-linear case:

$$J = -\left(\frac{\partial U}{\partial A}\right) = -\frac{1}{B}\left(\frac{\partial U}{\partial a}\right) \qquad \text{non-linear} \qquad (5.34)$$

Crack growth occurs when the energy release rate J is greater than or equal to the crack propagation resistance J_R. The criterion $J \geq J_\mathrm{R}$ is similar to the condition $G_\mathrm{I} \geq G_\mathrm{IC}$ in linear elastic fracture mechanics.

There are major problems in defining and measuring energy release rates when non-linearity in the force–deflection curves is due to extensive yielding and/or viscoelasticity, rather than to non-linear elastic behaviour. Under these conditions, energy dissipation may occur in areas remote from the crack tip, and have little connection with the essential process of crack growth. Even when yielding is confined to a region close to the crack tip, and energy absorption is directly determined by the area of new crack surface, there are problems. Because the plastic zone does not return to its original dimensions upon unloading, the crack cannot close completely, and some elastic energy therefore remains stored in the specimen. Consequently, a full unloading curve can no longer be used to determine $\mathrm{d}U_\mathrm{el}$, the decrease in elastic energy during crack growth. This path-dependence in the force–deflection curves, which is inherent in specimens undergoing elastic–plastic fracture, means that valid measurements relating to J can be made only when the loading is basically montonic, and there is only a limited amount of unloading, if any.

One method of overcoming these problems is to carry out force–deflection tests on a series of fresh specimens, each with a different initial crack length a_0, in order to produce a small amount of stable crack growth $\mathrm{d}a$. Referring to Figure 5.16, the total energy U absorbed in reaching a given crack length $a = (a_0 + \mathrm{d}a)$ is obtained from the areas under the **loading** curves, while the crack extension $\mathrm{d}a$ is determined by breaking open each specimen at low temperature and measuring the maximum crack length at the centre of the specimen. The energy release rate $-(\mathrm{d}U/\mathrm{d}a)$ at constant displacement h then defines the parameter J. When the degree of non-linearity is small, J becomes identical to the linear elastic strain energy release rate G_I, and the crack growth resistance J_R tends to G_IC.

Earlier standards attempted to define an initiation parameter J_IC by plotting J_R against Δa for true crack growth, and obtaining the intercept with the 'blunting line' constructed at $J = 2\Delta a \sigma_y$ in accordance with eqn 5.33. However, this procedure has now been dropped as invalid in many materials. Current standards define a pseudo-initiation parameter $J_{0.2}$, the crack resistance at $\Delta a = 0.2$ mm.

Standard test methods specify sharply-notched single-edge-notched bend (SENB) or compact tension (CT) specimens identical with those used in linear elastic fracture mechanics, except that blunt side-grooves are recommended to guide the crack. Grooves reduce experimental scatter when crack growth resistance is high. The total energy input to the specimen U is measured as a function of Δa, and J is calculated from the equation:

$$J = \frac{\eta U}{B_N(W - a_0)} \tag{5.35}$$

where a_0 is the initial crack length, B_N is the net thickness of the specimen after allowing for side-grooves, and η is a dimensionless constant. For SENB specimens $\eta = 2$, and for CT specimens $\eta = [2 - 0.522(1 - a_0/W)]$. The fracture resistance J_R is then plotted against crack growth Δa, using a simple power law to fit data for Δa greater than 0.05 mm (avoiding the crack-blunting zone) and less than $0.1(W - a_0)$ (beyond which eqn 5.35 is no longer applicable).

Example 5.10

Fracture mechanics tests were carried out on ABS polymer, using 6 side-grooved compact tension specimens with $W = 50$ mm, $a_0 = 25$ mm, $B = 15$ mm. The side grooves were 1.5 mm deep, with an included angle of $45°$ and a root radius of 0.25 mm. Measured values of Δa were 0.1, 0.3, 0.5, 0.7, and 0.9 mm. The corresponding work inputs were 0.41, 0.66, 0.81, 0.92, and 1.02 J. Calculate J_R for each valid data point, and hence obtain a value for $J_{0.2}$.

Solution

Net thickness $B_N = 15 - (2 \times 1.5) = 12$ mm.

$$J = \frac{\eta U}{B_N(W - a_0)} = \frac{(2 - 0.522 \times 0.5)U}{12 \times 10^{-3} \times 25 \times 10^{-3}} = 5797U$$

All data points are valid, in that Δa lies between 0.05 mm and $0.1(W - a_0)$. They give J_R values of 2.38, 3.83, 4.70, 5.33, and 5.91 kJ m^{-2}. On plotting J_R against Δa and interpolating, $J_{0.2}$ is found to be 3.25 kJ m^{-2}.

As illustrated in Example 5.10, J_R increases with Δa. The slope dJ_R/da, which decreases as the crack extends, defines the increase in the resistance of the material to (stable) crack propagation during the initial stages of tearing. At the same time, the increasing applied load causes a matching increase in the effective strain energy release rate J. Crack growth finally becomes unstable when $(\mathrm{d}J/\mathrm{d}a) > (\mathrm{d}J_\mathrm{R}/\mathrm{d}a)$. In polymers, instability is often associated with a sharp drop in fracture resistance, caused by the strain-rate dependence of deformation behaviour. Both G_IC and J_R are functions of crack speed, and in some cases the dynamic fracture resistance is substantially lower than the static value. This may result in 'stick-slip' fracture, in which blunting of the static crack tip alternates with unstable crack propagation.

At speeds of order $500 \mathrm{\ m\ s}^{-1}$, which in rigid polymers are close to the velocity of the transverse wave, the quasi-static stress analysis upon which fracture mechanics is based breaks down completely, and inertial effects become predominant. The strain energy released is almost completely converted to kinetic energy, and critical energy release rates are no longer related to the energy absorbed in forming new crack surface. Under these conditions, apparent values of G_IC tend asymptotically to infinity.

Elastic–plastic fracture mechanics has been extensively used in studies on plastics, but to data mainly by research laboratories. The plastics industry continues to rely on impact testing, and in particular the notched Izod test, as the principal method for assessing toughness. One of the most important contributions of fracture mechanics to polymer engineering has been to provide a theoretical basis for understanding these impact test data.

5.6   Brittle fracture of polymers

Predicting when a material will fail under a particular set of loading conditions presents a major problem to the engineer. The problem is less severe for materials such as low-density polyethylene, which have very low yield stresses and usually undergo general yielding before fracture; but for most engineering plastics the possibility of brittle fracture is very real. Whereas yielding is determined by the stresses acting throughout a section of the component (which can be calculated), fracture is controlled by cracks and other defects in the material, which are usually undetected and may, indeed, be undetectable. Fracture mechanics provides valuable help in addressing this problem.

Figure 5.17 shows the relationship between critical stress and crack length a for a simple case of a rectangular bar of width W under uniaxial tension. The calculation is based on eqn 5.30, using published data for the geometrical factor Y. As a first approximation, we can draw a straight line

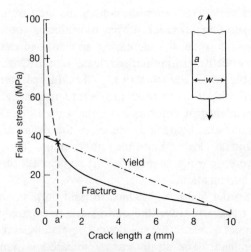

5.17 Relationship between crack length and type of failure in a bar of polymer of width $W = 10$ mm. The brittle fracture curve is calculated for a K_{IC} of 2.0 MPa m$^{0.5}$. Straight line shows ductile stress for $\sigma_y = 40$ MPa. Note that brittle fracture occurs before the material is able to yield, except at short crack lengths or very long crack lengths.

on the diagram relating the condition for general yielding to the area of the net section

$$\sigma_{max} = \sigma_y(W - a)/W. \tag{5.36}$$

It is clear that ductile failure occurs both at very short and at very long crack lengths, although the yield zone will be limited in extent in the latter case. The bar fails by brittle fracture at intermediate crack lengths. In this geometry, K_I increases with a, so there is no possibility that the crack will stop as a result of the tip becoming unloaded as it propagates; fracture is catastrophic.

Figure 5.17 provides the key to understanding the effects of test conditions and materials structure upon the fracture behaviour of polymers. All materials contain small defects of some kind, but whether they are large enough to cause brittle fracture depends upon the relative positions of the yield and fracture lines on the diagram, and therefore upon σ_y and K_{IC}.

The intersections of yield and fracture lines mark the transitions between modes. Most usually, the crack length a is a small fraction of the width of the bar. The critical crack length is then a' (see Figure 5.17). As stress is increased from zero, the event which occurs first, depending on the size a of crack present, is

yield   (i.e. ductile failure) if $a < a'$

or

fracture   (i.e. brittle failure) if $a > a'$.

First consider the effects on Figure 5.17 of varying the thickness B of the bar. We have already noted that plane-strain conditions cause reduced crack-tip plasticity compared with plane stress, and hence $G_{IC} < G_c$ and $K_{IC} < K_c$. Consequently, if bars of different thicknesses are compared, with increasing B the point of intersection of yield and fracture lines (starred) moves to the left (i.e. a' decreases), giving a greater tendency for brittle failure. A limiting value of a' is approached at large thicknesses, where plane strain conditions are fully developed and the fracture stress is determined by K_{IC}.

In some polymers, dramatic changes in toughness take place at room temperature, when the thickness is varied over the range likely to be encountered in practice. For example, under impact loading, **sharp cracks in thick sections can give rise to low-energy fractures even in polycarbonate**, which is normally a very tough polymer. Fracture mechanics measurements on polycarbonate show a decrease in toughness from $G_c = 10$ kJ m^{-2} under plane-stress conditions (in relatively thin specimens, 3 mm thickness) to $G_{IC} = 1.5$ kJ m^{-2} under plane-strain conditions (thick specimens).

Now consider the effects of strain rate. Experiments show that K_{IC} increases with increasing crack speed, as illustrated in Figure 5.18, but the resulting shift in fracture stress with increasing strain rate is usually smaller than the increase in yield stress. Consequently, a' decreases. As a result, impact and other forms of rapid loading tend to cause brittle failures.

A particularly awkward feature of fracture in polymers, as in some other materials, is that under certain conditions small cracks grow very slowly under **sub-critical** stresses. As a crack grows K_I increases. Finally, K_I may

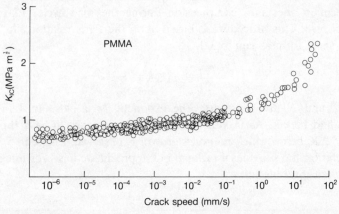

5.18 Relationship between K_{IC} and crack speed in PMMA. Data from many different authors (after W. Doell).

reach its critical value. There is then a sudden brittle fracture, which may occur after the component has been carrying load apparently safely for a long period of time (even several years).

The term 'static fatigue' is used to refer to slow crack growth under long-term steady loading. There appears to be a threshold value of K_I below which no crack growth occurs; above this limit, sub-critical crack growth can be represented by an empirical power law relating the applied K_I and (da/dt):

$$\frac{da}{dt} = \beta K_I^m \tag{5.37}$$

where β and m are constants. Times to failure can be calculated from this equation, either by integration or by iterative methods using a computer. Integration is convenient where Y varies only slowly with a. Final fracture occurs when K_I reaches K_{IC}. This type of fracture mechanics analysis of static fatigue has been of crucial importance in the development of polyethylene gas pipe materials, which must give safe operation over many years.

Dynamic fatigue is a more serious problem, because fluctuations in load accelerate sub-critical crack growth. One way to study this effect is to subject specimens to square-wave loading, in which they alternate between a fixed stress and zero. Comparisons can then be made between the fixed stress applied continuously for a certain period, and the same stress applied intermittently for the same total time. In practice, most experimenters employ sinusoidal loading, which gives lower fatigue crack propagation rates. A standard method for studying fatigue is to cycle specimens between load limits, and to make a log–log plot of crack growth rate da/dn against ΔK, where n is the number of fatigue cycles and ΔK is the difference between the maximum and minimum current values of K_I: if the specimen goes into compression during the load cycle, $K_I(\text{min})$ is taken as zero. This procedure is based upon the Paris relationship, which has the same form as eqn (5.37):

$$\frac{da}{dn} = \zeta (\Delta K)^N \tag{5.38}$$

where ζ and N are constants. The exponent N is close to 4 in many plastics and metals. **As in static fatigue, there appears to be a threshold value of ΔK below which no crack growth is observed.** Like eqn 5.37, the Paris relationship provides a rational but approximate basis for integration to calculate total lifetime.

Example 5.11
Rigid PVC is a polymer that conforms well to the Paris relationship.

Fatigue crack growth data at 20°C for PVC can be represented by the equation

$$\frac{\mathrm{d}a}{\mathrm{d}n} = 0.035\Delta K^{2.4},$$

where $\mathrm{d}a/\mathrm{d}n$ is in μm per cycle and ΔK is in MPa m$^{0.5}$. A compact tension specimen has $B = 6$ mm, $W = 50$ mm, and $a = 20$ mm. Calculate $\mathrm{d}a/\mathrm{d}n$ when the specimen is cycled between $F_{max} = 100$ N and $F_{min} = 50$ N.

Solution

Using the expression for Y given in eqn 5.31, at $a/W = 0.4$,

$$Y = 16.70 - 41.88 + 59.18 - 36.72 + 9.23 = 6.51;$$

$$\Delta K = Y(\pi a)^{\frac{1}{2}}(\sigma_{max} - \sigma_{min}) = Y(\pi a)^{\frac{1}{2}}\frac{(F_{max} - F_{min})}{BW},$$

$$= 6.51 \times (\pi \times 0.02)^{\frac{1}{2}}\frac{(100 - 50)}{0.006 \times 0.05}$$

$$= 0.272 \text{ MPa m}^{0.5};$$

$$\frac{\mathrm{d}a}{\mathrm{d}n} = 0.035\Delta K^{2.4}$$

$$= 1.54 \text{ nm per cycle}.$$

Dynamic fatigue is a complex process in which frequency, waveform, loading history, and the complete range of current K_I values play a part in determining crack growth rates, and no simple equation is likely to provide a satisfactory correlation of all fatigue data for a material. Equation 5.38 implies that the absolute values of K_I, $K_I(\max)$, and $K_I(\min)$ are of no significance, and that ΔK_I ($\equiv K_I(\max) - K_I(\min)$) alone matters, whereas experiment shows that this is rarely true. In some materials, a constant component of stress accelerates crack growth, essentially by adding a contribution due to static fatigue; in others, usually those with low yield stresses, the constant component slows the crack by blunting the tip. Heating at the crack tip appears to be largely responsible for the observed frequency effects, but again the result can be either faster or slower crack growth, depending upon the material.

Reducing temperature has a similar effect to increasing strain rate; both σ_y and K_{IC} increase, but the dominant factor is the increase in σ_y. Plastics therefore become more brittle at low temperatures (as a' decreases—see Figure 5.17), and more ductile at high temperatures (as a' increases). In

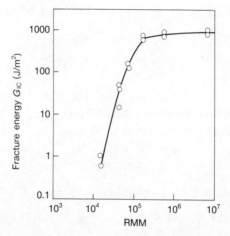

5.19 Relationship between G_{IC} and relative molar mass in PMMA (after R. P. Kusy and D. T. Turner).

the light of this familiar observation, it is perhaps rather surprising to find that G_{IC} and K_{IC} fall as the temperature is raised. The reason is that the thickness of material needed to form a craze is relatively independent of temperature, whereas the crazing stress falls on heating; crazing, like shear yielding, is an activated flow process driven by thermal energy and following the Eyring equation. Consequently, the work done in forming a unit area of craze on a fracture surface decreases with increasing temperature. Crazing and shear yielding are affected differently by temperature and strain rate because of differences in activation energy and activation volume, and in the more ductile polymers the most obvious effect of raising the temperature is to accelerate shear yielding.

The fracture resistance of a thermoplastic depends critically upon its relative molar mass, essentially because G_{IC} and K_{IC} fall rapidly when the chains are too short to form effective entanglements. This point is illustrated in Figure 5.19. Very short chains are held together only by van der Waals forces, and are unable to form stable crazes.

Of more practical interest are variations in the RMM of commercial materials, which occur for two reasons. Firstly, injection moulding grades cannot have very high RMM if they are to be processed satisfactorily, whereas extrusion and thermoforming grades do not suffer from the same kind of limitation; the differences in fracture resistance are especially apparent in dynamic fatigue tests (5.N.8). Secondly, degradation in chain length during processing or subsequent use of the polymer can result in a reduction in fracture toughness. Degradation may be purely thermal, but is accelerated by a trace of water in polymers containing hydrolysable linkages: careful drying is therefore necessary before moulding polyesters,

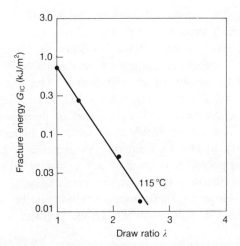

5.20 Fracture energy of polystyrene at 23°C after drawing above the glass transition at 10^{-2} s^{-1}, showing effect of molecular orientation. The crack plane is parallel to the draw direction (after L. J. Broutman and F. J. McGarry).

polyamides, and polycarbonates. Provided that suitable stabilizers and processing agents are added to the polymer, few problems are encountered in moulding fresh material, but difficulties are likely to arise when large amounts of reground polymer are used.

Molecular orientation is one of the most important factors affecting the fracture resistance of moulded or formed polymeric products. Biaxial stretching in the melt state makes a polymer much more resistant to crazing and cracking under the action of in-plane stresses, and is used to improve the properties of film, sheet, and other products (see Chapters 2 and 7). The hot-drawing operation rotates chain segments into the plane, and biaxially oriented sheet is therefore anisotropic in its strength properties: resistance to cracks propagating parallel to the surface of the sheet is reduced by the stretching. Anisotropy is a more obvious problem in **uniaxially** drawn material. The effect can be demonstrated in disposable coffee cups, which are thermoformed from HIPS. The cup slits easily along the draw direction, but is very resistant to tearing at right angles to this direction. A more quantitative experiment is illustrated in Figure 5.20, which shows that uniaxial drawing of polystyrene at 115°C (just above T_g) reduces G_{IC} by almost a factor of 100 for cracks **propagating along the draw direction**. Measurements are difficult to make transverse to the draw direction, but such data as are available show a substantial **increase** in G_{IC} for cracks propagating normal to this direction.

5.7 Rubber toughening

A process for adding rubbers to rigid plastics in order to increase their fracture resistance was first used commercially in 1948, with polystyrene as the matrix. The early success of 'high-impact' polystyrene (HIPS) led to the development of similar blends based on other rigid polymers, with the result that rubber-toughened grades are now available for most commercial plastics and thermosets of any significance. The main exceptions fall into two categories: soft polymers, notably LDPE and flexible PU, which already contain elastomeric domains as part of their intrinsic structure; and thermosets that are so highly cross-linked as to lack the minimum ductility necessary to benefit from rubber toughening. The other major plastics, including polymers as tough as polycarbonate, all show worthwhile improvements in fracture resistance on blending with 5 to 20% of a suitable rubber. However, these improvements must be balanced against the inevitable drop in modulus when a rigid polymer is blended with a rubber.

The added rubber forms a fine dispersion of particles which are approximately spherical in the relaxed state, but become distorted into ellipsoidal or other shapes during melt processing. Some particles consist simply of rubber, while others contain rigid polymeric inclusions, as illustrated in Figure 5.21. Particle sizes range from 50 nm to 10 μm; optimum values depend upon the properties of the matrix.

Experiments have shown that almost any type of elastomer can be made to work as a toughening additive, although not all have the exact combination of chemical reactivity, ease of processing, low T_g, and moderate cost that is required to make the product successful commercially. Furthermore, almost any standard process may be employed to mix the matrix polymer with the rubber, including: (a) simple melt blending; (b) controlled formation of 'core-shell' rubber particles by sequential polymerization in emulsion, followed by drying and melt blending with the matrix; and (c) dissolving the rubber in the liquid monomer(s) and polymerizing with stirring to produce the preferred two-phase morphology. In all three of these processes, it is possible both to cross-link the rubber and to bond it to the matrix polymer by means of a grafting reaction. Crosslinking stabilizes the particles against being broken up during melt processing, and grafting ensures good strength at the interface.

Toughening involves the matrix and rubber particles in complex interactions, which change and develop throughout the deformation process. One obvious effect of adding rubber to a rigid polymer is to lower its yield stress, and in some cases this alone is sufficient to produce a marked improvement in fracture resistance. The polymers that show the greatest direct benefit from a reduction in yield stress are ones that have low impact energies at room temperature, but become significantly tougher under slightly different loading conditions: higher temperature, lower loading rate, or modified geometry (e.g. thinner specimens, blunter

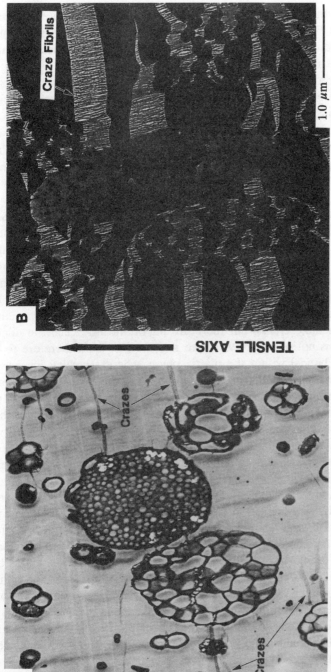

5.21 Transmission electron micrographs of HIPS: (A) osmium-stained thin section showing 'salami' structure, with polystyrene sub-inclusions embedded in (black) rubber phase, in a matrix of polystyrene; (B) unstained thicker section stretched on the microscope stage. The unloaded specimen (A) has largely recovered, and shows compressed crazes. The stressed sample (B) shows extended craze fibrils and fibrillated rubber. (Micrographs by courtesy of R. C. Cieslinski, Dow Chemical USA.)

notches). In polycarbonate, which shows pronounced thickness effects over the range $B = 3-6$ mm in notched Izod tests, a modest amount of toughening can be achieved simply by introducing a few well-dispersed air bubbles to reduce notch-tip constraints.

Polycarbonate is exceptional in this respect. Most polymers are embrittled by large voids or debonded rubber particles, which act as nuclei for crazes and cracks, whereas the same polymers are toughened by small well-bonded rubber particles, which generate similar stress concentrations, and are also capable of forming voids, by cavitating under stress. The key factor appears to be the size of the voids, which determines whether they nucleate crazes and cracks (see Sections 5.3 and 5.4), or promote shear yielding.

In both metals and polymers, void formation followed by dilatational shear yielding occurs preferentially at crack tips, where it overcomes the geometrical constraints imposed on plastic deformation. In rubber-toughened plastics, the process is initiated by cavitation of the rubber particles in response to the triaxial stress field, and allows the polymer to form planar **dilatation bands** in the crack plane, at stresses well below those required for yielding in the solid material. Superficially, dilatation bands resemble crazes, but there are important differences: in dilatation bands, it is not necessary to create new surfaces (and therefore to break chemical bonds) within the matrix phase, whereas crazing involves both types of damage to the matrix. Furthermore, the voids within dilatation bands are discrete, in contrast to the connected void structure in a craze. Dilatational shear yielding is an effective mechanism of rubber toughening in polymers that are relatively resistant to crazing.

Cavitation of the rubber particles is also an important part of the toughening mechanism in polystyrene and other polymers that have relatively high shear yield stresses and a low resistance to crazing. In these materials, craze formation initiated by particle cavitation is inevitable, and the material is designed to yield by multiple craze formation. The rubber particles are large enough (typically 1-5 μm) to cavitate at low stresses, and have a complex 'salami' morphology, which forces the rubber phase to fibrillate, as shown in Figure 5.21. Large numbers of rubber fibrils bridge the gaps between neighbouring polystyrene sub-inclusions, acting as strain-hardening elements in the structure, thereby preventing the associated craze fibrils from becoming overloaded. As in the case of dilatational shear deformation, cavitation of the rubber particles is the first stage in a process that enables the polymer to deform and strain-harden under the constraints imposed at a crack tip.

A simple energy-balance model can be used to analyse the factors affecting cavitation in homogeneous spherical rubber particles. Very little energy is required to produce small shear strains in rubber, because their shear moduli are so low; but the energy $U(0)$ required to generate a small

volume strain Δ_V in an intact particle of radius R, in response to a mean normal stress (negative pressure) σ_m, is quite large:

$$U(0) = \tfrac{4}{3}\pi R^3 \left(\frac{\sigma_m \, \Delta_V}{2} \right) = \tfrac{2}{3}\pi R^3 K_r \, \Delta_V^2 \tag{5.39}$$

The bulk modulus K_r of a typical rubber is 2 GPa, only about half the value for a thermoplastic. If the volume strain applied to the **particle** is held constant, and a spherical void of radius r is formed within the rubber, the volume strain of the rubber phase decreases by r^3/R^3, and its strain energy decreases to

$$U(r) = \tfrac{2}{3}\pi R^3 K_r \left(\Delta_V - \frac{r^3}{R^3} \right)^2 \tag{5.40}$$

The energy released, represented by $U(0)\text{-}U(r)$, is required for the formation of new surface and for stretching the rubber membranes surrounding the void. The energy of void formation is then

$$U(\text{void}) = 4\pi r^2 \Gamma + 2\pi r^3 G_r F(\lambda_f) \tag{5.41}$$

where Γ is the surface energy of the rubber, G_r is its shear modulus, and the term $2\pi r^3 G_r F(\lambda_f)$ is obtained by integrating shear strain energy terms around the void, to a maximum extension ratio of λ_f (see Problem 5.26). Void formation becomes possible when $U(0)\text{-}U(r) > U(\text{void})$. Application of these equations shows that the critical volume strain at cavitation decreases with increasing particle size, and also with decreasing rubber shear modulus. In general, both mechanically imposed stresses and thermal contraction stresses contribute to the dilatational strains in the rubber particle.

The effects of rubber toughening on the glassy copolymer SAN (polystyrene–acrylonitrile) are illustrated in Figure 5.22. Toughened SAN is known as ABS (from the initials of the constituent monomers: acrylonitrile–butadiene–styrene—see 5.N.3): the grade chosen for this comparison contains 20 wt% of well-bonded polybutadiene rubber particles. Under the conditions of the notched Izod test, SAN is brittle over the whole temperature range, forming at the notch tip a small number of crazes, which then fail catastrophically. Crack speeds reach approximately 500 m s^{-1}, a figure determined by the velocity of the transverse wave. If we overlook the bluntness of the notch, and consider the test from a fracture mechanics point of view, we can say that SAN exhibits linear elastic fracture with a low G_{IC}. Indeed, Charpy impact tests on sharply notched SENB bars over the same range of temperatures give the same pattern of behaviour as that shown in Figure 5.22.

Below the T_g of the rubber particles, at about $-80°C$, the ABS behaves

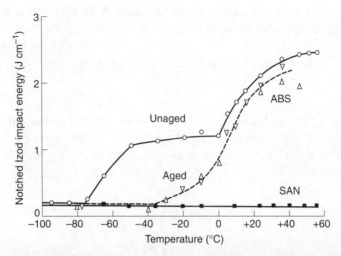

5.22 Notched Izod impact test results for SAN (0% rubber) and ABS (20% rubber), showing effects on ABS of 3 months' exposure to sunlight during the English summer.

in a manner similar to neat PSAN. The rubber particles are rigid, and are consequently unable either to cavitate or to act as effective stress concentrators. However, between −75 and −10°C, the impact resistance of ABS increases steadily, as the various rubber-toughening mechanisms come into play. Instrumented tests show that all of the energy is absorbed before the load maximum: subsequent crack propagation is rapid and unstable, and yielding is restricted to a stress-whitened region near the notch tip. Impact behaviour in this intermediate temperature range reflects a steady increase in G_{IC}. Above −10°C, the fracture behaviour changes again: there is a rapid increase in impact energy, much of the energy is absorbed **after** the load maximum, and the whole fracture surface becomes whitened. In this upper temperature region, the increase in fracture resistance with increasing crack length compensates for the increase in strain energy release rate. In fracture mechanics terms, J_R is high, and crack propagation becomes stable: crack speeds are typically below 10 m s^{-1}.

Electron microscopy reveals that stress whitening in ABS is due to two complementary mechanisms: fibrillation of the rubber phase in the complex particles, and multiple crazing in the SAN matrix. These processes allow the ABS to develop dilatational plasticity at the tip of the notch, ahead of the propagating crack. At lower temperatures, deformation in the matrix is dominated by crazing, but as the temperature increases towards the T_g of SAN, shear yielding, also accompanied by cavitation of the rubber, makes an increasing contribution. Raising the temperature accelerates shear yielding in the matrix, and also reduces the relaxation time of the rubber phase.

There is a sharp increase in the impact energy of ABS above $-10°C$, coinciding with the transition from unstable to stable crack propagation. The best explanation of these effects is that they reflect a change in material behaviour: below $-10°C$, the dynamic fracture resistance drops steeply at high velocities, before rising again as the crack approaches its terminal velocity, whereas above $-10°C$, the polymer exhibits high dynamic fracture resistance at all attainable velocities, up to the terminal velocity. This type of transition from low to high dynamic toughness is frequently observed in the fracture of polymers. In ABS, as in other rubber-toughened plastics, the transition is related to the relaxation behaviour of the rubber. In order to initiate yielding, the particles must achieve a low modulus (of order 1 MPa) before the matrix polymer reaches its critical fracture strain (of order 0.01). During initial loading, while the crack tip is stationary, the time scale for relaxation is about 1 ms, and toughening is observed above $-75°C$. On the other hand, during crack propagation at speeds of up to 500 m s^{-1}, the time scale available for relaxation at the crack tip is much smaller, of order 0.1 μs, and the transition is shifted to $-10°C$.

A high dynamic fracture toughness, which extends over all attainable velocities, is helpful in counteracting the effects of surface embrittlement, as illustrated in Figure 5.22. Results are given for notched ABS bars exposed to natural sunlight for a period of 3 months, causing surface embrittlement to a depth of about 60 μm. The ultraviolet component of sunlight causes oxidative degradation of both polybutadiene and SAN, a process known as 'u.v. ageing'. Consequently the COD at crack initiation is low. Below $-10°C$ this results in a drastic fall in impact resistance, but at higher temperatures the propagating crack slows as it passes from the degraded surface layer into the unaffected ABS beyond: because of the high dynamic fracture resistance in the non-degraded region of the specimen, the total energy absorbed is only a little lower than in bars that have not been aged. Several methods can be adopted to limit the effects of u.v. ageing, including the use of oxidation-resistant rubbers, addition of antioxidants, and compounding with carbon black to screen out the u.v. light.

Undoubtedly the most difficult polymer-based materials to toughen are the high-performance composites, which are usually based on highly cross-linked resins, because these have a high T_g (typically about 200°C). Some success has been achieved in toughening moderately cross-linked epoxy resins and other thermosets, using dissolved rubbers that precipitate as small particles during resin cure. Under the triaxial stresses generated at notch tips, the rubber particles cavitate to form dilatation bands, in a similar manner to toughened thermoplastics. These toughened thermosets are used as adhesives and in medium-grade composites. However, they are unsuitable for high-performance composites for two reasons: the rubber causes an unacceptable reduction in stiffness, and resins that are ductile enough to benefit from rubber modification typically have a relatively low T_g. Both problems can be overcome by replacing the rubber with a higher concentration of a ductile thermoplastic having a high T_g. At thermoplastic concentrations above about 20 vol.%, both components form continuous phases. Co-continuous resin–thermoplastic blends of this type, in which the fracture energy is absorbed mainly by the ductile thermoplastic, are now established as matrix materials for high-performance composites, with applications including primary structures in commercial airliners.

Notes for Chapter 5

5.N.1
Shoppers who are accustomed to carrying heavy loads in polyethylene bags will be familiar with cold drawing in the handle.

5.N.2
The activation volume should not be thought of as a physical volume. It is simply the product of two terms: an area over which the stress is acting, and a distance moved by the flow segment.

5.N.3
Crazing caused in early plastic picnic ware by the cream layer on milk led to the replacement of polystyrene by the more expensive SAN copolymer. Margarine containers are made from ABS (rubber-toughened SAN) for similar reasons. Recent concerns about solvent crazing have centred mainly upon the effects of fuel and oils on plastic components in cars.

5.N.4
It will help the reader in thinking about this statement to consider the following approximate, elementary, and conceptually useful argument. Let us assume that the effect of the crack is to 'drain off' completely the strain energy from a cylinder of diameter $2a$ centred on the crack, as indicated in Figure 5.23, and to leave the strain energy outside the cylinder at $+(\bar{\sigma}^2/2E)$ per unit volume. The introduction of the crack has then reduced the strain energy by

$$\text{Volume of cylinder} \times \left(\frac{\bar{\sigma}^2}{2E} \right) = (\pi a^2 \times B) \times \frac{\bar{\sigma}^2}{2E} = \frac{\bar{\sigma}^2 \pi a^2 B}{2E}.$$

This argument is dimensionally correct but is, of course, a very rough

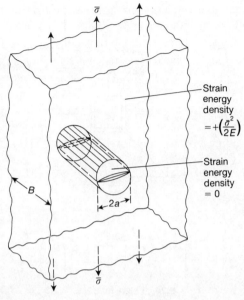

5.23 Figure for 5.N.4.

approximation. It differs from the precise, but highly involved, computation by a factor of a half.

5.N.5

An equation similar to eqn 5.21 was first derived by Griffith. The Griffith equation for a wide thin sheet is

$$\bar{\sigma}_\mathrm{F} = \left(\frac{2\gamma E}{\pi a} \right)^{\frac{1}{2}}.$$

Griffith assumed $G_\mathrm{c} = 2\gamma$: that is, the fracture surface has characteristics the same as those of a free surface. Griffith found this by experiment to be true for silica glass, but in general it is found that it is not true, and that $G_\mathrm{c} \geqslant 2\gamma$.

5.N.6

The significance of K is that it provides a measure of the magnitude of the stresses close to the tip of a **sharp** crack. Linear elastic stress analysis gives the following solution for the stresses acting at a point with polar coordinates r, θ relative to the tip of such a crack in a wide sheet (see Figure 5.24).

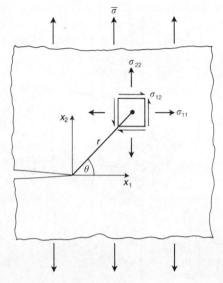

5.24   Figure for 5.N.6.

$$\sigma_{11} = \frac{K}{(2\pi r)^{\frac{1}{2}}} \cos \frac{\theta}{2} \left[1 - \sin \frac{\theta}{2} \sin \frac{3\theta}{2} \right] + \cdots$$

$$\sigma_{22} = \frac{K}{(2\pi r)^{\frac{1}{2}}} \cos \frac{\theta}{2} \left[1 + \sin \frac{\theta}{2} \sin \frac{3\theta}{2} \right] + \cdots$$

$$\sigma_{12} = \frac{K}{(2\pi r)^{\frac{1}{2}}} \sin \frac{\theta}{2} \cos \frac{\theta}{2} \cos \frac{3\theta}{2} + \cdots$$

(5.N.6.1)

$\sigma_{33} = 0$ (in plane stress).

$\sigma_{13} = \sigma_{23} = 0$ (in plane stress).

In this case, $K = \bar{\sigma}(\pi a)^{\frac{1}{2}}$. It can be seen that, in this extreme situation of a **perfectly sharp crack**, the stresses are predicted to become infinite right at the crack tip ($r \to 0$). This is as expected from the limiting case of an elliptical crack ($b \to 0$, eqn 5.16), but clearly, in reality, infinite stresses cannot be carried by the material. The very large strains around the tip will lead to some blunting of the crack, and stresses will fall to values which the material can carry. Nevertheless, the assumption of a perfectly sharp crack, leading to stresses given by eqns 5.N.6.1, is always used in the engineering treatment of linear elastic fracture mechanics (LEFM): it represents the **worst case**. If a component is designed not to fail in the presence of a perfectly sharp crack, one can be certain it will not fail in the presence of a real (slightly blunt) crack.

It is apparent from eqns 5.N.6.1 that K is a convenient factor which scales the magnitudes of all the stresses. It is for this reason that it is called the **stress intensity factor**.

The interesting feature of eqns 5.N.6.1 is that they in fact apply not only to the case considered here (simple tension of a wide sheet), but to plane stress mode I loading of any body containing a sharp crack. In passing from one loading geometry to another, all that changes is K. It is convenient to express this fact as follows:

$$K = Y\bar{\sigma}(\pi a)^{\frac{1}{2}},$$

where Y is a dimensionless parameter depending only on the geometry of the stressed body.

We have here an alternative mental picture of the criterion for crack propagation. The magnitude of the stress field around the crack tip is specified by K (fixed by geometry and loading conditions). The crack can be assumed to propagate when K reaches some critical value K_c (a material property).

5.N.7

Two alternative modes of fracture have been defined, both involving shear on the crack plane. In mode II, shear is parallel to the crack propagation direction. In mode III, shear is normal to the propagation direction. These types of fracture are important in long-fibre composites and in adhesives, because of anisotropy. They are of minor importance in thermoplastics. Cracks in isotropic materials tend to turn in a direction normal to the tensile stress, giving mode I fracture, whatever the initial orientation of the crack plane.

5.N.8

The need to compromise between melt viscosity and strength particularly affects speciality plastics that have high glass transitions (e.g. PC, PES, and PSF), as these are usually moulded just below the temperature at which thermal degradation becomes significant.

Problems for Chapter 5

5.1 An HDPE tensile bar yields at a nominal stress of 28 MPa at 8% extension, beyond which the stress falls to 22 MPa as a neck forms. The neck stabilizes with a cross-sectional area equal to 26% of the original area. Sketch a graph of true stress against percentage elongation.

5.2 Viscoelastic creep in a polypropylene homopolymer at 20°C can be fitted to the equation

$$\varepsilon = 0.066\sigma_t + 3.0 \times 10^{-6} t^{\frac{1}{3}} \sinh(0.6\sigma_t)$$

where ε is the percentage extension after time t (s) under true stress σ_t (MPa).

(1) Plot σ_t against λ at fixed times $t = 1$ s and 1000 s, and hence determine the nominal yield stress for each time.

(2) Calculate an equivalent strain rate corresponding to each yield stress by dividing yield strain by time to yield.

[Hint: Take $\sinh(0.6\sigma_t) = 0.5\exp(0.6\sigma_t)$.]

5.3 Calculate the activation volume of the creep process described in Problem 5.2.

5.4 The yield stress σ_y of an ABS at $\dot{\varepsilon} = 10^{-4}\mathrm{s}^{-1}$ is 44.0 MPa at 20°C, and 69.0 MPa at -60°C. At $\dot{\varepsilon} = 10^{-2}$ s^{-1}, σ_y is 53.2 MPa at 20°C, and 75.7 MPa at -60°C. Calculate ΔH and V^*, and hence find σ_y at 0°C for $\dot{\varepsilon} = 0.1$ s^{-1}.

5.5 Yield stress measurements were made at 20°C on three polypropylene materials, each containing different volume fractions ϕ of rubber particles. Table 5.1 gives values of yield stress σ_y as a function of $\dot{\varepsilon}$ and ϕ. Calculate V^* for each polymer, and suggest why V^* varies with ϕ.

Table 5.1   Yield stress (MPa) data for polypropylene

$\dot{\varepsilon}$ (s^{-1})	10^{-5}	10^{-4}	10^{-3}	10^{-2}
$\phi = 0$	23.1	26.7	30.2	33.8
$\phi = 0.1$	19.7	22.3	24.9	27.0
$\phi = 0.2$	15.5	17.6	19.7	21.8

5.6 By careful specimen preparation and testing, it is possible to obtain shear yielding in PS under tension. A PS gave $\sigma_y = 72.5$ MPa in tension and 90.4 MPa in compression. Assuming that PS obeys a pressure-dependent von Mises criterion, whereby C increases linearly with p in eqn 5.14, plot a yield envelope for PS under biaxial stress.

5.7 The shear yield stress σ_{sy} of PET at 22°C and $\dot{\varepsilon} = 0.002$ s^{-1} increases with hydrostatic pressure p according to the equation

$$\sigma_{sy} \text{ (MPa)} = 31 + 0.09p \text{ (MPa)}.$$

Assuming that yielding follows a pressure-dependent von Mises criterion, calculate σ_y in uniaxial tension and in uniaxial compression.

5.8 Crazing in PMMA follows a criterion which for fixed time under biaxial stress ($\sigma_3 = 0$) can be written

$$|\sigma_1 - \sigma_2| = C_A - \frac{C_B}{(\sigma_1 + \sigma_2)},$$

where C_A and C_B are constants and $(\sigma_1 + \sigma_2) > 0$. Note that σ_1 and σ_2 are **in-plane** principal stresses. Measurements at 50°C show crazing to occur within 1000 s when: (a) $\sigma_1 = 31$ MPa, $\sigma_2 = \sigma_3 = 0$; and (b) $\sigma_1 = \sigma_2 = 30$ MPa, $\sigma_3 = 0$. Plot the failure envelope for crazing according to the above criterion, and show that it is asymptotic to the pure shear line.

5.9   An HIPS tensile specimen extends to a strain of 0.56 by the multiple crazing mechanism. If the average extension ratio $\lambda = 4.2$ in the fibrils, and the craze thickness is 1.2 μm in the full extended state, calculate the number of crazes per mm length of the stressed bar at fracture. What fraction of the HIPS is converted into crazes?

5.10   A wide sheet of PC, containing a central sharp crack 25 mm in length, fractures at a stress of 12.1 MPa. Calculate (1) K_{IC} for PC under these conditions, and (2) the fracture stress for a wide sheet containing a 50 mm crack.

5.11   A notched rectangular bar of POM with dimensions $a = 10$ mm, $W = 20$ mm, $S = 80$ mm, and $B = 8$ mm undergoes plane-strain fracture in three-point bending at 20°C at an applied force $F = 390$ N. For this geometry $(S/W = 4)$,

$$Y = 1.09 - 1.73(a/W) + 8.20(a/W)^2$$
$$- 14.17(a/W)^3 + 14.55(a/W)^4.$$

Calculate K_{IC}, and find the applied force needed to fracture a compact tension specimen having $a = 25$ mm, $W = 50$ mm, and $B = 10$ mm.

5.12   Single-edge-notched bending tests were carried out on rectangular PC bars with dimensions $B = 5$ mm, $W = 10$ mm, and $S = 80$ mm. The PC has $E = 3.2$ GPa, $\nu = 0.40$ $\sigma_y = 64$ MPa, and $G_{IC} = 1.5$ kJ m^{-2} in plane strain. (1) Calculate the force required to fracture the bars when $a = 5$ mm. (2) Calculate the minimum a to initiate brittle fracture before first yield. For this geometry $(S/W = 8)$, use the expression for Y given in Example 5.8.

5.13   The data in Table 5.2 on force to fracture F_c were obtained from single-edge-notched bending tests on sharply notched bars having $a = 5$ mm, $W = 10$ mm, $B = 6$ mm, and $S = 80$ mm. Calculate an apparent K_{IC} for each material, and determine which of the tests give valid plane-strain results. Use the expression for Y given in Example 5.8 for bars having $S/W = 8$.

Table 5.2   Information for Problem 5.13

Material	F_c (N)	σ_y (MPa)
Epoxy resin	16.4	100
Toughened epoxy	63	65
Dry PA 6.6	85	75
Saturated PA 6.6	210	35
PC	60	63
PES	33	85
POM	156	62
PS	30	50
SAN	33	80

5.14   A highly cross-linked epoxy resin has a coefficient of linear thermal expansion $\alpha = 5 \times 10^{-5}$ K^{-1}, $G_{IC} = 120$ J m^{-2}, $E = 3.2$ GPa, and $\nu = 0.35$. A thick layer of resin is cured and is firmly bonded to an aluminium part ($\alpha = 2.5 \times 10^{-5}$ K^{-1}) at 180°C. Calculate the minimum defect size needed to initiate cracking in the resin on cooling to 20°C. Take $Y = 2/\pi$ for semi-circular edge-cracks of radius a in a wide sheet.

5.15   A rectangular strip of PMMA has length $L = 100$ mm, $B = 3$ mm, and $W = 10$ mm. A 1 mm edge crack is formed in the strip. Calculate the work to break in tension if $G_{IC} = 675$ J m^{-2}, $E = 3.0$ GPa, and $\nu = 0.40$. Take $Y = 1.12$ at $a/W = 0.1$, and neglect the small increase in specimen compliance due to the crack.

5.16   Elastic fracture mechanics can be applied to rubbers provided that allowance is made for their non-linear stress–strain behaviour. The steps are: (a) replacing $\sigma^2/2E$ in LEFM by U_∞, an experimentally determined strain energy density remote from the crack; and (b) replacing a_0, the crack length in the unstrained state, by $a = a_0/\lambda^{\frac{1}{2}}$ in order to allow for the shortening of the crack at high extension ratio λ. Verify these expressions, and obtain an equation for G_{IC} in wide sheets of rubber by substitution in eqn 5.19. Hence calculate G_{IC} from the data in Table 5.3 for wide strips of natural rubber in uniaxial tension.

Table 5.3   **Information for Problem 5.16**

a_0 (mm)	4.5	5.0	8.6	11.0	16.0
Critical λ	2.21	2.00	1.82	1.60	1.62
U_∞ (MJ m^{-3})	0.33	0.25	0.17	0.09	0.095

5.17   A cross-linked natural rubber has $G_{IC} = 9.4$ kJ m^{-2}. The threshold for slow crack growth is at $G_I = 40$ J m^{-2}. The strain energy density remote from the notch is given by

$$U_\infty = (\lambda^2 + 2/\lambda - 3)/8 \ (\text{MJ m}^{-3}) \quad (\text{see eqn 3.34}).$$

Using the principles outlined in Problem 5.16, calculate: (1) the minimum value of λ needed to cause crack growth over a long period in a wide sheet containing a 1 mm edge cut (take $Y = 1.12$); and (2) the value of λ at which a 10 mm central cut in a sheet of the same rubber causes catastrophic tearing.

5.18   The threshold for slow crack growth in a HDPE occurs at $K_I = 0.2$ MPa m$^{0.5}$. Above this threshold, $\log(da/dt)$ under static load increases linearly with $\log K_I$ until catastrophic fracture occurs at $K_{IC} = 1.6$ MPa m$^{0.5}$. Specific values of da/dt are 1 nm s^{-1} at 0.4 MPa m$^{0.5}$ and 40 nm s^{-1} at 1.0 MPa m$^{0.5}$. Assuming that HDPE contains defects equivalent to 100 μm cracks, and that $Y = 1.15$

independent of crack length, calculate the time to failure t_f, in bars subjected to a constant tensile stress of 10 MPa.

5.19 The yield stress σ_y of the HDPE described in Problem 5.18 is given by

$$\sigma_y \text{ (MPa)} = 23.5 - 1.43 \log_{10} t \text{ (s)}.$$

Plot a graph of failure stress against $\log t$ over the range $t = 1$ to 10^8 s, and hence find the stress and time at which a transition from ductile to brittle failure occurs.

5.20 The hoop stress at the outer surface of a pipe of internal diameter $2R_1$ and external diameter $2R_2$ under internal pressure P is given by

$$\sigma = \frac{2PR_1^2}{(R_2^2 - R_1^2)}.$$

When $a/W < 0.05$, $Y = 1.12$. A toughened rigid PVC water pipe has $R_1 = 147.5$ mm and $R_2 = 167.5$ mm. Tests show $K_{IC} = 3.1$ MPa m$^{0.5}$ and $\sigma_y = 40$ MPa. The largest defects are equivalent to 100 μm cracks in the outer surface. (1) Determine whether the pipe will fail by yield or fracture in a pressurization test, and find the pressure to cause failure. (2) Find the maximum value of yield stress $\sigma_y(\text{max})$ that will permit ductile failure in the pipe as strain rate increases, assuming that K_{IC} remains constant. (3) Calculate $\sigma_y(\text{max})$ for poorly processed PVC pipe, for which $K_{IC} = 1.5$ MPa m$^{0.5}$, assuming that a remains unchanged at 100 μm.

5.21 Experiments on PP homopolymer show that $K_{IC} = 4.7$ MPa m$^{0.5}$ in plane-strain at all temperatures between $-100°C$ and $25°C$. Tensile yield stresses at $25°C$ intervals over this range are 90, 85, 75, 50, 35, and 30 MPa at $\dot{\varepsilon} = 10^{-3}$ s^{-1}. (1) Use these data to calculate for each temperature the minimum crack length required to initiate fracture before yielding in a tensile bar (take $Y = 1.12$ for small a/W). (2) Plot a graph of critical crack length against temperature, and hence estimate the ductile–brittle transition temperature for PP specimens containing 1 mm, 2mm, and 4 mm notches and tested at this strain rate.

5.22 Compact tension tests on PMMA specimens give a value of 325 J m^{-2} for G_{IC} at 23°C, for a crosshead speed of 5 mm min^{-1}. The critical crack tip opening displacement δ_c, determined using an optical interferometry technique, is 4.2 μm. Calculate the surface stress acting on the craze formed ahead of the crack tip.

Most materials, including polymers, show an increase in toughness with increasing temperature and decreasing strain rate. However, measurements of G_{IC} in PMMA and other glassy polymers show the opposite trends. Suggest an explanation for this paradox.

5.23 Between the fatigue threshold at $\Delta K = 0.16$ MPa m$^{0.5}$ and $K =$

$K_{IC} = 2.2$ MPa m$^{0.5}$, fatigue crack growth in an ABS follows the equation

$$da/dn \ (\mu m \text{ per cycle}) = 0.18 \, \Delta K^{3.5} \quad (K \text{ in MPa m}^{0.5}).$$

A bar of the ABS has $a = 1$ mm, $B = 6$ mm, and $W = 10$ mm, and is subjected to tension–compression cycling between ± 300 N. Assuming $Y = 1.2$ for $0.1 < a/W < 0.3$, calculate the number of cycles for the crack to grow to 3 mm.

5.24   Fatigue tests on the rigid PVC described in Problem 5.20 show crack growth given by

$$da/dn \ (\mu m \text{ per cycle}) = 0.035 \, \Delta K^{2.4} \quad (K \text{ in MPa m}^{0.5}).$$

A PVC pipe of the dimensions given in Problem 5.20 is subjected to cyclic variations in internal pressure P. For this geometry W is the wall thickness. Noting that $Y = 1.12$ for small a/W, calculate the pressure variation ΔP at which an initial 100 μm equivalent crack would grow to 1.0 mm over 500 000 cycles.

5.25   More extensive fatigue tests show that the pipe-grade PVC described in Problems 5.20 and 5.24 has a fatigue threshold at $\Delta K = 0.2$ MPa m$^{0.5}$. Using $a = 100$ μm, and pipe dimensions given in Problem 5.20, calculate the minimum ΔP necessary to initiate fatigue crack growth in the pipe.

5.26   A spherical rubber particle of radius R is subjected to a small dilatational strain, and cavitates to form a void of radius r. Show that a rubber obeying eqn 3.29 gives the following expression for the function $F(\lambda_f)$ in eqn 5.41:

$$F(\lambda_f) = \rho \int_{\lambda=1}^{\lambda_f} \frac{2\lambda^4 - 3\lambda^2 + \lambda^{-2}}{(\lambda^3 - \rho)^2} \, d\lambda$$

where ρ is the ratio of densities in the rubber phase before and after cavitation (to avoid a singularity at $\lambda = 1$ when $\rho = 1$), and λ_f is the maximum attainable value of λ in biaxial tension (to avoid $\lambda = \infty$ at the void surface).

5.27   The T_g of an ethylene–propylene copolymer rubber is $-50°$C. When the EPR is blended with PA6.6, the fracture resistance of the blend in slow bending tests on notched bars first shows a significant increase at $-50°$C. In this test, the time taken to reach peak load is 1s. Toughening occurs when the rubber particles have time to stress-relax and to initiate plastic deformation in the PA6.6 matrix. Calculate the minimum temperature required to achieve toughening in the Charpy impact test. Use the WLF equation,

$$\log_{10} a_T = -17.4(T - T_g)/(51.6 + T - T_g)$$

(see Section 4.3.3 for definition of a_T). In the Charpy test, the time to peak load is 1 ms.

6   Reinforced polymers

6.1   Introduction

The major problem in the application of polymers in engineering is their low stiffness and strength when compared with metals; the moduli are ~100 times lower and strengths ~5 times lower. Two methods are used to offset these deficiencies:

- the ingenious use of shape, designing into a component the necessary stiffness and strength by means of ribs, box sections, etc. (see Chapter 8); and
- the addition of reinforcing particles or fibres to the resin to form a **composite material**, the subject of this chapter.

Three highly successful applications of reinforced polymers are shown in Figure 6.1. Fibre-reinforced plastics have long been the favoured materials for small boats; a good example is the competition kayak shown in Figure 6.1(a). It is constructed from epoxy resin in which are embedded continuous Kevlar and carbon reinforcing fibres (see Section 6.2). All three constituents play vital roles. The thermoset polymer matrix provides economical construction for small batches, excellent corrosion resistance in water, and lightness. The Kevlar and carbon fibres provide the essential strength and rigidity, at very little cost in extra weight. The result is a kayak of outstanding performance: very fast, manoeuvrable, and light. This construction is considered by many to be the best currently available for the rigorous conditions of competitive 'white-water' canoeing. In most small boats, however, performance is less critical and a low price assumes more importance. The preferred material is then glass fibre reinforced plastic (GRP), in which the plastic is a glassy polyester (see Section 6.2).

Another imaginative application of a reinforced polymer is shown in Figure 6.1(b). It is a carbon-fibre reinforced nylon tennis racket. The designers chose a polymer matrix for two reasons: economical manufacture in large batches by injection moulding; and low density. They exploited the ability of injection moulding to produce complex shapes economically. The racket frame is hollow, with moulded-in string holes. This is achieved by moulding the polymer around a low melting-point

6.1 Successful application of reinforced polymers.

  (a) Competition kayak constructed from epoxy resin reinforced by carbon and Kevlar fibres (photo courtesy of Pyranha Mouldings Ltd).

  (b) Tennis racket frame injection moulded from carbon–fibre reinforced nylon 6.6 (photo by courtesy of Dunlop Slazenger International Ltd).

  (c) Car tyre. The rubber is reinforced by carbon black, polymeric fibres, and steel (photo by courtesy of The Goodyear Tyre and Rubber Co. (Great Britain) Ltd).

metal alloy core, which is later removed by melting out. A racket made from the pure polymer could not withstand the combination of large static load from string tension and impulsive load from striking a ball at high speed (over 100 km h^{-1} for a top-class tennis player), without breaking or distorting. Consequently, a grade of material was selected for which the polymer matrix (nylon 6.6) is reinforced by chopped carbon fibres. The fibre lengths lie mostly in the range 0.1–1 mm: long enough to provide adequate rigidity and strength, yet not so long as to impede injection moulding.

The third example is the rubber car tyre (Figure 6.1(c)). The side-wall of the tyre must be highly flexible and resilient for repeated deformation in bending, in order to absorb shock loads applied between the axle and road. A pure rubbery polymer would not have sufficient rigidity or strength for deformation in other modes, or sufficient wear resistance. For this reason, **a tyre is reinforced at several different levels**. At the microscopic level, there are particles of carbon black mixed with the polymer (see Section 6.5). They adhere to the polymer molecules to provide increased stiffness, strength and wear resistance. On a large scale, the tyre is reinforced by rigid cords (e.g. drawn polyester fibres or steel wires) to provide strength and stiffness under tensile forces in radial and circumferential directions which are due to inflation and cornering.

These examples illustrate the important principles of polymer reinforcement. A good reinforcing additive has the following attributes:

- it is stiffer and stronger than the polymer matrix;
- it has good particle size, shape, and surface character for effective mechanical coupling to the matrix; and
- it preserves the desirable qualities of the polymer matrix.

Of course, the best reinforcement in any particular application is the one that achieves the designer's objectives at lowest net cost.

We examine next the mechanism of reinforcement. Consider the case of a single cylindrical reinforcing particle embedded in a block of polymer matrix, and perfectly bonded to it (see Figure 6.2(a)). The particle is of length l and diameter d. Situations before and after a tensile load is applied are sketched in Figure 6.2(b). Horizontal lines are imagined to be drawn in the block before load is applied (see Figure 6.2(a)), in order to demonstrate the strain distribution which develops under load (see Figure 6.2(b)). The particle is stiffer than the matrix and deforms less, causing the matrix strain to be reduced overall, especially in the vicinity of the particle. Note that the particle achieves its restraining effect on the matrix entirely via the particle–matrix interface. This highlights the critical role played by

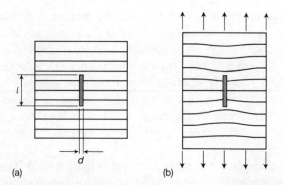

6.2 A cylindrical reinforcing fibre in a polymer matrix: (a) in the undeformed state; (b) under a tensile load. The fibre 'pinches' the polymer in its vicinity, reducing the strain and increasing stiffness.

the interface. The strength of the composite depends on the strength of the bond between particle and matrix: the more interface there is, the more effective is the reinforcement.

A useful parameter for characterizing the effectiveness of a reinforcement is the ratio surface area of reinforcement to volume of reinforcement. If the surface area of a particle is A and its volume is V, we require the surface-to-volume ratio A/V to be as high as possible. For a cylindrical particle

$$A = \frac{\pi d^2}{2} + \pi dl, \tag{6.1}$$

and

$$V = \frac{\pi d^2 l}{4},$$

and consequently

$$\frac{A}{V} = \frac{2}{l} + \frac{4}{d}. \tag{6.2}$$

This can be rewritten in terms of V and the **aspect ratio** of the cylinder $a = l/d$:

$$\frac{A}{V} = \left(\frac{2\pi}{V}\right)^{\frac{1}{3}} (a^{-\frac{2}{3}} + 2a^{\frac{1}{3}}). \tag{6.3}$$

A plot of A/V against a is shown in Figure 6.3.

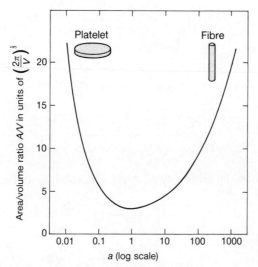

6.3 Surface-area-to-volume ratio A/V of a cylindrical particle of given volume, plotted versus aspect ratio $a = l/d$.

It will be seen from Figure 6.3 that the predicted optimum shape for the cylindrical reinforcement—to maximize A/V—is

- $a \gg 1$ (a fibre), and
- $a \ll 1$ (a platelet).

It will come as no surprise, therefore, that the two main classes of reinforcement are fibres and platelets. These shapes maximize the particle-matrix interaction through the interface.

Examples of fibrous reinforcements are glass fibres and carbon fibres. Examples of platelet[1] reinforcements are mica and talc.

Almost as critical in commercial practice as the effects of reinforcement on properties are the effects of reinforcement on the cost of the material and on its processing. The perceived effect of material cost depends on whether the decisive factor is cost per unit mass or cost per unit volume. Since the additive normally has a density considerably different from that of the polymer matrix, the density of the composite differs from that of the polymer. Consider the fibre-reinforced polymer shown schematically in Figure 6.4. A mass m of composite occupies a volume v. It contains a mass m_f of fibres occupying a volume v_f, and a mass m_m of matrix occupying a

---

1   When platelet reinforcements are discussed elsewhere, it is common practice to redefine the aspect ratio as d/l so that numbers much larger than 1 are obtained.

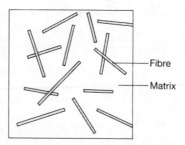

6.4   Block of fibre-reinforced polymer.

volume v_{m}. Then, by additivity of mass and volume, and assuming there are no voids,

$$m = m_{\mathrm{f}} + m_{\mathrm{m}} \tag{6.4(a)}$$

and

$$v = v_{\mathrm{f}} + v_{\mathrm{m}}. \tag{6.4(b)}$$

We express the proportions of fibre and matrix in the composite by the fractions of the total volume that they occupy.

$$\phi_{\mathrm{f}} = \frac{v_{\mathrm{f}}}{v} \quad \text{and} \quad \phi_{\mathrm{m}} = \frac{v_{\mathrm{m}}}{v}. \tag{6.5}$$

Note that from eqn 6.4(b) these are related through

$$\phi_{\mathrm{m}} = 1 - \phi_{\mathrm{f}}, \tag{6.6}$$

e.g. if the composite is 30% by volume fibre then it must be 70% by volume matrix.

Dividing eqn 6.4(a) by v yields an expression for the density of the composite ρ, in terms of the densities of matrix ρ_{m} and of fibres ρ_{f}, using eqns 6.5 and 6.6:

$$\rho = \phi_{\mathrm{f}} \rho_{\mathrm{f}} + (1 - \phi_{\mathrm{f}}) \rho_{\mathrm{m}}. \tag{6.7}$$

Reinforced polymers generally have low densities. For example, the density of an epoxy reinforced with 70% carbon fibres is only 1700 kg m^{-3}.

Equation 6.7 expresses a simple additivity of properties (in this case densities) that applies to some, but not all, of the other physical properties of the composite. This is known as the rule of mixtures.

Example 6.1

Composite materials are to be prepared with the following compositions (by volume):

- polypropylene reinforced by 20% glass fibre; and
- epoxy reinforced by 25% carbon fibre and 25% Kevlar fibre.

In each case, find the mass of each constituent required per unit mass of composite. The densities (in kg m^{-3}) are: glass fibre 2540; polypropylene 900; carbon fibre 1790; Kevlar fibre 1450; and epoxy 1300.

Solution

(i) First find the density ρ of the composite. From eqn 6.7

$$\rho = 0.2 \times 2540 + 0.8 \times 900 = 1228 \text{ kg m}^{-3}.$$

In 1.0 m^3 of composite there is 0.2 m^3 of glass. Hence the mass of glass in 1 m^3 of composite $= 0.2 \times 2540$ kg $= 508$ kg. The mass of glass in 1 kg of composite is therefore $\frac{508}{1228} = 0.414$ kg.

In 1 kg of composite there are 0.414 kg of glass and 0.586 kg of polypropylene.

(ii) Again, find the density of the composite.

$$\rho = 0.25 \times 1450 + 0.25 \times 1790 + 0.5 \times 1300 = 1490 \text{ kg m}^{-3}.$$

The mass of carbon in 1 m^3 of composite $= 0.25 \times 1790 = 447.5$ kg. Hence the mass of carbon in 1 kg of composite $= \frac{447.5}{1460} = 0.307$ kg. Similarly, the mass of Kevlar in 1 kg of composite $= \frac{362.5}{1460} = 0.248$ kg.

In 1 kg of composite there are 0.307 kg of carbon, 0.248 kg of Kevlar and 0.445 kg of epoxy.

The cost of a composite material has two parts: the cost of its constituent materials; and the cost of compounding them (that is, incorporating one of the constituents (fibres) into the other (polymer matrix)). Suppose the cost per unit mass of fibres is C_f and that of the matrix is C_m and the cost of incorporation per unit mass of composite is C_i. The total cost per unit mass of the composite is C and the cost for mass m is then

$$mC = C_f m_f + C_m m_m + C_i m. \tag{6.8}$$

It follows from eqns 6.4 to 6.6 that

$$C = \phi_f \frac{\rho_f}{\rho} C_f + (1 - \phi_f) \frac{\rho_m}{\rho} C_m + C_i. \tag{6.9}$$

In practice, composite materials contain voids which comprise trapped air, or solvents, etc. A void is a source of weakness. A void content greater than 2% indicates poor fabrication; a void content below 0.5% indicates high-class, 'aircraft quality', fabrication. Most voids are spherical, but a very thin, flat bubble spread over a large area is a serious point of weakness if transverse properties are under consideration. Given the exact composite density—and with ϕ_f, ρ_f, and ρ_m known—the fraction of voids can be calculated.

6.2 Reinforced plastics

6.2.1 Polymer matrices

The first reinforced plastics were all based on thermoset polymers. For many years the most popular have been the family of polymers known as **thermoset polyesters** (6.N.1).[1] These are versatile, inexpensive polymers, used extensively with glass-fibre reinforcement, often in substantial plastic components (such as storage tanks, pipes, boat hulls, and seating for public places). Other thermosets are in competition with them, the foremost being the epoxies (6.N.2). These are preferred to polyesters in demanding applications, where their superior mechanical performance (especially greater toughness) justifies their higher cost.

Thermoset polymers have some great virtues when used as matrices in reinforced plastics:

- low viscosity of the precursor liquids, prior to cross-linking, facilitates thorough wetting of reinforcing particles by the polymer;
- economical forming is possible for large components (see Section 6.3); and
- high softening points can be achieved in materials of only moderate cost.

Recent years, however, have seen rapid growth in the use of reinforced thermoplastic polymers. The semi-crystalline polymers polypropylene and nylon are especially popular as matrices. A major advantage of a thermoplastic matrix is that **forming is possible by normal injection moulding or extrusion techniques**. These are the most economical processes when cheap and precise manufacture of very large quantities of components is required (see Chapter 7). Allowance must be made for the effect of the reinforcing particles on the flow of molten plastic during forming; the viscosity, for example, is significantly increased. As a result, some modifications to tooling and process parameters are usually necessary.

6.2.2 Fibrous reinforcement

Fibrous reinforcements take diverse forms: continuous bundles of fibres, woven fabrics, chopped fibres, and numerous others. But whatever the

---

1 These should not be confused with **thermoplastic polyesters**—see 1.N.10.

form in which they are used, they are initially manufactured as bundles of continuous filaments. Each filament normally has a round cross-section with diameter in the range 5–15 μm, and the bundle (known as a 'roving' or a 'tow') consists of a large number of filaments (1000–10 000). The size of the bundle is sometimes specified by the number of filaments it contains, but more usually by its mass per unit length, or **linear density**. The standard unit of linear density is now the **tex** (1 tex = 1 g km^{-1} or 10^{-6} kg m^{-1}). Thus, a typical glass-fibre roving, containing, about 7000 filaments of diameter 13 μm, is said to be of 2400 tex.

A crucial feature of fibres used for reinforcement is the coating of **size** that is applied to the fibre surface during manufacture. The size is designed to:

- hold the fibres together as a coherent bundle before incorporation into the plastic matrix;
- protect the fibre surface from mechanical and environmental damage; and
- provide chemical bonding between fibres and matrix in the final composite.

The last function is achieved by 'tailoring' the size molecule so that one end reacts chemically (and bonds to) the fibre surface and the other end cross-links into the polymer network of the matrix.

Three types of reinforcing fibre dominate the reinforced plastics scene. These are: glass fibres, carbon fibres, and oriented polymeric fibres (the best known is the Du Pont company's 'Kevlar'). Their tensile stress–strain curves are shown in Figure 6.5(a), which includes mild steel for comparison. It can be seen that they are all linearly elastic right up to fracture, but they differ in their tensile strengths and moduli, although these are all comparable with, or superior to, those of steel. Because of the low densities of the fibres, these mechanical properties seem especially attractive when expressed per unit mass. This is done by replotting Figure 6.5(a) as **specific stress** σ/ρ (6.N.3) versus strain, as in Figure 6.5(b). Their specific moduli (slopes of specific stress versus strain curves) and specific strengths (σ/ρ at failure) are generally greatly in excess of those of the common metals. This fact is the cause of increasing interest in replacing metals by fibre-reinforced plastics in light-weight structures, most notably in transportation. Mechanical properties of reinforcing fibres are summarized in Table 6.1.

Glass fibre is by far the most widely used fibre reinforcement for plastics. Its success stems from an excellent balance of moderate price (about the same per kilogram as the common plastics) and desirable properties for reinforcement. The term 'glass' here refers to a range of

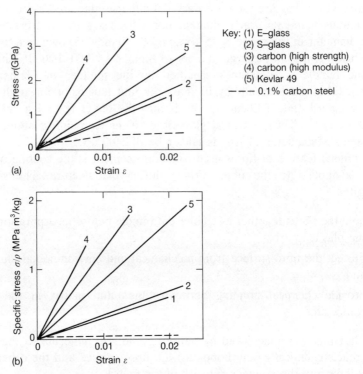

6.5 Stress–strain curves for reinforcing fibres compared with 0.1% plain carbon steel: (a) stress versus strain; and (b) specific stress versus strain.

inorganic glasses. They are all based on silica (SiO_2), but also contain smaller quantities of other inorganic oxides. Various compositions are available, each identified by a code letter. Thus, the fibre most commonly used in reinforced plastics is E-glass. Its main constituents are SiO_4 (54%), CaO (17.5%), Al_2O_3 (14.0%), B_2O_3 (8.0%), and MgO (4.5%). In order to achieve a different mix of properties, the recipe is varied. Examples are: S-glass with higher modulus and strength; and C-glass with improved resistance to attack by water and acids.

Table 6.1   Typical properties of some reinforcing fibres

		E-glass	S-glass	Carbon (high strength)	Carbon (high modulus)	Aramid (Kevlar 49)
Density	(kg m^{-3})	2540	2490	1790	1860	1450
Axial tensile modulus	(GPa)	76	86	230	340	124
Axial tensile strength	(GPa)	1.5	1.9	3.2	2.5	2.8

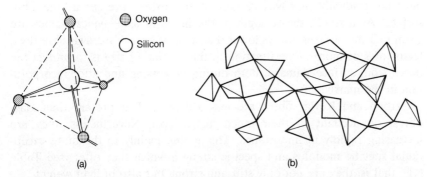

6.6 Structure of silica glass: (a) the silicon tetrahedron; and (b) non-crystalline network of silicon tetrahedra.

Glass fibres are manufactured by extruding molten glass at high linear velocity through a large number (100–1000) of holes in a platinum plate known as a 'bushing'. The resulting filaments are then wound up at an even faster rate as they cool through their glass transition and solidify. This has the effect of drawing them down to a much smaller diameter than when they left the bushing. On route, the continuous filaments are coated in size. Bundles of them from several bushings are then combined together to constitute a roving.

The structure of glass is dominated by the large proportion of silica present. Like carbon, silicon has a valency of four, and an atom of silicon bonds covalently to four neighbouring atoms (Figure 6.6(a)) in such a manner that they are placed at the four corners of a tetrahedron, with the silicon atom itself at the center. In silica, the tetrahedron has an oxygen atom at each corner, and each of these is shared with another silicon atom. The entire structure consists of a non-crystalline network of silicon tetrahedron linked together by oxygen atoms. Figure 6.6(b) shows a sketch of the network. Other oxides are interspersed among the SiO_2 groups. **There must be no crystallinity**; the fibres are cooled rapidly through T_g during drawing in order to ensure this. Crystals act as stress raisers in glass, and would greatly weaken the fibres.

Three further characteristics of glass fibre are advantageous in certain applications:

- resistance to high temperatures—the softening point is about 850°C;
- transparency to visible light—a composite therefore takes the colour of its matrix; and
- isotropy—for example thermal expansion is identical in axial and radial directions.

A disadvantage is that glass is very susceptible to surface damage. This can be caused by rubbing the surface or by the action of moisture, which seeks out and gradually dissolves certain of the oxides present at the fibre surface. As a result, the strength of the fibres in use—typical values are given in Table 6.1—is always lower than that of a freshly drawn fibre (by a factor of 2 or more). One function of the coating of size is to protect the fibres from surface damage, both during processing and after incorporation in the matrix.

Carbon and Kevlar fibres are less widely used at present than glass fibres, on account of their much higher cost. Nevertheless, they are increasing rapidly in importance. This is due mainly to the quite exceptional specific modulus and specific strength which they offer (see Table 6.1); that is, they are not only stiff and strong but also of light weight.

The best carbon fibres are prepared from polyacrylonitrile (PAN). PAN is converted into graphite through a sequence of carefully controlled heat-treatment operations. The key to the success of the process is that the high degree of orientation of the initial polymer fibre is preserved, ensuring that the final graphite structure has a similar high degree of crystal orientation.

Graphite has a highly anisotropic sheet-like structure, with strong covalent bonding between hexagonally arranged carbon atoms within the sheets or layer planes, but much weaker bonding between them (see Figure 6.7(a)). As a result, the microstructure of a carbon fibre resembles a stack of playing cards standing on end (Figure 6.7(b)). The axial mechanical properties are critically dependent on the extent to which: (a) layer planes lie parallel to the fibre axis; and (b) voids and other flaws are absent. The graphitic crystal structure of commercial carbon fibres is always somewhat disordered and microvoided, as can be seen from their relatively low density (Table 6.1) compared with that of a single crystal of graphite ($\rho = 2200$ kg m^{-3}). Both structure and properties are sensitive to

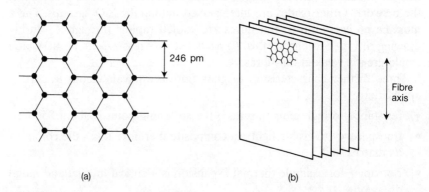

(a)   (b)

6.7 Graphitic structure of carbon fibre: (a) hexagonal arrangement of carbon atoms within a layer plane; and (b) structural unit of carbon fibre (a stack of layer planes aligned parallel to the fibre axis).

the details of manufacturing, and this is exploited commercially. Fibres are produced with varying properties, suited to different applications. Table 6.1 shows two examples: (a) fibres with the highest modulus (but lower breaking strain and strength); and (b) fibres with the highest strength (but lower modulus).

Other features of carbon fibre offer great advantages in some applications:

- chemical inertness—they are highly resistant to moisture and most common chemicals;
- high electrical and thermal conductivity along the fibre axis; and
- dimensional stability—axial thermal expansion is extremely low, and negative.

A disadvantage of carbon fibres for consumer products is that, being composed of graphite, they are black and impart this colour to the composite. In engineering design, allowance must be made for the high anisotropy of carbon fibres, which reflects the anisotropy of graphite (see Section 6.4). Transverse to the fibre axis, the tensile modulus is much lower and the thermal expansion much higher (and positive) than their respective values in the axial direction.

Oriented polymers clearly have great potential as reinforcing fibres, if molecules can be well-aligned parallel to the fibre axis. The stiff, strong covalent bonding along the molecular backbone can then be exploited. The problem is that most polymer molecules are so flexible (owing to bond rotation) that a complete molecular alignment is only partially achieved (see Chapter 2).

To date, only one class of polymer has been found in which this problem can be overcome and highly successful reinforcement fibres produced. These are the **aramid polymers**, so-called because they contain both 'aromatic' and 'amide' groups in the molecular chain. Indeed, it is this combination which is the key to their success. Kevlar 49 is currently the most widely used aramid fibre for reinforcement of plastics, and it illustrates the point well. It consists of the polymer poly(paraphenylene terephthalamide):

$$
\left[\begin{array}{c} \text{C} \\ \| \\ \text{O} \end{array} - \bigcirc - \begin{array}{c} \text{C} \\ \| \\ \text{O} \end{array} - \begin{array}{c} \text{H} \\ | \\ \text{N} \end{array} - \bigcirc - \begin{array}{c} \text{H} \\ | \\ \text{N} \end{array} \right]_n
$$

(XXIV)

The molecule is rigid because of the dominance of rigid benzene rings (the aromatic component). It cannot turn back on itself and form chain-folds.

Moreover, the rod-like molecules pack together like pencils and bond firmly to their neighbours because of the amide groups present:

$$-\underset{\underset{O}{\parallel}}{C}-\underset{\overset{H}{|}}{N}-\,.$$

These allow large numbers of hydrogen bonds to form between adjacent molecules:

$$\diagdown C\!\!=\!\!O\text{----}H\!\!-\!\!N\diagup\,,$$

providing an excellent 'glue'.

The net result is as desired. When Kevlar fibres are extruded from solution and stretched in order to align the modules parallel to the fibre axis, a remarkably high degree of orientation is achieved. The orientation is not disrupted by periodic disordered regions along the fibre, and it is resistant to heating. The structure is that of a disordered crystal, **without discrete amorphous regions**. The properties of Kevlar fibres show two drawbacks: (1) weakness in axial shear; and (2) yellow coloration (it imparts this colour to its composites).

As a reinforcement, Kevlar 49 is in intense competition with carbon fibre. They are both much more costly than E-glass, and are considered only where their outstanding mechanical properties are really needed. Not surprisingly, there is currently increasing interest in 'hybrid' composites. A combination of two or more types of reinforcement is used to produce an optimum profile of cost and properties for a given application.

In many applications, the long-term 'in-use' properties of a loaded composite are of great importance. The required properties include:

- fatigue resistance;
- elevated temperature resistance (both instantaneous and long-term);
- chemical resistance; and
- weathering resistance.

One of the most important attributes of carbon fibre composites is their outstanding fatigue resistance. The order of excellence is

carbon > aramid > glass.

Carbon will endure many millions of cycles at up to 80% of static strength; aramid at up to 40%; and glass up to 25%.

6.2.3 Platelet reinforcement

The platelet reinforcements in common use are all minerals. Two in particular are notably successful: talc and mica. Talc in a magnesium silicate, while mica is an aluminium silicate.

Talc: $3MgO \cdot 4SiO_2 \cdot H_2O$.

Mica: $K_2O \cdot 3Al_2O \cdot 6SiO_2 \cdot 2H_2O$ (muscovite form)

Typically of naturally occurring materials, they are never obtained in pure form. The product of one mine always differs from that of another, because of the presence of varying proportions of contaminating minerals.

Talc and mica are both crystalline, with similar structures of a layered type. SiO_2 tetrahedra are firmly linked together in layers, sandwiching other oxides, but with only weaker bonding between the layers. It is the weakness of this inter-layer bonding that allows crystals to be readily split along the layers by the mechanical actions of crushing and grinding. The resulting particles therefore have the shape of small plates, $10-1000$ μm across and $1-5$ μm in thickness.

The success of mineral platelet reinforcements is due to their desirable combination of cost and properties:

- low price—price per unit mass is typically less than one-fifth that of the common plastics; and
- stiffness and strength are greater than those of plastics.

As a result, when they are used to reinforce plastics, significant increases in modulus (with talc and mica) and strength (with mica only) can be obtained **at little or no increase in cost**. There is also a further advantage over fibres: when platelets are aligned parallel to each other (as they are in the surface layers of an injection moulding) they provide reinforcement in all directions in their plane, not merely in one direction as with uniaxially aligned fibres.

6.3 Forming of reinforced plastics

When a component is manufactured from a reinforced plastic, the forming process does more than simply shape the component; it also positions the reinforcing particles, and fixes their orientation. This results in an important characteristic of a reinforced polymer: its structure usually has directionality, and it shows anisotropy in its physical properties. Moreover, the material is usually inhomogeneous, as the particle orientation and even degree of reinforcement may vary from point to point throughout a component.

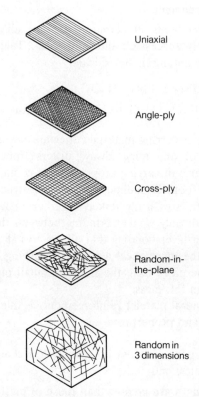

Uniaxial

Angle-ply

Cross-ply

Random-in-
the-plane

Random in
3 dimensions

6.8   Patterns of fibre reinforcement in thermosets.

In the forming of critical load-bearing products with continuous filament, these effects are exploited fully by the manufacturer. Reinforcement
is positioned and aligned where it produces optimum mechanical effect. In
other applications, inhomogeneity and anisotropy arise adventitiously, as in
the processing of fibre-reinforced thermoplastics. Here the short fibres are
distributed and aligned by the viscous flow of the surrounding molten
polymer. Clearly, the engineer must know what pattern of reinforcement is
possible with each forming process, and the control he can exert over it.

Consider first the important class of fibre-reinforced thermosets. Figure
6.8 shows the range of common fibre arrangements,[1] and Table 6.2
indicates the forming processes which lead to them. The salient features of
the processes are as follows.

---

1 Figure 6.8 shows idealizations which are not realized exactly in practice. For example, no
commercial process produces truly random orientation of fibres in three dimensions. Fibres
are always acted on by forces due to viscous flow during forming.

Table 6.2   Fibre arrangements possible in fibre-reinforced thermoset polymers

Forming method	Uniaxial	Angle-ply	Cross-ply	Random-in-the-plane	Random in 3-D
Pultrusion	*				
Filament winding	*	*	*		
Hand lay-up	*	*	*	*	
Hand spray-up				*	
Compression moulding	*	*	*	*	*
Reinforced reaction injection moulding					*

6.3.1 Pultrusion (see Figure 6.9(a))

Continuous fibre rovings are hauled through a bath of resin (polymer precursor) and then through a shaping die. The resin then cross-links and hardens, producing a long prismatic component with fibres aligned uniaxially parallel to its long axis.

6.3.2 Filament winding (see Figure 6.9(b))

Continuous fibre rovings are pulled through a bath of resin and are then wound on to a driven 'mandrel'. With the arrangement shown, a tubular component results when the resin has hardened and the mandrel is

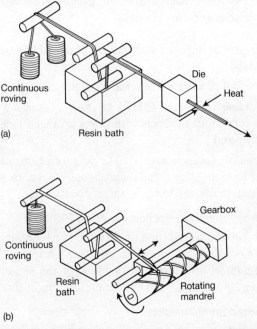

6.9 Two processes for the forming of reinforced thermosets: (a) pultrusion; and (b) filament winding.

withdrawn. The fibres are in an angle-ply arrangement. The method is clearly well suited to producing pipes and cylindrical vessels. With other winding geometries, however, it is possible to produce other shapes and either uniaxial or more complex fibre arrangements.

6.3.3 Hand lay-up

Fibre reinforcement is laid down by hand in the required arrangement and shape, and resin is applied with a brush. Reinforcement can be in the form of rovings, a woven roving fabric, or a mat of randomly arranged chopped fibres (chopped stand mat—CSM). The method is very versatile, but is also labour-intensive. It is therefore used for relatively short production runs, for example in the building of small boats.

6.3.4 Hand spray-up

Rovings are fed to a spray gun, which chops the fibres and then sprays them at the panel where reinforcement is needed. Resin is again applied with a brush. Fibres tend to be randomly arranged in the plane of the panel.

6.3.5 Compression moulding

Fibre reinforcement and resin are premixed, and other additives are added prior to compression moulding in a heated mould which triggers the hardening of the resin. Many forms of premix are available, making a variety of fibre arrangements possible:

- resin-impregnated tape of continuous fibres of uniaxial or angle-ply arrangements;
- resin-impregnated woven fabric for cross-ply arrangements;
- sheet moulding compound (SMC) (a sheet containing chopped fibres and resin) for moulding sheets with fibres arranged essentially random-in-the-plane; and
- dough moulding compound (DMC) (a dough-like mix of chopped fibres and resin) for moulding three-dimensional objects in which the fibres are arranged nearly randomly in three dimensions.

6.3.6 Reinforced reaction injection moulding (RRIM)

Liquid precursors are mixed (one of them already containing short fibres) as they are injected into a closed mould (see Chapter 7). The resulting fibre arrangement is influenced by the flow of the simultaneously cross-linking polymer, but approximates to being random in three directions.

6.3.7 Reinforced thermoplastics

Forming is usually carried out using the normal process of extrusion or injection moulding (see Chapter 7). Alignment of the reinforcement is caused by drag exerted on the particles by the flowing viscous polymer melt. It forces reinforcing particles to follow the flow of the melt. Where the melt is stretched (elongational flow, see Chapter 7) fibres align parallel to the direction of the stretching. Since the flow field varies throughout an extrudate or moulding, an inhomogeneous pattern of reinforcement results. Careful attention must be paid to this in design (see Chapter 8). An example can be seen in Figure 8.3, where cross-sections are shown through a glass-fibre reinforced polypropylene injection moulding. As is typical of fibre-reinforced thermoplastics, close to the surface of the moulding, fibres are preferentially aligned in the direction of mould filling. A similar pattern is observed in platelet-reinforced thermoplastics: near the moulding surface, platelets align parallel to it.

Another important effect is the damage that is done to reinforcing particles during passage through the thermoplastic compounding and forming machinery. In the extrusion or injection moulding of glass-fibre reinforced thermoplastics, for example, the mean fibre length is reduced to ~0.5 mm, irrespective of its initial value. Similarly, in mica-reinforced thermoplastics the mica platelets are broken down to a width of about 50 μm. In all cases, as the particle aspect ratio is changed to a value closer to 1, the effectiveness of reinforcement is impaired.

In recent years, thermoplastic composites known as GMTs (glass mat reinforced thermoplastics) have taken a big stride into the automotive market. GMTs are stampable sheets of glass fibre in a thermoplastic matrix. They can be processed by techniques similar to those used for sheet metals. The fibres can be randomly distributed, or unidirectional (and continuous), or a mixture of each. For example, in some Peugeot cars, the bumper is hot-pressed from a polypropylene GMT with 20% unidirectional and 20% random glass fibres. GMT technology is in its infancy and has tremendous potential: in the metals replacement area in particular. The technology can be used with the more expensive fibres (such as carbon) and thermoplastic matrices.

6.4 The mechanics of fibre reinforcement

Predicting accurately the mechanical properties of a composite material is not an easy task. Differences between the properties of reinforcing particles and the matrix cause complex distributions of stress and strain at the microscopic level, when load is applied. Reasonably accurate predictions can be made for fibre composites, however, by employing simplified assumptions about the stress and strain distributions.

6.4.1 Continuous fibres
First, consider the case of fibres which are so long that the effects of their

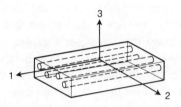

6.10 Definition of axes in aligned fibre composite.

ends can be ignored. A block of polymer reinforced with uniaxially aligned continuous fibres is sketched in Figure 6.10. Three orthogonal axes 1, 2, and 3 are defined within it. When loads are applied to the block we can assume that it deforms **as if** the mechanical coupling between fibres and matrix took special simple forms, in which stress and strain are uniform within each component.

If a stress σ_1 acts parallel to axis 1, fibres and matrix are approximately coupled together in 'parallel', as shown in Figure 6.11(a). Fibres and matrix elongate equally in the direction of σ_1. Their axial strains ε_{f_1} and ε_{m_1}, respectively, equal the strain ε_1 in the composite:

$$\varepsilon_{f_1} = \varepsilon_{m_1} = \varepsilon_1. \tag{6.10}$$

Fibres extend the entire length of the block, so that at any section the area fractions occupied by fibres and matrix equal their respective volume fractions. The total stress σ_1 must then equal the weighted sum of stresses in fibres and matrix σ_{f_1}, and σ_{m_1}, respectively:

$$\sigma_1 = \phi_f \sigma_{f_1} + (1 - \phi_f) \sigma_{m_1}. \tag{6.11}$$

Fibres and matrix are assumed to carry pure axial tension, with no stress in the 2–3 plane (6.N.4):

$$\sigma_{f_2} = \sigma_{m_2} = \sigma_{f_3} = \sigma_{m_3} = 0. \tag{6.12}$$

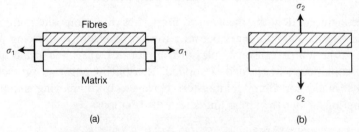

(a)   (b)

6.11 Fibre matrix coupling in aligned fibre composite: (a) coupling for tensile stress parallel to axis 1; and (b) coupling for tensile stress parallel to axis 2.

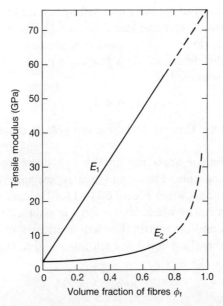

6.12 Predicted tensile moduli for aligned glass-fibre reinforced epoxy.

We neglect here the viscoelasticity of the polymer matrix, and treat fibres and matrix as being linearly elastic. Under tension parallel to axis 1, Hooke's law then relates stresses to strains for fibres, matrix, and composite as a whole:

$$\sigma_{f_1} = E_f \varepsilon_{f_1}, \quad \sigma_{m_1} = E_m \varepsilon_{m_1},$$

and

$$\sigma_1 = E_1 \varepsilon_1, \tag{6.13}$$

where E_f, E_m, and E_1 are the respective moduli.[1] Substituting the stresses from eqn 6.13 into eqn 6.11 and dividing through by ε_1 yields a prediction of the axial tensile modulus of the composite:

$$E_1 = \phi_f E_f + (1 - \phi_f) E_m. \tag{6.14}$$

An example of the application of eqn 6.14 is shown in Figure 6.12. The graph is drawn for the case of glass-fibre reinforced epoxy. At high fibre fractions E_1 can be seen to approach values typical of metals (e.g. aluminium with $E = 70$ GPa).

---

1 Care must be exercised to use the correct fibre modulus E_f in the case of anisotropic fibres such as carbon and Kevlar. In eqn 6.14 it is the axial tensile modulus; in eqn 6.20 it is the transverse tensile modulus.

The linear additivity of moduli expressed by eqn 6.14 is another example of the rule of mixtures applying (see Section 6.1). The second term in the equation, however, makes only a small contribution, since $E_m \ll E_f$. Except at very low fibre fractions, E_1 is given to a good approximation by the fibres' contribution alone:

$$E_1 \approx \phi_f E_f. \tag{6.15}$$

The polymer matrix is then simply acting as a glue, holding the load-bearing fibres together.

By contrast, when a stress acts in the 2–3 plane, the matrix plays a crucial load-bearing role. Fibres and matrix are now coupled approximately in 'series', as shown in Figure 6.11(b) for the case of a tensile stress σ_2 parallel to axis 2. The whole tensile force is assumed to be carried fully by both the fibres and the matrix. The tensile stresses in fibres and matrix σ_{f_2} and σ_{m_2} are therefore equal to each other and to the overall stress in the composite:

$$\sigma_{f_2} = \sigma_{m_2} = \sigma_2. \tag{6.16}$$

Displacement in fibres and matrix parallel to σ_2 can be approximated as being additive. The total strain is therefore the weighted sum of strains in fibres and matrix:

$$\varepsilon_2 = \phi_f \varepsilon_{f_2} + (1 - \phi_f) \varepsilon_{m_2}. \tag{6.17}$$

Again, the stress in each component is approximated as being pure uniaxial tension: there is no stress in the 1–3 plane:

$$\sigma_{f_1} = \sigma_{m_1} = \sigma_{f_3} = \sigma_{m_3} = 0. \tag{6.18}$$

Lastly, we again invoke Hooke's Law for fibres, matrix and composite:

$$\sigma_{f_2} = E_f \varepsilon_{f_2},$$

$$\sigma_{m_2} = E_m \varepsilon_{m_2},$$

and

$$\sigma_2 = E_2 \varepsilon_2. \tag{6.19}$$

When strains from eqns 6.19 are substituted in eqn 6.17, an expression is obtained for the tensile modulus of the composite transverse to the fibres:

$$E_2 = \frac{E_f E_m}{(1 - \phi_f) E_f + \phi_f E_m}. \tag{6.20}$$

Equation 6.20 is included in Figure 6.12 for the glass–epoxy system. It can be seen that E_2 is predicted to be much lower than E_1, as is found in practice.

It is interesting to note that, since $E_f \gg E_m$, eqn 6.20 approximates to the simpler expression

$$E_2 \approx \frac{E_m}{1 - \phi_f}. \qquad (6.21)$$

The meaning of eqn 6.21 is that the fibres now play the subordinate role: their presence serves merely to reduce the amount of deformable matrix.

Although based on only simple models of the mechanical coupling between fibres and matrix, eqns 6.14 and 6.20 are found to give a reasonable fit to observed properties of composites. The same models can be used to derive other elastic constants of the composite; for example Poisson's ratio. We shall define ν_{12} as being Poisson's ratio applying to **free contraction parallel to axis 2, when a tensile stress is applied parallel to axis 1.** With similar reasoning as used above to find E_1 and E_2, one can show that ν_{12} obeys the rule of mixtures:

$$\nu_{12} = \phi_f \nu_f + (1 - \phi_f)\nu_m \qquad (6.22)$$

(see Problem 6.7). The other Poisson's ratio ν_{21} applying to the 1–2 plane does not equal ν_{12} but can be found from ν_{12}, E_1, and E_2 (6.N.5).

Example 6.2

A composite material consists of 40% (by volume) continuous, uniaxially aligned, glass fibres in a matrix of thermoset polyester. A tensile stress of 100 MPa is to be applied parallel to the fibres. Predict the strains which will result. Take the tensile modulus and Poisson's ratio of glass to be 76 GPa and 0.22, and of thermoset polyester to be 3 GPa and 0.38, respectively.

Solution

From the uniaxial symmetry of the material and the loading system, the only strains which will arise with respect to axes 1, 2, and 3 (Figure 6.10) are (1) axial tensile strain ε_1, and (2) negative tensile strain in all directions transverse to the fibres $\varepsilon_2 = \varepsilon_3 = -\nu_{12}\varepsilon_1$.

(1) The tensile modulus E_1 is obtained by assuming parallel coupling between fibres and matrix, thus (from eqn 6.14)

$$E_1 = 0.4 \times 76 + 0.6 \times 3 = 32.2 \text{ GPa}.$$

Hence

$$\varepsilon_1 = 100 \times 10^6 / 32.2 \times 10^9 = 3.11 \times 10^{-3}.$$

(2) In order to find ε_2, first find Poisson's ratio ν_{12}. This is found by assuming series coupling between fibres and matrix parallel to axis 2, thus (from eqn 6.22)

$$\nu_{12} = 0.4 \times 0.22 + 0.6 \times 0.38 = 0.316.$$

Consequently, tensile strain transverse to the fibres is given by

$$\varepsilon_2 = -0.316 \times 0.00311 = -9.83 \times 10^{-4}.$$

Example 6.3

A tensile stress of 15 MPa is applied to the composite material of Example 6.2, in a direction transverse to the fibres. Predict the strain which will result in the direction parallel to the fibres.

Solution

Let the stress be parallel to axis 2 (Figure 6.10). The required strain ε_1 is then to be found from $\varepsilon_1 = -\nu_{21}\varepsilon_2$ (note carefully the different meaning of ν_{21} in comparison to the ν_{12} employed in Example 6.2). The first step is to find tensile modulus E_2. The convenient approximate procedure is to assume series coupling between fibres and matrix parallel to axis 2. Thus eqn 6.20 becomes

$$E_2 = \frac{76 \times 3}{0.6 \times 76 + 0.4 \times 3} = 4.87 \text{ GPa}.$$

Hence the strain in this direction will be

$$\varepsilon_2 = 15 \times 10^6 / 4.87 \times 10^9 = 3.08 \times 10^{-3}.$$

The relevant Poisson's ratio ν_{21} must be found from ν_{12}, E_1, and E_2. Thus, taking ν_{12} and E_1 from Example 6.2, we obtain (eqn 6.N.5.5)

$$\nu_{21} = 4.87 \times 0.316 / 32.2 = 0.0478,$$

and hence

$$\varepsilon_1 = -0.0478 \times 0.00308 = -1.47 \times 10^{-4}.$$

When a shear stress τ_{12} acts parallel to the fibres, the composite deforms as if fibres and matrix were coupled in series. Hence the shear strain γ_{12} can be found from a relation similar in form to eqn 6.17, and the corresponding shear modulus is found to be

$$G_{12} = \frac{G_f G_m}{(1 - \phi_f) G_f + \phi_f G_m}. \qquad (6.23)$$

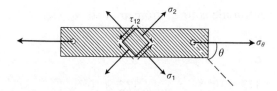

6.13 Tensile test on aligned fibre composite at angle θ to axis 1.

Once elastic constants are known relative to the axes 1, 2, and 3, Mohr's circle can be used to predict deformations resulting from stress applied in any direction. Suppose a tensile stress σ_θ is applied to the 1–2 plane, along a line at angle θ to axis 1 (Figure 6.13). The procedure is as follows.

- Resolve σ_θ into components σ_1, σ_2, and τ_{12} using Mohr's circle.
- Use known elastic constants to find ε_1, ε_2, and γ_{12}.
- Use Mohr's circle to resolve ε_1, ε_2, and γ_{12} into strain components in directions of interest.

Example 6.4

A composite material consists of 60% (by volume) continuous, uniaxially aligned, glass fibres in a matrix of epoxy. A tensile stress of 150 MPa is applied in a direction inclined at 30° to the fibres. Calculate the tensile strain which results parallel to the fibres. Take the tensile modulus and Poisson's ratio of glass to be 76 GPa and 0.22, and of epoxy to be 2.4 GPa and 0.34, respectively.

Solution

The applied stress results in stresses σ_1, σ_2, and τ_{12} relative to the axes 1, 2, and 3 (Figure 6.10). Of these, only σ_1 and σ_2 produce tensile strains parallel to the fibres. These stresses may be found by application of Mohr's circle as follows:

$$\sigma_1 = \tfrac{1}{2} \times 150(\cos(2 \times 30) + 1) = 112.5 \text{ MPa},$$

and

$$\sigma_2 = \tfrac{1}{2} \times 150(1 - \cos(2 \times 30°)) = 37.5 \text{ MPa}.$$

Evaluation of the strains which they cause parallel to the fibres requires knowledge of E_1, E_2, and ν_{21}. Following the methods of Examples 6.2 and 6.3,

$$E_1 = 0.6 \times 76 + 0.4 \times 2.4 = 46.6 \text{ GPa},$$
$$E_2 = \frac{76 \times 2.4}{0.4 \times 76 + 0.6 \times 2.4} = 5.73 \text{ GPa},$$
$$\nu_{12} = 0.6 \times 0.22 + 0.4 \times 0.34 = 0.268,$$

and

$$\nu_{21} = 5.73 \times 0.268/46.6 = 0.033.$$

Summing the contributions to ε_1 from σ_1 and σ_2 yields

$$\varepsilon_1 = \frac{\sigma_1}{E_1} - \nu_{21}\frac{\sigma_2}{E_2}$$

$$= \frac{113 \times 10^6}{46.6 \times 10^9} - 0.033 \times \frac{37.5 \times 10^6}{5.73 \times 10^9} = 2.21 \times 10^{-3}.$$

What if the fibres are not uniaxially oriented? Any misalignment of fibres causes E_1 to fall below the value predicted by eqn 6.14. Suppose E_{max} is defined as this maximum achievable modulus at any given fibre volume fraction. Two examples of other fibre arrangements are found to give tensile moduli as follows.

Random in two directions (in the 1–2 plane)

$$E_1 = E_2 \approx \tfrac{3}{8}E_{max} \tag{6.24(a)}$$

Random in three directions

$$E_1 = E_2 = E_3 \approx \tfrac{1}{5}E_{max}. \tag{6.24(b)}$$

Returning now to uniaxial fibre orientation, we can use the same models of mechanical coupling to predict thermoelastic properties of the composite. If the coefficients of linear expansion of fibres and matrix respectively are α_f^{1} and α_m, the coefficients α_1 and α_2 for the composite parallel to axes 1 and 2 can be shown to be

$$\alpha_1 = \frac{\phi_f E_f \alpha_f + (1 - \phi_f) E_m \alpha_m}{\phi_f E_f + (1 - \phi_f) E_m}, \tag{6.25(a)}$$

and

$$\alpha_2 = \phi_f \alpha_f (1 + \nu_f) + (1 - \phi_f)\alpha_m (1 + \nu_m) - \alpha_1 \nu_{12} \tag{6.25(b)}$$

(see Problem 6.13).

It is usually the case that $\alpha_f \ll \alpha_m$ and $E_f \gg E_m$. Equations 6.25(a) and (b) then predict widely differing linear thermal expansion parallel and perpendicular to the fibres. As an example, α_1 and α_2 (from eqns 6.25) are plotted versus ϕ_f in Figure 6.14 for epoxy reinforced by glass fibres. It

---

1   α_f is assumed here to be independent of direction within the fibre. In practice, carbon and Kevlar fibres are highly anisotropic in thermal expansion (see Table 6.3 below), and allowance for this must be made in calculating α_1 and α_2 (see Problem 6.14).

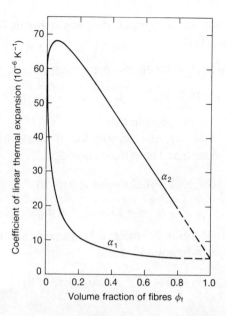

6.14 Predicted coefficients of linear thermal expansion for aligned glass-fibre reinforced epoxy.

is immediately clear that the fibres are highly effective at reducing thermal expansion in the direction in which they are aligned (axis 1). Meanwhile, at right angles (axis 2) thermal expansion is much greater, and may even exceed **both** α_f **and** α_m. A curious result of this anisotropy of the composite in thermal expansion is that a rectangular block of material cut with edges inclined to axes 1, 2, or 3 spontaneously deforms in shear when temperature is changed (Problem 6.16).

A further consequence of the difference between α_f and α_m is that microscopic thermal stresses are generated, which may significantly affect the strength of the composite (6.N.6).

Example 6.5

The composite material of Example 6.4 is subjected to a temperature rise of 100 K. Calculate its free thermal expansion in the direction inclined at 30° to the fibres. Take the coefficients of linear thermal expansion to be 5×10^{-6} K^{-1} and 60×10^{-6} K^{-1} for glass and epoxy, respectively.

Solution

The first step is to evaluate the coefficients α_1 and α_2 of thermal expansion parallel and perpendicular to the fibres. In order to obtain α_1, the assumption is made of parallel coupling between fibres and matrix. Thus (from eqn 6.25(a))

$$\alpha_1 = (0.6 \times 76 \times 5 \times 10^{-6} + 0.4 \times 2.4 \times 60 \times 10^{-6})/46.6$$

$$= 6.13 \times 10^{-6} \text{ K}^{-1}$$

(making use of E_1 from Example 6.4).

In order to obtain α_2, the assumption is made of series coupling between fibres and matrix. Thus (from eqn 6.25(b))

$$\alpha_2 = 0.6 \times 5 \times 10^{-6} \times (1 + 0.22) + 0.4 \times 60 \times 10^{-6} \times (1 + 0.34)$$

$$- 6.13 \times 10^{-6} \times 0.268 = 34.2 \times 10^{-6} \text{ K}^{-1}$$

(making use of ν_{12} from Example 6.4). Hence the thermally induced strains ε_1 and ε_2 can be found.

$$\varepsilon_1 = 6.13 \times 10^{-6} \times 100 = 6.13 \times 10^{-4}$$

and

$$\varepsilon_2 = 34.2 \times 10^{-6} \times 100 = 3.42 \times 10^{-3}.$$

Application of Mohr's circle to these strains then yields the thermal expansion at 30° to the fibres.

$$\varepsilon = 3.42 \times 10^{-3} + \tfrac{1}{2}(6.13 \times 10^{-4} - 3.42 \times 10^{-3}) \times (\cos(2 \times 30) + 1)$$

$$= 1.31 \times 10^{-3}.$$

We turn now to the important question of how the composite fails under load. Consider the case in which a tensile stress acts parallel to the fibres (axis 1). The sequence of events varies, depending on which of the two components is the more brittle: fibres or matrix (see Figure 6.15). Denoting fracture by an asterisk, the question is: which of ε_f^* and ε_m^* is the smaller? We consider the two cases in turn.

6.4.1.1 Case I: $\varepsilon_f^* < \varepsilon_m^*$ *(Figure 6.15(a))*   An example of this might be epoxy reinforced by carbon fibres. As before, we employ the simple assumption of parallel coupling between fibres and matrix, for stress parallel to axis 1. Furthermore, fibres and matrix in the composite are assumed to fail independently, at the same stresses and strains as when tensile tests are carried out on the pure materials.

The course of events is best followed by superimposing the fibre and matrix stress–strain curves, multiplied by their respective volume fractions.

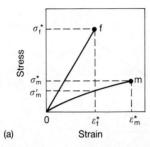
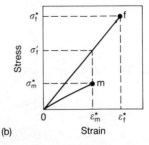

6.15 Schematic sketch of tensile stress–strain curves for fibres and matrix: (a) brittle fibres in ductile matrix; and (b) ductile fibres in brittle matrix.

The total stress $\sigma_1(\varepsilon_1)$ at strain ε_1 is then given from eqn 6.11 as the sum of these two (see Figure 6.16). As strain is increased, a point is reached around ε_f^* where fibres begin to break, and their contribution to the total stress falls to zero. What happens next depends on ϕ_f. At low ϕ_f (see Figure 6.16(a)) there is sufficient matrix to carry the tensile load even after all fibres have failed. The total stress at strain ε_1 is, from now on, just the contribution of the matrix alone, $(1 - \phi_f)\sigma_m$. On further straining, the composite fails when the matrix (now containing broken fibres) finally fails at $\varepsilon_1 = \varepsilon_m^*$. On the other hand, at higher ϕ_f, when the fibres break there is insufficient matrix to maintain the same stress. The point of failure of the fibres then marks the failure of the composite also. In this case, the stress just prior to failure is the sum of $\phi_f \sigma_f^*$ and $(1 - \phi_f)\sigma_m'$, where $\sigma_m' = \sigma_m(\varepsilon_f^*)$, the stress carried by the matrix **at the failure strain of the fibres** (see Figure 6.15(a)). To summarize, there are two criteria for failure:

$$\text{Low } \phi_f: \qquad \sigma_1^* = (1 - \phi_f)\sigma_m^*. \qquad (6.26(a))$$

$$\text{High } \phi_f: \qquad \sigma_1^* = \phi_f \sigma_f^* + (1 - \phi_f)\sigma_m'. \qquad (6.26(b))$$

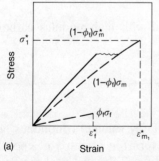
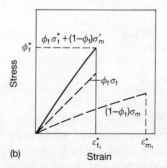

6.16 Schematic sketch of tensile stress–strain curves for aligned fibre composite containing brittle fibres in a ductile matrix. Contributions from fibres and matrix are shown as dashed lines: (a) low fibre volume fraction ϕ_f; and (b) high ϕ_f.

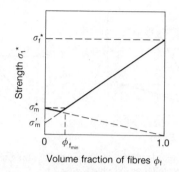

6.17 Schematic sketch graph of tensile strength σ_1^* versus fibre volume fraction ϕ_f for aligned fibre composite containing **brittle** fibres in a **ductile** matrix (case I).

Equations 6.26 give two estimates for σ_1^*, shown schematically in Figure 6.17 as two intersecting straight lines. The greater of the two represents the actual strength of the composite—shown as the full line. As we found before for other properties measured parallel to the fibres, except at very low fibre fractions tensile strength is dominated by the contribution from the fibres. This can be approximated well by the fibre contribution alone:

$$\sigma_1^* \approx \phi_f \sigma_f^* \qquad (6.27)$$

An important fact emerges from Figure 6.17. A minimum fibre fraction $\phi_{f_{min}}$ is required before the fibres exert a strengthening effect at all. Lower values of ϕ_f simply cause a weakening of the matrix and must be avoided. $\phi_{f_{min}}$ is clearly a critical parameter. It is obtained by equating σ_m^* with σ_1^* from eqn 6.26(b) and rearranging:

$$\phi_{f_{min}} = \frac{\sigma_m^* - \sigma_m'}{\sigma_f^* - \sigma_m'}. \qquad (6.28)$$

As an example, for epoxy reinforced with high-modulus carbon fibres, $\phi_{f_{min}} = 0.03$.

Example 6.6

A composite material consists of 50% (by volume) continuous, uniaxially aligned, carbon fibres (high-strength form) in a matrix of epoxy. Predict the tensile strength parallel to the fibres. Take the axial tensile strength and modulus of these fibres to be 3200 MPa and 230 GPa, and of epoxy to be 60 MPa and 2.4 GPa, respectively.

Solution

The failure strains of fibres and matrix are as follows

$$\varepsilon_f^* = 3200 \times 10^6/230 \times 10^9 = 1.39 \times 10^{-2},$$

and

$$\varepsilon_m^* = 60 \times 10^6/2.4 \times 10^9 = 2.5 \times 10^{-2}.$$

Therefore, the fibres fail before the matrix does so. When this occurs, the matrix carries a stress

$$\sigma_m' = E_m \varepsilon_f^* = 2.4 \times 0.0139 = 0.0334 \text{ GPa}.$$

Provided the composite fails when the fibres fail (a good assumption at this fibre volume fraction—see text), the stress carried by the composite at failure is given by the assumption of parallel coupling between fibres and matrix:

$$\sigma_1^* = 0.5 \times 3.20 + 0.5 \times 0.0334 = 1.62 \text{ GPa}.$$

6.4.1.2 *Case II:* $\varepsilon_m^* < \varepsilon_f^*$ *(Figure 6.15(b))*   This might apply to a thermoset polyester reinforced with glass fibres. It can be analysed in a similar manner to the previous case. The details are left as an exercise for the reader (Problem 6.18). The result, as before, is two criteria for failure of the composite. They are:[1]

$$\text{Low } \phi_f: \qquad \sigma_1^* = \phi_f \sigma_f' + (1 - \phi_f) \sigma_m^*. \qquad (6.29(a))$$

$$\text{High } \phi_f: \qquad \sigma_1^* = \phi_f \sigma_f^*. \qquad (6.29(b))$$

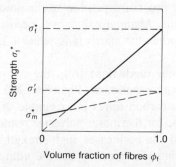

6.18 Schematic sketch graph of tensile strength σ_1^* versus fibre volume fraction ϕ_f for aligned fibre composite containing **ductile** fibres in a **brittle** matrix (case II).

---

1   By analogy with case I, σ_f' is the stress carried by fibres **at the failure strain of the matrix**.

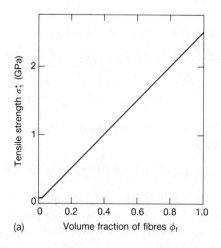

(a)

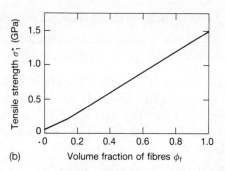

(b)

6.19 Tensile strength versus ϕ_f predicted for: (a) carbon-fibre reinforced epoxy (case I); and (b) glass-fibre reinforced polyester (case II).

These two estimates are shown as two intersecting straight lines in Figure 6.18. The larger of the two, corresponding to the predicted tensile strength, is shown as the full line. Again, a switch-over occurs. At low ϕ_f failure of the composite occurs when the matrix fails, while at high ϕ_f it occurs when the fibres fail.

Figure 6.19 illustrates predictions from the above analysis, which are found to agree well with experiment. Figure 6.19(a) shows predictions for epoxy reinforced with carbon fibres (an example of case I), while Figure 6.19(b) shows the same for thermoset polyester reinforced with glass fibres (an example of case II). In both cases, high strengths are obtained at high fibre fractions. For comparison, recall that the ultimate tensile strength of mild steel is about 0.45 GPa.

Example 6.7

A composite material consists of 50% (by volume) continuous, uniaxially aligned, E-glass fibres in a matrix of thermoset polyester. Predict the tensile strength parallel to the fibres. Take the tensile strength and modulus of E-glass fibres to be 1800 MPa and 76 GPa, and of thermoset polyester to be 55 MPa and 3 GPa, respectively.

Solution

The failure strains of fibres and matrix are as follows.

$$\varepsilon_f^* = 1800 \times 10^6 / 76 \times 10^9 = 2.37 \times 10^{-2},$$

and

$$\varepsilon_m^* = 55 \times 10^6 / 3 \times 10^9 = 1.83 \times 10^{-2}.$$

Therefore, the matrix fails before the fibres do so. When the matrix fails, the stress carried by the fibres is given by

$$\sigma_f' = E_f \varepsilon_m^* = 76 \times 1.83 \times 10^{-2} = 1.39 \text{ GPa},$$

and the stress carried by the composite is given by (assuming parallel coupling between fibres and matrix)

$$\sigma_1 = 0.5 \times 1.39 + 0.5 \times 0.055 = 0.723 \text{ GPa}.$$

Whether or not this constitutes failure of the composite depends on whether the composite can sustain a higher stress than this on further straining. The maximum stress which can be carried by the composite when only the fibres are contributing (i.e. when the matrix has failed) is given by

$$\sigma_1 = 0.5 \times 1.8 = 0.90 \text{ GPa}.$$

This exceeds the stress carried at the point of failure of the matrix. We deduce, therefore, that the composite can continue to carry stress up to the point where the fibres fail, and the tensile strength is given by

$$\sigma_1^* = 0.90 \text{ GPa}.$$

The high strengths predicted are realized only when loads are parallel to the fibres. The composite is much weaker under stress in other directions. This is because cracks seek out the easiest path along which to propagate. In a fibre-reinforced polymer, the easiest paths are through the matrix and along the fibre–matrix interface.

In consequence, when a tensile stress acts transversely to the fibres, fracture occurs as sketched in Figure 6.20(a), without the need to break any fibres. Indeed, they now serve as stress raisers and actually **reduce** the strength σ_2^* to below that of the pure matrix σ_m^*. Similarly, a low strength τ_{12}^* is obtained under shear parallel to the fibres (Figure 6.20(b)).

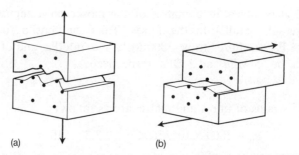

6.20 Schematic sketches of other failure modes in the aligned fibre composite: (a) tension parallel to axis 2; and (b) shear parallel to axis 1.

When a tensile stress is applied in an arbitrary direction in the 1–2 plane (Figure 6.13), the failure stress can be predicted with the aid of a simple assumption. Failures under stresses σ_1, σ_2, and τ_{12} are assumed to occur independently of each other. Consider the case shown, of tensile loading at angle θ to axis 1. The stress σ_θ can be resolved into components σ_1, σ_2, and τ_{12} by application of Mohr's circle.

$$\left.\begin{array}{l} \sigma_1 = \sigma_\theta \cos^2\theta \\ \sigma_2 = \sigma_\theta \sin^2\theta \\ \tau_{12} = \sigma_\theta \sin\theta \cos\theta \end{array}\right\} \qquad (6.30)$$

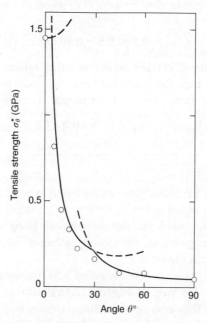

6.21 Angular dependence of tensile strength of aligned carbon-fibre reinforced epoxy ($\phi_f = 0.66$), as predicted and as measured.

Failure will occur immediately σ_1, σ_2, or τ_{12} reaches its limiting value. Three modes of failure are to be expected, each implying a different tensile strength.

Axial tensile failure $\qquad\qquad \sigma_\theta^* = \sigma_1^*/\cos^2\theta.$

Axial shear failure $\qquad\qquad \sigma_\theta^* = \tau_{12}^*/\sin\theta\cos\theta.$ $\qquad$ (6.31)

Transverse tensile failure $\qquad \sigma_\theta^* = \sigma_2^*/\sin^2\theta.$

They are shown plotted as dashed lines in Figure 6.21, for the case of an epoxy reinforced with 66% by volume carbon fibres. The composite is expected actually to fail at the lowest of the three predicted failure stresses —shown as a full line. Measured strengths lie close to this line, but reveal more gradual transitions between the competing modes of failure, as can be seen in Figure 6.21.

Example 6.8

A composite material consists of 55% (by volume) continuous, uniaxially aligned, S-glass fibres in a matrix of epoxy. Such a composite is found to have a tensile strength transverse to the fibres $\sigma_2^* = 25$ MPa and shear strength parallel to the fibres $\tau_{12}^* = 55$ MPa. The tensile strength and modulus of the fibres are 1900 MPa and 86 GPa, and of the matrix are 60 MPa and 2.4 GPa, respectively. The composite is to be subjected to tensile stress in a direction inclined at 20° to the fibre axes. Predict the stress at failure and determine the mode of failure.

Solution

We can consider there to be three possible modes of failure: axial tensile failure parallel to the fibres; shear failure parallel to the fibres; and transverse tensile failure perpendicular to the fibres. Adopting the approximation that each mode of failure occurs independently of the others (see text), identify failure stresses corresponding to each mode. Failure stresses σ_2^* and τ_{12}^* are given: it remains to find σ_1^*, by employing the methods of Examples 6.6 and 6.7. For axial tensile failure

$$\varepsilon_f^* = 1900 \times 10^6/86 \times 10^9 = 2.21 \times 10^{-2},$$

and

$$\varepsilon_m^* = 60 \times 10^6/2.4 \times 10^9 = 2.50 \times 10^{-2},$$

and hence (see Example 6.6)

$$\sigma_m' = 2400 \times 2.21 \times 10^{-2} = 53.0 \text{ MPa}.$$

The stress carried by the composite at the point of failure of the fibres is given (assuming parallel coupling) by

$$\sigma_1 = 0.55 \times 1.90 + 0.45 \times 0.053 = 1.07 \text{ GPa.}$$

After the fibres have failed, the maximum stress which can be carried by the matrix acting alone is

$$\sigma_1 = 0.45 \times 60 = 27 \text{ MPa.}$$

Consequently, the point of failure of the fibres corresponds to the failure of the composite, and the axial tensile strength $\sigma_1^* = 1.07$ GPa.

We may now predict the tensile failure stress at 20° to the fibre axes, by finding the strength corresponding to each potential mode of failure. Applying eqn 6.31 we obtain the following.

Axial tensile failure $\sigma_{20}^* = 1070/\cos^2 20° = 1212$ MPa;

Axial shear failure $\sigma_{20}^* = 55/\sin 20° \cos 20° = 171$ MPa;

Transverse tensile failure $\sigma_{20}^* = 25/\sin^2 20° = 213$ MPa.

It is clear that failure is predicted to occur by axial shear, at a tensile stress of 171 MPa.

A uniaxially aligned fibre composite is too weak in directions close to $\theta = 90°$ for most engineering applications. This problem is overcome by combining together fibres inclined in two or more directions within the same material. The result is a fibre arrangement which is angle-ply, cross-ply, or random-in-the-plane. The usual method of achieving this to to bond together one above the other a sequence of sheets, or **laminae**, of uniaxially aligned fibres, woven fabric, or chopped strand mat. The resulting multilayer sheet is known as a **laminate**. Its properties can be predicted by a statics analysis of the stack, taking due account of compatibility of strains in all laminae in the planes of the laminate (see Chapter 8).

6.4.2 Discontinuous fibres

Generally, the highest strength and stiffness are obtained with continuous fibre reinforcement. Discontinuous fibres are used only when manufacturing economics dictate the use of a process where the fibres must be in this form—for example injection moulding.

When considering discontinuous fibres it is necessary to take account of the fibre ends. These are weak points in the composite—sites of high

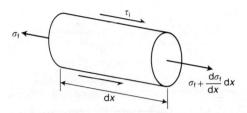

6.22 Stresses acting on an element of fibre embedded in a matrix.

stress concentration in the matrix. It must be assumed that negligible stress gets transferred to the fibres across their end faces. Stress builds up along each fibre from zero at its ends to a maximum at the centre. How is this stress transferred to the fibre? Examination of Figure 6.2 indicates that the matrix is severely sheared in the vicinity of the fibre. It is the shear stress at the fibre–matrix interface τ_i that does the vital job of transmitting tension to the fibre, in rather the same way as a hand pulling on a rope. By invoking static equilibrium of the forces acting on an element of the fibre of length dx, a relation is obtained between τ_i and the axial tension σ_f in the fibre (see Figure 6.22). Equilibrium requires

$$\left(\sigma_f + \frac{d\sigma_f}{dx} dx \right) \frac{\pi d^2}{4} - \sigma_f \frac{\pi d^2}{4} + \tau_i \pi d \, dx = 0,$$

and rearranging,

$$\frac{d\sigma_f}{dx} = -\frac{4}{d} \tau_i. \tag{6.32}$$

Predicting σ_f and τ_i involves stress analysis of the fibre–matrix composite under load. A simplified but highly successful solution is known as the **shear lag theory** (6.N.7). It assumes perfect bonding between fibres and matrix, and results in the following prediction of σ_f as a function of distance x along the fibre, measured from its centre (recall $a \equiv l/d$):

$$\sigma_f = E_f \varepsilon_1 \left\{ 1 - \left[\frac{\cosh\left(na \, \dfrac{2x}{l} \right)}{\cosh(na)} \right] \right\}, \tag{6.33}$$

where n represents a dimensionless group of constants,

$$n = \sqrt{\left[\frac{2G_m}{E_f \ln(2R/d)} \right]}, \tag{6.34}$$

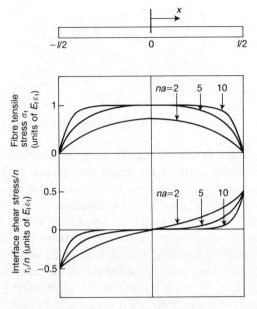

6.23 Distributions along a short fibre of tensile stress σ_f and interfacial shear stress τ_i, as predicted by eqns 6.33 and 6.35.

where $2R$ is the distance from the fibre to its nearest neighbour. Equation 6.33 is differentiated to obtain τ_i via eqn 6.32:

$$\tau_i = \frac{n}{2} E_f \varepsilon_1 \frac{\sinh\left(na\,\frac{2x}{l}\right)}{\cosh(na)}. \tag{6.35}$$

Equation 6.33 shows clearly that the critical parameter dictating how rapidly stress builds up along the fibre is the product na. This is illustrated in Figure 6.23, where graphs of σ_f and τ_i/n are plotted versus x/l for three values of na. The graphs show that for most efficient stress transfer to the fibres na should be as high as possible. This confirms the desirability of a high aspect ratio a, as mentioned before (in Section 6.1), but also shows that n should be high: the ratio G_m/E_f should therefore be as high as possible. Typical values encountered in practice are $a = 50$, $n = 0.24$, and hence $na = 12$ (for 30% by volume glass fibres in a nylon matrix).

The most important effects of the fibre stress falling away to zero at the ends are reductions in the axial tensile modulus and strength of the composite. The effect on modulus can be found as follows. Suppose the fibres are parallel, and a line is imagined drawn across the composite at right angles to them, as in Figure 6.24. The line intersects fibres at

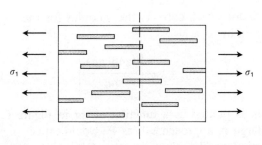

6.24   Polymer reinforced by aligned discontinuous fibres.

longitudinal positions chosen at random. It follows that the stress carried
by the composite is no longer given by eqn 6.11, but instead by

$$\sigma_1 = \phi_f \bar{\sigma}_f + (1 - \phi_f)\sigma_m, \tag{6.36}$$

where $\bar{\sigma}_f$ is the mean fibre stress intercepted:

$$\bar{\sigma}_f = \frac{1}{l} \int_{-l/2}^{l/2} \sigma_f \, dx. \tag{6.37}$$

Substituting from eqn 6.33 into eqn 6.37 and carrying out the integration
yields

$$\bar{\sigma}_f = E_f \varepsilon_1 \left\{ 1 - \left[\frac{\tanh(na)}{na} \right] \right\}. \tag{6.38}$$

Consequently, the axial tensile modulus obtained by dividing stress σ_1 by
strain ε_1 becomes

$$E_1 = \eta_1 \phi_f E_f + (1 - \phi_f)E_m, \tag{6.39}$$

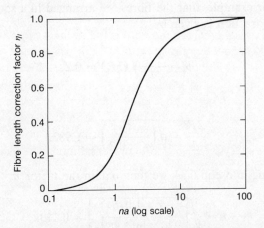

6.25   Fibre length correction factor for modulus plotted against the product *na*.

where η_l is a factor which corrects the modulus for the shortness of the fibres and is given by

$$\eta_l = 1 - \left[\frac{\tanh(na)}{na} \right]. \qquad (6.40)$$

The factor η_l is plotted as a function of na in Figure 6.25. When na becomes very large η_l approaches 1, as we would expect since this limit is the case of continuous fibres (see equation 6.14). But when na falls below about 10, η_l is reduced significantly below 1.

Example 6.9

An injection-moulded bar contains 20% (by volume) of short carbon fibres (high strength form) in a matrix of nylon 6.6. A tensile strain of 10^{-4} is applied along the axis of the bar. Determine the mean tensile stress carried by the fibres and the overall stress carried by the bar. Assume that the fibres are all of length 400 μm and diameter 6 μm, and are perfectly aligned along the axis of the bar and perfectly bonded to the matrix. Take the axial tensile modulus of the carbon fibres to be 230 GPa, and the tensile and shear moduli of nylon 6.6 to be 2.7 GPa and 1.015 GPa, respectively.

Solution

Since the fibres are short, we must take account of the variation of tensile stress along each fibre. This is determined by the aspect ratio $a = l/d = 400/6 = 66.7$, and the dimensionless factor n (from eqn 6.34). In order to evaluate n, we need the ratio $2R/d$ occurring in the shear lag theory. Assuming, for example, that the fibres are arranged in a square array, the fibre volume fraction is given by

$$\phi_f = \frac{\pi d^2}{4} \bigg/ (2R)^2 = 0.2.$$

Therefore

$$\frac{2R}{d} = \sqrt{\left(\frac{\pi}{4 \times 0.2} \right)} = 1.982.$$

By substituting into eqn 6.34, we then obtain the factor n:

$$n = \sqrt{\left[\frac{2 \times 1.015}{230 \times \ln(1.982)} \right]} = 0.1136.$$

With n and a both known, the fibre length correction factor η_l can be evaluated

$$\eta_l = 1 - \frac{\tanh(na)}{na} = 1 - \frac{\tanh(0.1136 \times 66.7)}{0.1136 \times 66.7} = 0.868.$$

The mean fibre stress is given by (from eqn 6.38)

$$\bar{\sigma}_f = E_f \varepsilon_1 \eta_l = 230 \times 10^3 \times 10^{-4} \times 0.868 = 20.0 \text{ MPa},$$

and the stress carried by the composite is obtained with the aid of the parallel coupling assumption (eqn 6.36):

$$\sigma_1 = 0.2 \times 20.0 + 0.8 \times 2.7 \times 10^3 \times 10^{-4} = 4.22 \text{ MPa}.$$

The interfacial shear stress τ_i is concentrated at the fibre ends (see Figure 6.23). With increase in strain, these are the sites where the interface first fails and **debonding** of fibres and matrix or **shear failure** of the matrix begins. This occurs when τ_i here reaches the **interfacial shear strength**. The magnitude of this strength is determined by three factors:

- the strength of the chemical bond between the fibres and the matrix;
- the friction between the fibres and the matrix, resulting from pressure exerted on the fibres by the matrix; and
- the shear strength of the matrix.

The pressure referred to here arises from differential thermal contraction during the cooling stage of manufacture, and from solidification shrinkage of the polymer (6.N.6).

With further straining, the debonding regions of the fibre slip within their hole in the matrix. Slippage is resisted by a constant friction stress τ_i^*, and the stress distributions switch from those shown in Figure 6.23 to the form shown in Figure 6.26. Since the fibre ends now carry a constant interfacial shear stress, the tensile stress here has a constant gradient (eqn 6.32): stress builds up linearly from each end.

To a good approximation, only the stresses shown shaded in Figure 6.26 need be considered. The length δ of the debond regions can then be found by applying eqn 6.32 to either one of them and rearranging:

$$\delta = \frac{E_f \varepsilon_1 d}{4\tau_i^*}. \tag{6.41}$$

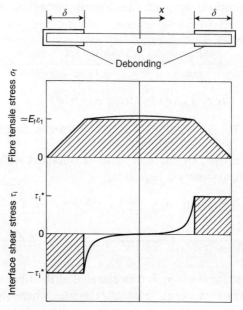

6.26 Schematic sketch of distributions of fibre tensile stress and interfacial shear stress after fibre–matrix debonding has commenced.

It is clear that, as straining proceeds, there is progressive growth of the debond regions: δ increases with ε_1. When the fibre eventually breaks at a strain $\varepsilon_1 = \varepsilon_f^* = \sigma_f^*/E_f$, the debond regions have lengthened to

$$\delta^* = \frac{\sigma_f^*}{4\tau_i^*}\, d. \tag{6.42}$$

However, this point can be reached **only if fibres are longer than a critical length l_c**

$$l_c = 2\delta^* = \frac{\sigma_f^*}{2\tau_i^*}\, d. \tag{6.43}$$

If the fibres are shorter than l_c, failure of the composite occurs after the debond regions have extended along the full length of the fibre. The entire fibre is then merely slipping within its hole in the matrix: it cannot be raised to its failure stress and so cannot be broken. When failure of the matrix finally occurs, the fibre simply slides out of the matrix, leaving behind an empty hole. In practice, then, l_c is a parameter of vital importance. Strenuous efforts are usually made to ensure that $l > l_c$ (6.N.8).

The axial tensile strength of the composite can be calculated by applying eqn 6.36 again. Take the most common case, in which the matrix is more ductile than the fibres: $\varepsilon_m^* > \varepsilon_f^*$ (see Figure 6.15(a)). As with continuous fibres, the sequence of events depends upon the fibre volume fraction ϕ_f.

At low ϕ_f, the strength actually **exceeds** that derived for continuous fibres (eqn 6.26(a)). The reason for this is that those fibres located with their ends within a distance δ^* of the failure plane do not fracture. Instead, they simply pull out of the matrix, and this is resisted by the friction stress τ_i^*. The proportion of fibres in this category is l_c/l, and the mean stress resisting pull-out is $\sigma_f^*/2$. The fibres therefore contribute a mean stress $\bar{\sigma}_f = \sigma_f^* l_c/2l$, and from eqn 6.36 the strength becomes

$$\sigma_1^* = \frac{l_c}{2l} \phi_f \sigma_f^* + (1 - \phi_f)\sigma_m^*. \tag{6.44}$$

At high ϕ_f, the strength is **reduced** by the shortness of the fibres. The mean fibre stress at failure is obtained from the trapezium-shaped distribution of tensile stress sketched in Figure 6.26, with $\varepsilon_1 = \sigma_f^*/E_f$ and $\delta = \delta^* = l_c/2$. It can readily be seen that

$$\bar{\sigma}_f = \sigma_f^*(1 - l_c/2l). \tag{6.45}$$

As with continuous fibres, the stress in the matrix is now σ_m'. Equation 6.36 then gives

$$\sigma_1^* = (1 - l_c/2l)\phi_f \sigma_f^* + (1 - \phi_f)\sigma_m'. \tag{6.46}$$

Equations 6.44 and 6.46 are shown plotted schematically in Figure 6.27 as dashed lines, with the greater of the two (corresponding to the predicted strength) shown as a full line. For comparison, the case of continuous

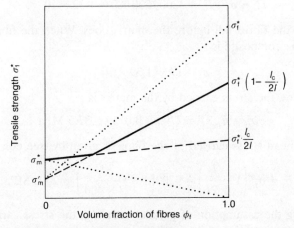

6.27 Schematic sketch graph of tensile strength σ_1^* versus fibre volume fraction ϕ_f for an aligned fibre composite containing **discontinuous** brittle fibres in a ductile matrix (dotted lines show the case of continuous fibres, for comparison).

fibres is shown dotted. What can be clearly seen is that reducing fibre length has two major effects on failure of the composite:

- strength is reduced (except at very low ϕ_f);
- there is a wider range of ϕ_f over which failure is matrix-controlled.

Example 6.10

The injection-moulded bar described in Example 6.9 is tested to failure in tension parallel to the axis of the bar. Predict the mean fibre stress when the bar fails, and the tensile strength of the bar. Take the tensile strengths of the carbon fibres and the nylon to be 3200 MPa and 70 MPa, respectively, and assume the shear strength of the carbon fibre–nylon interface to be 32 MPa.

Solution

We calculate the critical length l_c of these carbon fibres in a matrix of nylon. Substituting into eqn 6.43 we obtain

$$l_c = \frac{3200}{2 \times 32} \times 6 = 300 \ \mu m.$$

Therefore, in this moulded bar, the fibre length exceeds l_c. The carbon fibre–nylon case corresponds to brittle fibres in a more ductile matrix:

$$\varepsilon_m^* = \sigma_m^*/E_m = 70/2700 = 2.59 \times 10^{-2},$$

and

$$\varepsilon_f^* = \sigma_f^*/E_f = 3200/230\,000 = 1.39 \times 10^{-2}.$$

Therefore, the fibres fail before the matrix does. When the fibres fail the strain in the composite is

$$\varepsilon_1 = \varepsilon_f^* = 1.39 \times 10^{-2}.$$

At this point, the stress carried by the matrix is

$$\sigma_m' = E_m \varepsilon_f^* = 2700 \times 0.0139 = 37.5 \text{ MPa},$$

while the mean stress carried by the fibres is given by (see eqn 6.45)

$$\bar{\sigma}_f = \sigma_f^*\left(1 - \frac{l_c}{2l}\right) = 3200\left(1 - \frac{300}{2 \times 400}\right) = 2000 \text{ MPa}.$$

By invoking the assumption of parallel coupling, the stress carried by the composite at this point is obtained from

$$\sigma_1 = 0.2 \times 2000 + 0.8 \times 37.5 = 430 \text{ MPa}.$$

Whether or not this in fact corresponds to failure of the composite depends on whether a higher stress can be supported by the combination of the matrix and the slipping and broken fibres. The maximum stress which can be carried by the matrix and the slipping fibre ends is given by eqn 6.44:

$$\sigma_1 = \frac{300}{2 \times 400} \times 0.2 \times 3200 + 0.8 \times 70 = 296 \text{ MPa}.$$

We conclude that the tensile strength of the bar equals $\sigma_1^* = 430$ MPa, and that the mean fibre stress when this is reached equals 2000 MPa.

6.5 Reinforced rubbers

Polymers used in engineering in the rubbery state are often highly reinforced. Certain components—conveyor belts, tyres, inflatable boats—rely heavily for their strength on fibre reinforcement of the rubber. The mechanisms of reinforcement are then as described in Section 6.4.

The most widely used reinforcement for rubbers, however, is not a fibre but the black powder known as 'carbon black'. This is produced when hydrocarbons are burnt in an atmosphere deficient in oxygen. It consists of fine, spherical particles of carbon 0.01–1 μm in diameter. The best reinforcement is obtained when the particles are fused together into chain-like aggregates: such blacks are said to have 'structure' (a sketch is given in Figure 6.28). Within each particle the carbon atom arrangement is that of highly disordered, polycrystalline graphite similar to that in carbon fibres (see Figure 6.7), but lacking any overall alignment of layer planes.

The reinforcement is successful because rubbery polymer molecules bond well to carbon black particles through two mechanisms:

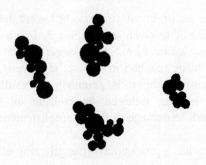

0.1 μm

6.28 Schematic sketch of the chain-like aggregates of carbon particles in carbon black.

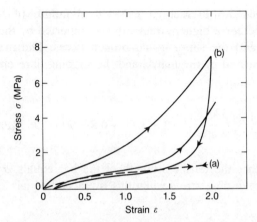

6.29 Tensile stress–strain curves for natural rubber: (a) without carbon black; and (b) with carbon black reinforcement (50 parts per 100 of rubber).

- adsorption of segments of polymer molecules at the disordered and somewhat porous surfaces of the particles; and
- chemical bonding between the polymer and residual hydrogen and oxygen-based chemical groups at the surfaces of the particles.

The dominant effects of carbon black on the tensile stress–strain curve of natural rubber can be seen in Figure 6.29. Loading/unloading curves are compared for unloading from a strain of 2. The **stiffening** action of carbon black is clearly in evidence. In addition, however, it accentuates two undesirable features:

- mechanical hysteresis; and
- strain softening (on **re**-loading, the new stress–strain curve lies below the initial curve).

These two features occur in all rubbers, but were ignored in the ideal theory of rubber elasticity given in Chapter 3. In **un**-reinforced rubbers they do not become significant except at larger strains (in the range 4–6) which are unlikely to be reached in service. Hysteresis, in a crystallizable rubber such as natural rubber, **is primarily a result of stress-induced crystallization on loading followed by melting on unloading.** Strain-softening corresponds to damage to the internal structure of the rubber on first loading.

Carbon black also has a pronounced strengthening effect. It is especially effective in those rubbers which do not crystallize on stretching (e.g. SBR, see Chapter 3), and which therefore do not benefit from the reinforcing action of crystals. Carbon black can increase their strengths by up to ×10 from an initial low value of about 2 MPa to about 20 MPa. Crystallizable rubbers, even unreinforced, normally have strengths in the range 10–30 MPa. These can be increased by a factor of up to ×1.5 by the addition of carbon black. Further advantages of carbon black are increases in abrasion resistance (very important in tyres) and fatigue resistance. In addition, its black coloration shows that it is an excellent absorber of light. It therefore absorbs most of the ultaviolet component of sunlight, which can otherwise initiate oxidative degradation of the rubber.

The main disadvantage of carbon black as a reinforcing additive in rubber is the greater mechanical hysteresis it introduces. Under high-speed oscillatory loading (e.g. in a rapidly revolving car tyre) excessive hysteresis leads to high heat generation. In practice, the volume fraction of carbon black is chosen so as to achieve a desirable compromise.

Notes for Chapter 6

6.N.1

Thermoset polyesters are, like thermoplastic polyesters (see Chapter 1), polymers resulting from condensation reactions between glycols of general formula

$$HO - R_g - OH, \qquad\qquad (XXV)$$

and dicarboxylic acids of general formula

$$HO - \overset{\overset{\displaystyle O}{\|}}{C} - R_a - \overset{\overset{\displaystyle O}{\|}}{C} - OH, \qquad\qquad (XXVI)$$

where R_g and R_a can represent numerous different chemical groups. A polyester monomer unit then has the form,

$$- \underset{\underset{\displaystyle O}{\|}}{C} - R_a - \underset{\underset{\displaystyle O}{\|}}{C} - O - R_g - O -. \qquad\qquad (XXVII)$$

The built-in distinguishing feature of a **thermoset** polyester, however, is that it is **unsaturated**. That is, prior to cross-linking, there are some R_a groups present in which there is a carbon–carbon double bond ($C=C$ bond) contained in the molecular backbone. The mixture for polymerization contains (XXVII) plus monomers such as styrene (or other unsaturated non-polymeric molecules) in liquid form. When catalyst is added and heat is applied the double bonds open and an addition reaction leads to the formation of a highly cross-linked network, with polyester chains interconnected by many short segments of polystyrene. Since the resulting polymer has its T_g well above room temperature, it is a hard, glassy polymer.

6.N.2

Epoxy resins are a family of thermoset polymers of increasing importance in engineering. They are glassy at room temperature, and are popular as matrix materials for high-performance fibre-reinforced plastics. In addition they are valuable engineering adhesives, bonding well to metals. Their characteristic feature is the presence in the liquid precursor of the epoxide three-membered ring. Thus, one of the common forms is a diepoxide of structure

$$CH_2\!\!-\!\!CH\!\!-\!\!R\!\!-\!\!CH\!\!-\!\!CH_2, \qquad\qquad\text{(XXVIII)}$$

where R can have various forms, but is commonly

$$-CH_2\!\!-\!\!O\!\!-\!\!\bigcirc\!\!-\!\!\underset{\underset{CH_3}{|}}{\overset{\overset{CH_3}{|}}{C}}\!\!-\!\!\bigcirc\!\!-\!\!O\!\!-\!\!CH_2- \qquad\qquad\text{(XXIX)}$$

Active groups on the curing agent, or 'hardener', molecules open the reactive epoxide rings to produce crosslinks. For example, a diamine hardener

$$H_2N\!\!-\!\!R'\!\!-\!\!NH_2 \qquad\qquad\text{(XXX)}$$

(where R' can take various forms) provides four active hydrogen atoms, each capable of being a site for attachment of an epoxide molecule, e.g.

$$\text{(XXXI)}$$

6.N.3

The significance of specific stress is as follows. Suppose a tensile force F is to be carried by a bar of length l and cross-sectional area A (Figure 6.30). The mass W of the bar is $Al\rho$. Hence the mass required to carry the force can be written in terms of the stress σ,

$$W = Al\rho = \frac{Fl}{(\sigma/\rho)}. \qquad\qquad\text{(6.N.3.1)}$$

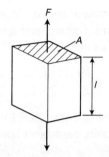

6.30 A bar to carry a force F and to be designed to have minimum weight.

If the bar is to be designed to have minimum mass (e.g. as part of an aircraft), the designer must seek a material which can support a high value of σ/ρ without failure.

6.N.4

In practice, there must be some transverse stress. Since Poisson's ratios of matrix and fibres normally differ, in the absence of transverse stress there would be incompatible transverse strains—fibres would no longer fit in their holes in the matrix. Often, Poisson's ratio of the fibres is less than that of the matrix, e.g. for glass fibres $\nu_f \approx 0.2$, and for most plastics $\nu_m = 0.3$–0.4. This has an advantageous result. Under tension parallel to the fibres, the matrix exerts a pressure on the fibres which assists in maintaining adhesion between them.

6.N.5

If tensile stresses σ_1 and σ_2 act parallel to axes 1 and 2, respectively, the in-plane strains are

$$\left.\begin{aligned} \varepsilon_1 &= \frac{\sigma_1}{E_1} - \nu_{21}\frac{\sigma_2}{E_2} \\ \varepsilon_2 &= -\nu_{12}\frac{\sigma_1}{E_1} + \frac{\sigma_2}{E_2} \end{aligned}\right\} \tag{6.N.5.1}$$

The strain energy (free energy of deformation) per unit volume is given by

$$A = \tfrac{1}{2}(\sigma_1\varepsilon_1 + \sigma_2\varepsilon_2), \tag{6.N.5.2}$$

and hence by substitution from above,

$$A = \frac{1}{2}\left[\frac{\sigma_1^2}{E_1} - \left(\frac{\nu_{21}}{E_2} + \frac{\nu_{12}}{E_1}\right)\sigma_1\sigma_2 + \frac{\sigma_2^2}{E_2}\right]. \tag{6.N.5.3}$$

Since the composite is assumed to be linearly elastic, A equals the complementary energy per unit volume, and strains are obtainable from A by differentiation. Thus, for example,

$$\varepsilon_1 = \frac{\partial A}{\partial \sigma_1} = \frac{\sigma_1}{E_1} - \frac{1}{2}\left(\frac{\nu_{21}}{E_2} + \frac{\nu_{12}}{E_1}\right)\sigma_2. \qquad (6.N.5.4)$$

Since eqn 6.N.5.4 and the first of eqns 6.N.5.1 must apply for any values of σ_1 and σ_2, the elastic constants must be related through

$$\frac{\nu_{21}}{E_2} = \frac{\nu_{12}}{E_1} \qquad (6.N.5.5)$$

6.N.6

The usual case is $\alpha_m > \alpha_f$. Therefore, **unrestricted** cooling from the forming temperature would result in more contraction of matrix than of fibres. Because in reality displacement of both must be equal at the interface, internal stresses are spontaneously generated on cooling, to expand the matrix and compress the fibres. The residual stress on the matrix is therefore tensile (see Figure 6.31(a)) and, if excessive, can cause premature failure of the matrix. The residual stress on the fibres is compressive (see Figure 6.31(b)), and provides a useful strengthening of

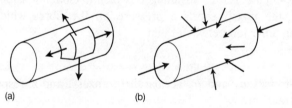

(a)   (b)

6.31 Residual thermal stresses after cooling from the forming temperature: (a) stresses on the matrix; and (b) stresses on the fibre.

the bond between fibre and matrix. In practice, these stresses are often enhanced by the matrix shrinking by a few per cent as it changes structure during solidification: cross-linking if it is a thermoset; crystallizing if it is a semi-crystalline thermoplastic.

6.N.7

The linear elastic stress analysis of a short fibre composite by the shear lag theory is due originally to H. L. Cox. The real composite (see Figure 6.32(a)) is assumed to deform in the vicinity of any particular fibre as if it were the model system shown in Figure 6.32(b). In the model, the fibre (of diameter d) is surrounded by a cylinder of matrix of radius R, embedded in a homogeneous block of material which deforms as the composite as a

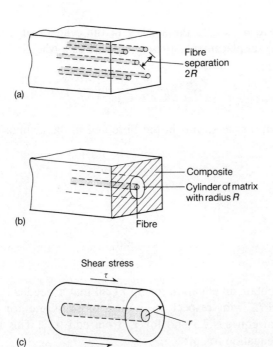

6.32 Fibre stress σ_f is evaluated with the shear lag theory. (a) The mean fibre separation is labelled $2R$. (b) The composite in the vicinity of the fibre is modelled as a cylinder of matrix with radius R surrounding the fibre, embedded in a block which deforms as the composite as a whole. (c) Within the cylinder of matrix, a shear stress τ acts at radius r.

whole. R is chosen to be half the mean centre-to-centre separation between nearest-neighbour fibres in the real composite (Figure 6.32(a)). Consider a cylinder of radius r cut concentrically from within the cylinder of matrix with radius R. The shear force per unit length carried by its surface is $2\pi r\tau$. This must be transmitted through the cylinder to the surface of the fibre without diminution. Hence the shear stress τ at radius r is given by

$$2\pi r\tau = \pi d\tau_i. \qquad (6.N.7.1)$$

Let u be the axial displacement at radius r and axial position x; then $\partial u/\partial r$ is the shear strain in the cylinder of matrix and must equal τ/G_m. Substituting for τ from eqn 6.N.7.1 yields

$$\frac{\partial u}{\partial r} = \frac{d}{2r}\frac{\tau_i}{G_m}.$$

This equation is integrated with respect to r from the fibre surface at

$r = d/2$, where $u = u_f$, to the surface of the cylinder of matrix at $r = R$, where $u = u_1$ (displacement of the composite as a whole):

$$u_1 - u_f = \frac{d}{2} \frac{\tau_i}{G_m} \ln\left(\frac{2R}{d}\right). \tag{6.N.7.2}$$

Now the axial tensile strains in the fibre and in the composite as a whole are given by

$$\varepsilon_f = \frac{\sigma_f}{E_f} = \frac{du_f}{dx},$$

and

$$\varepsilon_1 = \frac{du_1}{dx}. \tag{6.N.7.3}$$

In order to obtain an equation for $\sigma_f(x)$, all that is needed is to differentiate eqn 6.N.7.2 with respect to x, and to substiiute for du_1/dx and du_f/dx from eqn 6.N.7.3 and for τ_i from eqn 6.32. This results in the differential equation for σ_f:

$$\left(\frac{d}{2n}\right)^2 \frac{d^2\sigma_f}{dx^2} - \sigma_f = -E_f \varepsilon_1, \tag{6.N.7.4}$$

where

$$n = \sqrt{\left[\frac{2G_m}{E_f \ln(2R/d)}\right]}.$$

Equation 6.N.7.4 has a solution of standard form

$$\sigma_f = E_f \varepsilon_1 + C \sinh\left(\frac{2nx}{d}\right) + D \cosh\left(\frac{2nx}{d}\right),$$

where C and D are constants of integration. The assumed boundary conditions are that no axial tensile stress is transmitted to the fibre across its ends, i.e. $\sigma_f = 0$ at $x = -l/2$ and $x = l/2$. These conditions require that

$$C = 0,$$

and

$$D = \frac{-E_f \varepsilon_1}{\cosh\left(\dfrac{nl}{d}\right)}.$$

Finally, by recalling that the fibre aspect ratio a is l/d, we obtain the solution for σ_f:

$$\sigma_f = E_f \varepsilon_1 \left[1 - \frac{\cosh\left(na \dfrac{2x}{l} \right)}{\cosh(na)} \right].$$

6.N.8

In the great majority of cases, where debonding is to be minimized in order to maximize tensile strength, it is essential that $l > l_c$. In a minority of applications, however, composites are required to absorb as much **energy** as possible in failing: that is, to have a high toughness (see Chapter 5). In such cases, energy absorption due to debonding and fibre pull-out are exploited, by using short fibres and weak interfaces so that $l \lesssim l_c$.

Problems for Chapter 6

Use data from Table 6.3.

Table 6.3 Information for Chapter 6 problems

	Density ρ (kg m^{-3})	Tensile modulus E (GPa)	Poisson's ratio ν	Tensile strength σ^* (MPa)	Coefficient of linear thermal exp. α (10^{-6} K^{-1})
E-glass fibre	2540	76	0.22	1800	5
Carbon fibre (high modulus)	1860	340 (7)	0.35	2500	-1.2 (27)
Carbon fibre (high strength)	1790	230 (13)	0.25	3200	-1.2 (27)
Kevlar fibre	1450	124 (5)	0.35	2800	-2 (59)
Nylon 6.6	1140	2.7	0.33	70	90
Epoxy	1300	2.4	0.34	60	60
Thermoset polyester	1280	3.0	0.38	55	75

(Note: Where the fibres are anisotropic, values given refer to the fibre axis direction, except values in parenthesis which refer to the radial direction. The data all refer to 20°C, and E, ν, and σ^* were obtained in tensile tests at constant strain-rates of approximately 10^{-1} min^{-1}.)

6.1   Glass particles for reinforcement are available in various forms. Rank the following in order of their expected efficiencies as reinforcing particles. (Hint: consider the area/volume ratio as a measure of efficiency.)

(1) Glass flakes: diameter = 1 mm, thickness = 4 μm.
(2) Glass fibres: diameter = 12 μm, length = 3 mm.
(3) Glass fibres: diameter = 7 μm, length = 1 mm.
(4) Solid glass spheres: diameter = 44 μm.
(5) Hollow glass spheres: diameter = 75 μm.

6.2   Show that the theoretical maximum volume fraction for identical continuous fibres with circular cross-section in a composite material is 0.91.

6.3   A certain materials supplier sells grades of nylon 6.6 reinforced with the following mass fractions of high-strength carbon fibre: 20%, 30%, and 40%. Find the corresponding fibre volume fractions.

6.4   The supplier referred to in Problem 6.3 also sells a blend containing the following (percentages given are mass fractions): nylon 6.6 (55%), PTFE (15%) (see 2.N.1), and high strength carbon fibre (30%). Find its density and the fibre volume fraction. This blend excels in components where there is sliding contact between parts (e.g. a stationary bush on a rotating shaft). Explain this. (Take the density of PTFE to be 2200 kg m^{-3}.)

6.5   Pultruded rods, reinforced with aligned continuous fibres, are produced as follows (percentages given are fibre volume fractions):

(1) 50% E-glass fibres in epoxy;
(2) 60% carbon fibres (high modulus form) in epoxy; and
(3) 60% Kevlar fibres in epoxy.

Predict the axial tensile modulus and specific modulus for (1)–(3).

6.6   Estimate the tensile moduli of the rods referred to in Problem 6.5, in the radial direction.

6.7   For a composite material reinforced with aligned continuous fibres, show that Poisson's ratio ν_{12} may be estimated by applying the rule of mixtures:

$$\nu_{12} = \phi_f \nu_f + (1 - \phi_f)\nu_m.$$

Assume that the fibres are isotropic. (Hint: use assumptions of parallel and series coupling for strains in axial and transverse directions, respectively.)

6.8   An E-glass fibre–thermoset polyester composite tube is carefully constructed with the objective of achieving maximum axial tensile modulus. The glass fibres are continuous and highly aligned parallel to the tube axis. The tube contains 70% by volume glass, and has dimensions: length 1 m, diameter 50 mm, and wall thickness 1 mm. Predict the following deformations:

(1) the changes in length and diameter when the tube carries an axial tensile load of 10 kN;

(2) the change in diameter when the tube is internally pressurized (with closed ends) to a pressure of 0.4 MPa; and

(3) the relative rotation of the tube ends when it is subjected to an axial torque of 30 N m.

6.9 A sheet of fibre reinforced plastic consists of E-glass chopped-strand mat (30% mass fraction) in a matrix of thermoset polyester. The glass fibres are randomly arranged in the plane of the sheet. Estimate the in-plane tensile modulus of the sheet. (In CSM the fibres are sufficiently long—e.g. 10 to 50 mm—that they may be considered to be continuous as far as elastic stress transfer is concerned.)

6.10 An E-glass fibre/epoxy thin-walled tube, of length 2 m and diameter 100 mm, has a fibre volume fraction of 60%. The fibres are continuous and parallel, and are wound spirally around the tube at 45° to the tube axis. A tensile stress of 50 MPa is applied parallel to the tube axis. Predict the strains ε_1, ε_2, and γ_{12} which result in the wall of the tube, referred to axes parallel and perpendicular to the **fibres**. (Hint: first find the elastic constants of tube wall with respect to these axes.)

6.11 Continue the solution to Problem 6.10, to predict (a) the change in length, (b) the change in diameter, and (c) the relative rotation of the ends of the tube, which result from the action of the stress.

6.12 A sheet of composite material consists of aligned continuous fibres in an isotropic matrix. A tensile specimen is cut from the sheet at angle θ to the fibres. Show that the tensile modulus E of the specimen will vary with θ as follows:

$$\frac{1}{E} = \frac{\cos^4\theta}{E_1} + \frac{\sin^4\theta}{E_2} + \sin^2\theta\cos^2\theta\left(\frac{1}{G_{12}} - \frac{2\nu_{12}}{E_1}\right).$$

(Hint: remember that $1/E = \partial\varepsilon/\partial\sigma$, where ε and σ are tensile strain and stress in the direction to which E refers.)

6.13 Show that the coefficients of linear thermal expansion for a composite material reinforced with aligned, continuous isotropic fibres may be estimated from

$$\alpha_1 = \frac{\phi_f E_f \alpha_f + (1 - \phi_f)E_m \alpha_m}{\phi_f E_f + (1 - \phi_f)E_m},$$

and

$$\alpha_2 = \phi_f \alpha_f(1 + \nu_f) + (1 - \phi_f)\alpha_m(1 + \nu_m) - \alpha_1\nu_{12}.$$

(Hint: use assumptions of parallel and series coupling between fibres and matrix, as in Problem 6.7.)

6.14 For the following composite materials, consisting of aligned, continuous fibres in a polymer matrix, estimate the axial and transverse coefficients of linear thermal expansion (percentages given are fibre volume fractions)

(1) 50% E-glass fibres in thermoset polyester.
(2) 60% carbon fibres (high strength form) in epoxy.
(3) 65% Kevlar fibres in epoxy.

6.15 For the following fibre composite systems, find the fibre mass fraction at which the axial coefficient of linear thermal expansion becomes zero (assume fibres to be aligned and continuous):

(1) carbon fibres (high modulus form) in epoxy; and
(2) Kevlar fibres in epoxy.

6.16 For the spirally wound fibre composite tube of Problem 6.10, estimate the coefficients of linear thermal expansion parallel to and perpendicular to the fibres. If the temperature of the tube were raised from 20°C to 50°C, show that it would spontaneously twist, and predict the relative rotation of the ends of the tube. Also predict the changes in length and diameter.

6.17 For each of the pultruded rods described in Problem 6.5, predict the axial tensile strength.

6.18 For a composite material consisting of aligned, continuous, ductile fibres in a relatively more brittle matrix, show that the axial tensile strength can be estimated from the greater of the following two predictions:

$$\sigma_1^* = \phi_f \sigma_f' + (1 - \phi_f)\sigma_m^*,$$

and

$$\sigma_1^* = \phi_f \sigma_f^*,$$

where σ_f' is the axial tensile stress carried by fibres, at the failure strain of the matrix.

6.19 For unidirectional fibre composites with each of the following compositions, predict the strengths and describe the failure processes under axial tensile stress.

(1) 10% E-glass fibre in thermoset polyester.
(2) 50% E-glass fibre in thermoset polyester.

(Percentages are fibre volume fractions.)

6.20 The spirally wound E-glass–epoxy tube of Problem 6.10 is loaded to failure in axial tension. Predict the failure stress and describe the mode of failure. (The wall of the tube may be taken to have a tensile strength perpendicular to the fibres $\sigma_2^* = 40$ MPa and a shear strength $\tau_{12}^* = 60$ MPa.)

6.21   A composite sheet consists of 50% (by volume) aligned continuous carbon fibres (high-strength form) in an epoxy matrix. In use, the sheet is to be subjected to tension at various angles θ to the fibre direction. Describe how the mode of failure will vary with θ, and predict the ranges of θ over which each failure mode will be observed. (Take $\sigma_2^* = 40$ MPa and $\tau_{12}^* = 70$ MPa.)

6.22   For each of the following short-fibre composite materials, tested in tension parallel to the fibres, predict the tensile modulus. (Percentages given are fibre volume fractions.)

   (1) 30% aligned E-glass fibres in a matrix of nylon 6.6 (interface shear strength $\tau_i^* = 20$ MPa, fibre diameter $d = 12$ μm, and length $l = 1$ mm).

   (2) 40% aligned E-glass fibres in a matrix of epoxy ($\tau_i^* = 10$ MPa, $d = 12$ μm, and $l = 2$ mm).

6.23   For each of the composites described in Problem 6.22, predict the tensile stress on the composite at which fibre–matrix debonding will begin.

6.24   For each of the composites described in Problem 6.22 predict the tensile stress at which failure will occur and give further details of the anticipated process of failure.

7 Forming

7.1 Introduction

Of the many factors which have caused the rapidly increasing use of polymers in engineering, two are dominant. First, of course, there are the interesting intrinsic properties of polymers such as lightness, cheapness, toughness, and corrosion resistance, etc. Second, and of no less importance, are the simplicity, versatility, and rapidity of the shaping operations by which the raw polymer is transformed into useful objects.

For plastics the dominant forming operations are **injection moulding** and **extrusion**. They both follow the same sequence of steps, characteristic of polymer melt forming processes:

- heating the polymer into the molten state;
- pumping the melt to the forming unit;
- forming the melt into the required shape;
- cooling and solidification.

The simplest technique is extrusion. An extruder (Figure 7.1) is essentially a large, hot motor-driven 'meat grinder'. The cold polymer granules are fed into the hopper; a gravity feed supplies granules to the screw which rotates within the heated barrel. Granules are advanced along the barrel between the flights of the screw and the hot wall of the barrel. As the

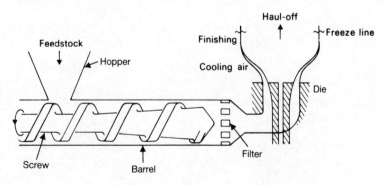

7.1 Extruder and die in a film blowing unit (after Cogswell).

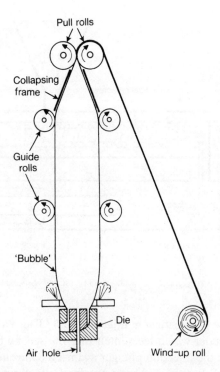

7.2 Die, 'bubble', and take-off equipment in a film blowing unit.

polymer moves along the extruder it is turned into a liquid: in the crystalline polymers by melting; and in amorphous polymers by passing up through the glass transition. The liquid, and now pressurized, polymer then moves from the extruder into a die, the first part of the forming system. Shown here (in Figures 7.1 and 7.2) is a system for producing polyethylene film. The liquid flow is turned through a right angle (so that it is moving vertically), separated into an annulus, and then forced upwards through a thin circular slit, finally emerging into the atmosphere as a thin-walled continuous tube. The tube is then, and very rapidly

- hauled upwards by the pull rolls,
- expanded by internal pressure (inflated into a sausage-shaped bubble), and at the same time
- acted upon by cooling air jets which cause it to solidify at a 'freeze-line,' some tens of centimetres above the shaper lips.

After these operations the bubble is collapsed and the film is rolled on to a wind-up roll. The stability of the bubble is a matter of great importance to the film fabricator.

The operation of an injection moulding machine is illustrated in Figure

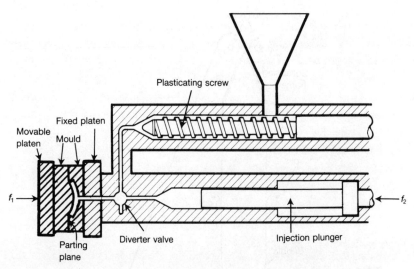

7.3 An injection moulding machine with mould; screw preplasticator injection unit (after ICI plc).

7.3: this is a **screw preplasticator injection unit**. The plasticating screw is essentially an extruder which feeds melt into the second barrel. As it does this it pushes back the injection plunger against a controlled back pressure. When the required shot size has been obtained, the diverter valve connects the injection barrel to the mould and disconnects it from the plasticating barrel. The injection plunger then rams forward, injecting the liquid into the mould. Because the polymer liquid is highly viscous, a large force f_2 is required to move the ram rapidly, thus forcing the liquid through the many channels in the mould. An even larger force f_1 is required to hold the mould closed during injection. The diverter valve is then turned again, thus isolating the mould, and the plasticating screw commences to refill the injection barrel. Whilst this continues, the shot injected into the mould cools; the movable platen is then retracted to the left, thus opening the mould; the moulding is then removed and the mould closed again preparatory to the next shot. The plasticating barrel can either be parallel to the injection barrel (as in Figure 7.3) or at an angle. The cycle time depends on various factors (see Section 7.5), but is normally in the range of several seconds (for the smallest mouldings) to several minutes. The alternative (and more common) machine—the reciprocating screw—is described in Section 7.5.

The dominant properties of the polymer which are of interest in these operations are summarized in the following questions:

• What are the shear viscosity of the liquid polymer and its temperature and rate dependence?

- What are the elongational viscosity and elongational strength of the liquid polymer? Note that in the polyethylene film unit the tubular film exists as a liquid as it emerges from the shaping unit. **It must here sustain tensile stresses** and the questions of interest concern the liquid response to these tensile stresses.[1]

- If the liquid film, with free surfaces, can sustain tensile stresses, it then follows that tensile (as well as shear) stresses may be generated in polymer liquids flowing in closed vessels. What effect will this have on polymer flow?

- What factors control the rate of cooling? This is often the rate-determining step in manufacture of polymer products.

- When the liquid polymer solidifies in the mould or in the film, it does so under unusual conditions. That is, the liquid has been stressed by shear and by tensile stresses which may not relax to zero. What effect does crystallization under these conditions have on the microstructure and properties of the product?

Before examining these vital points in order, we make one comment of general validity. The film process produces not merely a film, but one with specific properties. The haul-off and inflation of the bubble and the rapid and simultaneous crystallization produce the correct shape (a film of the designed thickness) **combined with a property-enhancing microstructural texture**. This is true generally: the technology of polymer processing involves the handling of melts so as to produce the correct shape and, in addition, the most advantageous texture. An incorrect procedure can produce the wrong shape or a texture which weakens the product, or in other ways lowers it value.

7.2 The flow properties of polymer melts

We consider three types of flow: elongational flow, simple shearing flow, and first, but quite briefly, the behaviour of melts under hydrostatic pressure. This latter behaviour is particularly significant in injection moulding in which liquid flows may be generated by hydrostatic pressures $\sim 10^3$ atmospheres (1 atmosphere $= 10^5$ Pa).

7.2.1 Bulk deformation

The bulk modulus K of polymer liquids is normal in the sense that:

---

1 Note that normal liquids cannot sustain tensile stresses, so this question cannot arise for them. It is for this reason that the reader may not have considered the question of tensile stress in liquids before.

- it has a value of order 1 GPa (typical of liquids);
- it is **not** dependent on relative molecular mass; and
- it is not particularly viscoelastic (the response of an applied stress is 95% complete within 1 second).

Pressures up to $\sim 10^3$ atmospheres are encountered in injection moulding, and this leads to a fractional volume decrease of

$$\frac{\Delta V}{V} = \frac{P}{K} = \frac{10^3 \times 1.01 \times 10^5}{10^9} \tag{7.1}$$
$$= 0.101.$$

This is to say a 10% change in volume. This volume decrease raises the shear viscosity, as described below. Note also that if a mould is filled at 1000 atmospheres it will contain $\sim 10\%$ more polymer than if it were filled at 1 atmosphere; this fact plays a central role in injection moulding.

7.2.2 Elongational flow

This type of flow is of much greater importance in polymer liquids than in normal liquids. It is named variously: elongational, tensile, stretching, or free-surface flow. Elongational flow is important not only in film processes but also in fibre forming as well as in blow moulding and vacuum forming. Consider a fibre of liquid being pulled continuously and at a constant temperature out of a long tube, as indicated in Figure 7.4. The force F which generates the flow could be applied, for instance, by haul-off rolls. The volume rate of flow Q (m^3 s^{-1}) past a plane is a constant. Because the cross-section decreases, the tensile stress increases with increasing z (see Figure 7.4). Consider a small element of liquid of length l at z below the outlet. At this point it moves downwards with velocity v; a time dt later on, it has moved to $z + dz$:

$$dz = v \, dt.$$

During its passage from z to $z + dz$, its cross-section changes from A to $A + dA$ and its length from l to $l + dl$. Since its volume remains constant, it follows that

$$Al = (A + dA)(l + dl), \tag{7.2}$$

$$\therefore \qquad \frac{dA}{A} = -\frac{dl}{l}. \tag{7.3}$$

The volume rate of flow Q is independent of z:

$$Q = vA = \text{constant}, \tag{7.4}$$

and therefore

$$\frac{dA}{A} = -\frac{dv}{v}. \tag{7.5}$$

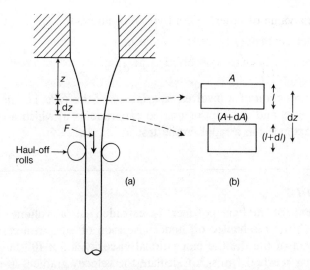

7.4 (a) A fiber of liquid polymer is pulled continuously from a reservoir by a force F. (b) At z below the outlet a small element of length l and area A; in time dt it moves to $z + dz$, where $dz = v\,dt$; its volume remains constant but the dimensions change to $l + dl$ and $A + dA$.

The change in logarithmic strain as the element moves from z to $z + dz$ is

$$d\varepsilon_t = \frac{dl}{l} = \frac{dv}{v}, \tag{7.6}$$

from eqns 7.3 to 7.5, so that

$$\frac{d\varepsilon_t}{dt} = \frac{1}{dt}\left(\frac{dv}{v}\right). \tag{7.7}$$

But since $dt = dz/v$, it follows that

$$\frac{d\varepsilon_t}{dt} = \frac{dv}{dz}. \tag{7.8}$$

Now, according to Trouton's definition of the elongational viscosity λ, a liquid instantaneously elongating at $d\varepsilon_t/dt$ generates a tensile stress σ_t $(= F/A)$,

$$\sigma_t = \lambda(d\varepsilon_t/dt).$$

It follows that

$$\lambda = \frac{F/A}{(d\varepsilon_t/dt)}$$

$$= \frac{F/A}{dv/dz}. \tag{7.9}$$

From eqn 7.5,

$$\lambda = -\frac{F/v}{\mathrm{d}A/\mathrm{d}z}. \tag{7.10}$$

The determination of λ involves experiments of this type, i.e. the measurement of F, v, and $\mathrm{d}A/\mathrm{d}z$, or related experiments in which a cylinder of liquid is extended in a rapid tensile test.

Example 7.1

A filament of molten polymer is extruded at a volume flow rate 10^{-6} m³ s⁻¹ and is hauled off under a tension of 3 N. At a certain point downstream of the die the elongational viscosity is 5×10^5 Pa s and the velocity has reached 1 m s⁻¹. Calculate the velocity gradient at this point.

Solution

The velocity gradient $\mathrm{d}v/\mathrm{d}z$ may be identified with the true strain rate $\mathrm{d}\varepsilon_t/\mathrm{d}t$ (eqn 7.8), which in turn may be found from the elongational viscosity λ and the true tensile stress σ_t acting on the melt. The stress is given by

$$\sigma_t = \frac{F}{A} = \frac{Fv}{Q} = \lambda \frac{\mathrm{d}\varepsilon_t}{\mathrm{d}t}.$$

At any point along the extrudate, therefore, the following equation applies:

$$\frac{Fv}{Q} = \lambda \frac{\mathrm{d}v}{\mathrm{d}z}.$$

Rearranging, the velocity gradient is found as follows:

$$\frac{\mathrm{d}v}{\mathrm{d}z} = \frac{1}{\lambda}\frac{Fv}{Q} = \frac{1}{5 \times 10^5} \times \frac{3 \times 1}{10^{-6}} = 6 \, \text{s}^{-1}.$$

Measurements of λ are shown in Figure 7.5 for two polymers: linear and branched polyethylene. It will be seen that λ depends on stress for both polymers:[1] **the elongational viscosity is non-Newtonian.** At low stresses λ

---

1 Relative molecular mass controls the magnitude of λ, but has no effect on the general pattern of behaviour (the stress dependence of λ).

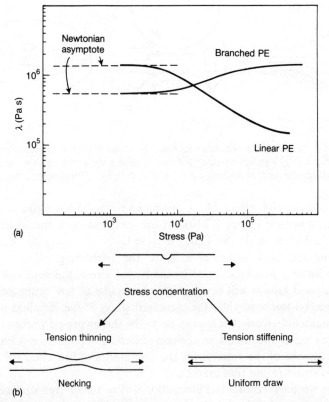

(a)

(b)

7.5   (a) The dependence of λ, the elongational viscosity, on tensile stress for branched and linear polyethylene. At stresses below 10^3 Pa, both polymers approach Newtonian behaviour (λ independent of σ). Non-Newtonian behaviour occurs above 10^3 Pa: branched PE tension-stiffens; linear PE tension-thins.

(b) Illustration of the way in which tension-stiffening liquids may be drawn without necking: the higher stress at the incipient neck (an adventitious small reduction in cross-section) is neutralized by λ increasing at that point owing to the stress increase (after Cogswell).

becomes more or less Newtonian, independent of stress, and approaches asymptotically a constant value for each polymer. At stresses over 10^3 Pa:

- λ increases with stress for branched PE; and
- λ decreases with stress for linear PE.

Branched PE is said to 'tension-stiffen' and linear PE to 'tension thin'. One way of rationalizing this effect, or at least committing it to memory, is to invoke the action of the side-branches in branched PE acting as 'hooks', increasing the resistance to flow when the molecules slide by each other as the liquid elongates.

The ability of a few materials to draw without necking can be examined

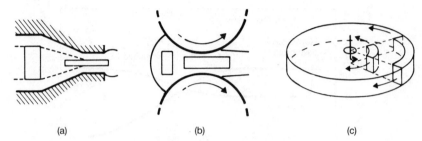

7.6 Tensile deformation in constrained flow: (a) tapering section of extrusion die; (b) calendering; and (c) injection moulding fed from a sprue gate; as the liquid expands radially into the mould, elongational stresses are set up in th direction of the tangent (after Cogswell).

by considering the effect of an incipient neck. This produces a stress concentration (see Figure 7.5(b)). At the incipient neck the higher stress imposes a higher ε_t. If the liquid tension-thins, the viscosity falls at this point and the neck is unstable, since the perturbing effect of stress-induced lower λ is compounded by the higher stress; the neck will remain localized, and failure will occur relatively rapidly at low extension. If the liquid tension-thickens, then the increased stress at the incipient neck can be counteracted—completely or in part—by the increased viscosity. In this way, drawing is facilitated in tension-stiffening liquids (see Chapter 5). This fact is one of the reasons for the dominance of branched PE over all other polymers in the film market.

So far we have considered elongational flow under free-surface conditions: this is intuitively the simplest case. Significant elongational effects occur in flowing polymer liquids whenever there is a change of cross-section (see Figure 7.6). Work must be done elongating the liquid as it passes from one cross-section to another. In Section 7.2.3 we will examine the shear-flow behaviour of polymer liquids, but it will be noted that elongational effects are normally present with shear effects, since rarely do flows occur without some change in section. **The pressure drop generating the combined flow thus drives both shear and elongational flow simultaneously.**

The presence of elongational flows is an important factor in limiting the speed of processing. The reason is that, in a similar manner to a solid, a polymer melt fractures if elongated under too high a stress. In extrusion, for example, this can occur either within the die (see Figure 7.6) or during draw-down of the extrudate (see above). It leads either to complete failure of the product or at least to an unacceptably distorted product, and must be prevented. Processing equipment is carefully designed with this in mind. The case of an extruder die is considered in 7.N.1.

7.2.3 Shear flow

The most important flow process in polymer liquids is shear flow. Polymer

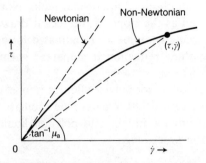

7.7 Observed dependence of shear stress τ on $\dot{\gamma}$ for a polymer liquid. The apparent viscosity at $(\tau, \dot{\gamma})$ is $\mu_a = \tau/\dot{\gamma}$; μ_a decreases with increasing τ.

liquids differ from simple liquids, first in that the shear viscosity is invariably extremely large, and second in that Newton's empirical equation giving a linear relationship between shear stress τ and shear strain rate $\dot{\gamma}$ with constant shear viscosity μ,

$$\tau = \mu\dot{\gamma}, \tag{7.11}$$

does not apply. The τ versus $\dot{\gamma}$ plot for a non-Newtonian polymeric liquid has the form indicated in Figure 7.7: as τ increases, the liquid appears to 'yield'. Of the several non-Newtonian characteristics of polymers this is the most important for polymer processing. It is known as **pseudoplasticity** or **shear thinning**. It is clearly an advantageous property. The opposite effect,

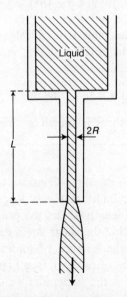

7.8 Capillary flow apparatus for determining μ_a. Note the swelling of the extruded polymer liquid as it leaves the capillary.

shear thickening, is not observed in polymer melts. Note that the apparent viscosity μ_a is defined as the slope of the secant drawn as indicated. **The conditions in processing equipment are adjusted to make full use of this large order-of-magnitude reduction in apparent shear viscosity with increasing stress.**

The determination of μ_a for polymer liquids is most commonly obtained by capillary flow (see Figure 7.8). We examine capillary flow for this reason and because it will provide insight into polymer liquid flow in the other cross-sections used in processing.

Example 7.2

A layer of molten poly(methyl methacrylate) at 190°C is of uniform thickness 3 mm, and is sandwiched between two flat, parallel plates. A shear stress of 100 kPa is applied to the melt. Find the relative sliding velocity of the plates, using data for apparent viscosity from Figure 7.13.

Solution

From the data of Figure 7.13, the apparent viscosity of poly(methyl methacrylate) at 190°C and at a shear stress of 100 kPa is 3.9×10^4 Pa s. The shear stress is uniform through the layer of melt and hence, even for this non-Newtonian fluid, the shear strain-rate $\dot{\gamma}$ will be uniform with a value

$$\dot{\gamma} = \tau/\mu = 10^5/(3.9 \times 10^4) = 2.56 \text{ s}^{-1}.$$

The shear strain-rate in the melt is identically equal to the transverse velocity gradient. Hence, the relative sliding velocity v of the plates is given by

$$v = 3 \times 2.56 = 7.68 \text{ mm s}^{-1}.$$

Flow of this type in simple shear is known as 'drag flow.'

7.2.3.1 Stress distribution in a capillary Consider a continuous and steady flow of liquid down a long cylindrical tube of length L and radius R. Over an infinitesimal segment of length dz, let the pressure change by dP (see Figure 7.9). We require to find the shear stress distribution along a radius of the capillary. Consider the balance of forces acting upon the liquid within the cylindrical surface of radius r (see Figure 7.9): the shear stress everywhere on this surface is the same, and we denote it $\tau(r)$. The total traction on the cylindrical surface is then $2\pi r \, dz \, \tau(r)$ and it acts in

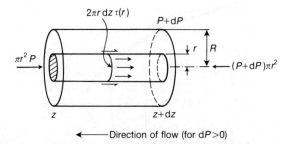

7.9   Laminar flow in an element of a capillary. The force $\pi r^2 \, \mathrm{d}P$ acting to push the cylinder of liquid of radius r in the direction of flow is balanced by the surface traction $2\pi r \, \mathrm{d}z \, \tau(r)$.

the direction indicated in Figure 7.9. Since the flow is steady (non-accelerating) this traction must be balanced by the net force $\pi r^2 \, \mathrm{d}P$ acting in the opposite direction. It follows therefore that

$$2\pi r \, \mathrm{d}z \, \tau(r) = \pi r^2 \, \mathrm{d}P,$$

$$\tau(r) = \frac{r}{2}\frac{\mathrm{d}P}{\mathrm{d}z}. \qquad (7.12)$$

Thus the shear stress is zero in the centre of the capillary and increases linearly with r, reaching a maximum value τ_w at the capillary wall

$$\tau_w = \frac{R}{2}\frac{\mathrm{d}P}{\mathrm{d}z}. \qquad (7.13)$$

This result is independent of the stress dependence of μ: that is, it holds both for Newtonian and non-Newtonian liquids. Since for given experimental conditions $(R, \mathrm{d}P/\mathrm{d}z)$ we know τ_w, it remains to find $\dot{\gamma}_w$ in order to determine μ_a. This is difficult for a non-Newtonian fluid, but is quite straightforward for a Newtonian fluid.

7.2.3.2 Newtonian flow in a circular pipe   We review the description of laminar flow in a circular pipe of radius R for a Newtonian fluid with constant viscosity μ flowing under an imposed negative pressure gradient $\mathrm{d}P/\mathrm{d}z = -\Delta P/\Delta L$.

- The dependence of the velocity of flow on r is parabolic:

$$v = \frac{1}{4\mu}\left(\frac{\Delta P}{\Delta L}\right)(R^2 - r^2). \qquad (7.14)$$

The cylindrical lamina in contact with the wall has $v = 0$: the maximum velocity occurs in the centre (see Figure 7.10 for the case $n = 1$).

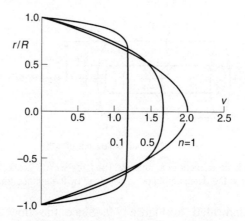

7.10 Velocity profiles for laminar flow of shear-thinning power-law fluids down a capillary of circular cross-section of radius R; velocity v is as calculated in Example 7.4, in units of $Q/\pi R^2$. The Newtonian case is the parabolic profile represented by $n = 1$.

- The liquid in the annulus between r and $r + \mathrm{d}r$ flowing at velocity v delivers a throughput per second

$$\mathrm{d}Q = \frac{\pi}{2\mu}\left(\frac{\Delta P}{\Delta L}\right)[R^2 r - r^3]\,\mathrm{d}r. \tag{7.15}$$

The throughput from the whole cross-section is

$$Q = \frac{\pi R^4}{8\mu}\left(\frac{\Delta P}{\Delta L}\right). \tag{7.16}$$

- From eqn 7.14,

$$\frac{\mathrm{d}v}{\mathrm{d}r} = -\frac{r}{2\mu}\left(\frac{\Delta P}{\Delta L}\right),$$

and since $\mathrm{d}v/\mathrm{d}r = \dot{\gamma}$,

$$\dot{\gamma}_\mathrm{w} = \frac{4Q}{\pi R^3}, \tag{7.17}$$

using eqn 7.16.

Example 7.3

A molten polymer flows through a cylindrical tube of diameter 2 mm at a volume flow rate 10^{-8} m^3 s^{-1}. Assuming that under these conditions the polymer is a Newtonian fluid with viscosity 10^2 Pa s, find the pressure drop per unit length along the tube and the shear stress at the wall of the tube.

Solution

From eqn 7.16,

$$\frac{\Delta P}{\Delta L} = \frac{8\mu}{\pi} \frac{Q}{R^4} = \frac{8}{\pi} \times 10^2 \times \frac{10^{-8}}{(10^{-3})^4} = 2.55 \times 10^6 \text{ Pa m}^{-1}.$$

The wall shear stress τ_w is obtained from $\Delta P/\Delta L$ by balancing tractions on the cylinder of melt moving through the tube (eqn 7.13):

$$\tau_w = \frac{R}{2} \frac{\mathrm{d}P}{\mathrm{d}z} = -\frac{R}{2} \frac{\Delta P}{\Delta L} = -\frac{10^{-3}}{2} \times 2.55 \times 10^6 = -1.27 \times 10^3 \text{ Pa}.$$

7.2.3.3 Non-Newtonian flow in a circular pipe   The particular $\dot{\gamma}$ established at radius r is determined by τ:

$$\dot{\gamma} = f(\tau).$$

This function is unknown; however, it is often of some use to assume the liquid may be represented by a 'power-law fluid':

$$\tau = K\dot{\gamma}^n, \tag{7.18}$$

in which K and n are constants and for shear-thinning polymer melts $n < 1$. By rearrangement, eqn 7.18 can be shown to lead to a straight-line graph of log μ_a versus log τ with negative slope of magnitude $n^{-1} - 1$ (an exercise for the reader). Data given below in Figures 7.13 and 7.14 will show that this is a reasonable approximation over only a limited range of shear stress: but it is used, nevertheless, as a convenient aid to calculation.

We saw that satisfaction of equilibrium, whatever the fluid, requires shear stress to vary linearly across the capillary channel (eqn 7.12). It follows that in a shear-thinning polymer melt, the velocity profile must deviate from the Newtonian parabolic law, eqn 7.14. High shear-rate becomes more concentrated near the capillary wall, leaving a wider region of low shear at the centre of the channel. This can be seen clearly in Figure 7.10. Velocity profiles are shown for power-law fluids with various values of the exponent n, as calculated using the expression for velocity derived in Example 7.4. In the extreme case of a highly shear-thinning fluid (very low n) there is 'plug flow': the melt passes down the capillary almost as a plug moving at uniform speed, as if it were sliding against the channel wall.

Example 7.4
A certain polymer melt behaves as a power-law fluid, conforming to eqn

7.18. It flows through a cylindrical tube of radius R. Derive an expression for the volume flow-rate Q which results from a pressure drop per unit length $\Delta P/\Delta L$.

Solution

Since the pressure gradient $dP/dz = -\Delta P/\Delta L$, equilibrium of forces acting on a cylinder of fluid of radius r requires (from eqn 7.12) that

$$\tau(r) = \frac{r}{2}\left(-\frac{\Delta P}{\Delta L}\right).$$

Substituting for τ into the power law (eqn 7.18), we obtain

$$\frac{dv}{dr} = \dot{\gamma} = \left(\frac{\tau}{K}\right)^{1/n} = -Cr^{1/n},$$

where

$$C = \left(\frac{1}{2K}\frac{\Delta P}{\Delta L}\right)^{1/n}.$$

Integrating yields the velocity profile

$$v = -\left(\frac{Cn}{1+n}\right)r^{(1+(1/n))} + A,$$

where the integration constant A is determined from the boundary condition: $v = 0$ when $r = R$. Hence

$$v = \frac{Cn}{1+n}[R^{(1+(1/n))} - r^{(1+(1/n))}].$$

From the velocity profile, the volume flow rate may be obtained

$$Q = 2\pi\int_0^R vr\,dr = \frac{n\pi C}{(3n+1)}R^{(3+(1/n))}$$

$$= \frac{n\pi}{(3n+1)}R^{(3+(1/n))}\left(\frac{1}{2K}\frac{\Delta P}{\Delta L}\right)^{1/n}.$$

Comment

This expression reduces to the Newtonian form (eqn 7.16) when $n = 1$, as expected.

If the function $f(\tau)$ is unknown and the Newtonian result (eqn 7.17) cannot be assumed, how may the wall shear-rate $\dot{\gamma}_w$ be found experimentally? Fortunately the unknown dependence of v on r controls both Q and $dQ/d\Delta P$, and it is possible to show that the wall shear rate is

$$\dot{\gamma}_w = \frac{1}{\pi R^3}\left[3Q + \Delta P\left(\frac{dQ}{d\Delta P}\right)\right]. \tag{7.19}$$

This is the Rabinowitsch equation. Thus, by determining Q and $dQ/d\Delta P$ as a function of ΔP, it is possible to determine $\dot{\gamma}_w$, the shear rate at the wall (eqn 7.19); τ_w, the shear stress at the wall is also known (from eqn 7.13). It is then possible to plot **wall shear stress** versus **wall shear rate** and to determine the apparent viscosity.

It is usual, in practice, to ignore the Rabinowitsch equation and to assume that the velocity profile for the polymer liquid is parabolic to within the required error. With this assumption the wall shear rate is taken to be given by the Newtonian value, so that, from eqns 7.13 and 7.17, the apparent viscosity is

$$\mu_a = \frac{\tau_w}{\dot{\gamma}_w}$$

$$= \left(\frac{\pi R^4}{8Q}\right)\left(\frac{\Delta P}{\Delta L}\right).$$

This is the value of μ_a observed at the particular wall shear stress τ_w. All the data shown in this chapter have been obtained in this way.

The apparent viscosity determined from eqns 7.13 and 7.17 is at most 15% greater than the apparent viscosity determined using the Rabinowitsch method (eqns 7.13 and 7.19). For practical purposes, this error (although a systematic error) is considered to be acceptable. This is a reasonable procedure in quality control when polymers are to be compared, and no absolute quantities are required.

There is a second systematic error in the capillary flow determination of μ_a which was first described by Bagley. As the polymer liquid converges from the barrel into the capillary, the converging flow requires an abrupt pressure drop due to the elongational flow viscosity λ. The drop in pressure along the barrel and the capillary is sketched in Figure 7.11. According to Bagley's correction, the pressure gradient in the capillary is

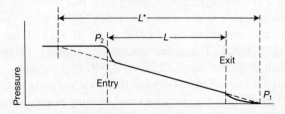

7.11 The variation of pressure along the flow length. An abrupt drop at the entry is due to the establishment of a converging flow at this point as the liquid is constrained to pass from the large diameter barrel into the much smaller capillary.

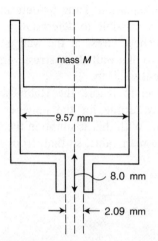

7.12   Illustration of a melt-flow indexer and the significant parameters.

not $(P_2 - P_1)/L$ but $(P_2 - P_1)/L^*$, in which the effective length L^* is determined by L and R:

$$L^* = L + mR.$$

The Bagley correction factor m has to be determined in experiments with capillaries of varying L/R. Its value falls in the range $1 < m < 15$.

7.2.3.4 Melt flow index   The melt flow index test is an extremely widely used, highly simplified variant of the capillary flow experiment. The rate of extrusion of a polymer liquid is determined through a given capillary in a closely defined apparatus (ASTM, D1238–62T, Figure 7.12). It consists, in outline, of a heated barrel of diameter 9.57 mm into which a capillary of diameter 2.09 mm and length 8.0 mm is fixed. Polymer granules are fed into the barrel and, after being heated to temperature T, forced through the capillary by a pressure induced by a piston of mass M. The flow is obtained by cutting off the length of extrudate which has flowed through the orifice in a certain time. For polyethylene, this flow rate, measured in grams per ten minutes, is the melt flow index (MFI) when the test is run according to the conditions $T = 190°C$ and $M = 2.160$ kg (yielding a pressure ~ 0.30 MPa). There are altogether another 12 specified test conditions with T in the range $125°C$ to $275°C$ and pressure in the range 0.045 MPa to 3.0 MPa. The reason for this wide range is to enable resins of widely different apparent viscosities and melting points to be indexed. It will be seen that the melt index is a 'single-point' test; it is nevertheless extremely widely used in quality control and product specification and exhibits good reproducibility (to within 3%).

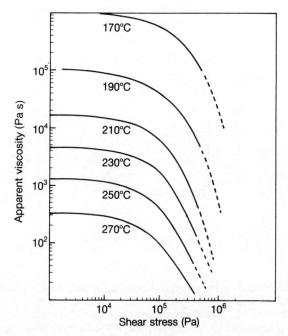

7.13 Stress dependence of the apparent viscosity μ_a at temperatures between 170°C and 270°C for a moulding grade of poly(methyl methacrylate). Measurements by capillary flow (after Cogswell).

7.2.3.5 Dependence of apparent viscosity on temperature and relative molecular mass The dependence of μ_a on stress for different temperatures between 170°C and 270°C is shown in Figure 7.13 for PMMA. The shear thinning characteristics of the curves are the same: that is, if the curves are shifted vertically (superposed at constant stress) they superpose. Note the large change in viscosity with temperature. As the temperature is lowered towards T_f ($\sim 100°C$) the low-stress temperature dependence of μ_a increases dramatically; this is a commonly observed effect in all glassy systems and is rationalized by the theory of Williams, Landel, and Ferry (7.N.2). For most liquids of very low relative molecular mass, and for polymers at temperatures more that 100 K above T_g, the Arrhenius equation (4.N.6) is a good fit for the temperature dependence of viscosity.

Above all other parameters, it is the relative molecular mass of a polymer which determines the apparent shear viscosity (7.N.3). Figure 7.14 shows the stress dependence at 170°C of four BPE polymers with relative molecular masses ranging from extremely high (MFI 0.2) to extremely low (MFI 200). Note that the shape of the curves (the shear thinning characteristics) is little changed by variations in relative molecular mass.

7.2.3.6 Viscoelasticity in polymer liquids Polymer liquids are invariably viscoelastic. The most important manifestation of viscoelasticity is die swell

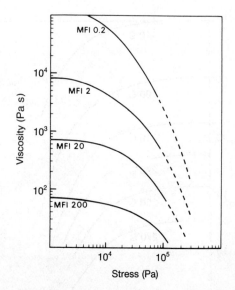

7.14 Stress dependence of the apparent viscosity μ_a at 170°C for four branched polyethylene resins ranging from a very high relative molecular mass, MFI = 0.20, to a very low relative molecular mass, MFI = 200. Measurements were made by capillary flow. Note the thousand-fold change in μ_a produced by changes in relative molecular mass (after Cogswell).

or post-extrusion swelling. For example, in the apparatus of Figure 7.8, the liquid extrudate swells as indicated. The cause is as follows. The liquid is forced under pressure from the barrel into the capillary: a change in area of, say, 100 to 1. The stresses so generated in the liquid do not relax to zero as it passes down through the capillary. At the instant of extrusion into the air, the physical restraint of the capillary wall is removed and the liquid responds to the unrelaxed internal stress and so increases in diameter: it, so to speak, attempts to return to the diameter of the barrel. The critical parameter is the time the polymer liquid is constrained within the capillary: if this is decreased (by decreasing the capillary length or by increasing the wall shear rate) then the **swell ratio** (extrudate diameter divided by capillary diameter) must increase. This is the observed behaviour, as indicated, for example, in Figure 7.15. The polymer liquid as it leaves the capillary 'remembers' the large cross-section in the barrel and tries to return to it by swelling.

Melt viscoelasticity is caused by physical entanglements of the molecules: the stresses induced are of rubber elastic origin. The rate of stress relaxation is of viscous origin. Thus, low temperatures (or high relative molecular masses) promote long memories and hence large swell ratios; high temperatures (or low relative molecular masses) promote short memories and hence small swell ratios.

Polymer liquid memory plays an important part in determining orientation, anisotropy and warping in injection moulding. In extrusion, the

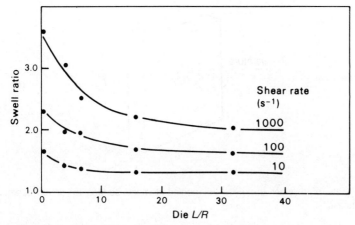

7.15 Dependence of the swell ratio of a polypropylene extruded at 210°C through a die of length L and radius R on the ratio L/R. Data taken at the three wall shear rates indicated (after Cogswell).

swelling which occurs as the polymer liquid leaves the die exit is counteracted by draw-down from pull rollers. The change in section can be large, as will be seen in Section 7.4.

7.3 Cooling and solidification

In polymer forming processes, the production rate is often limited by the time for cooling and solidification. For example, in manufacture of a typical small injection moulding, the cooling time required may be 30 seconds, while filling of the mould cavity itself might take only 1 second. Processing machines are designed to achieve the most rapid cooling possible, consistent with product quality. In injection moulding, the hot polymer melt is cooled by being pressed against the walls of a cold steel mould, which is usually cooled by circulating water or oil. In extrusion of say a pipe, the molten extrudate is cooled by passage through a bath of flowing cold water (see Figure 7.24). In both cases there is rapid heat transfer away from the surface of the polymer. The rate of internal heat transfer by conduction becomes the limiting process. Polymers are poor conductors of heat, as can be seen from Table 7.1 giving the thermal conductivities of some common polymers. For comparison, a typical commercial copper has a thermal conductivity of 390 $W\,m^{-1}\,K^{-1}$.

Table 7.1   Typical values of thermal conductivity ($W\,m^{-1}\,K^{-1}$)
for some common polymers

HDPE	PA6.6	PMMA	POM	PP	PTFE
0.43	0.33	0.18	0.31	0.21	0.25

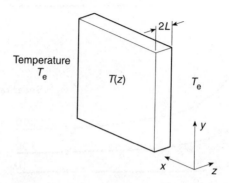

7.16 A plate of polymer thickness 2L, cooled at each surface by an environment of temperature T_e.

Most polymer products approximate to series of flat or curved plate-like elements joined together. This simplifies the analysis of heat flow by reducing it to a one-dimensional problem—see Figure 7.16. The wall of the component is of half-thickness L and is cooled by an environment at temperature T_e. Within a polymer of thermal conductivity k, density ρ, and specific heat capacity c_p, the variation of temperature T is then governed by the equation of one-dimensional heat conduction[1]:

$$\frac{\partial T}{\partial t} = \frac{1}{\rho c_p} \frac{\partial}{\partial z}\left(k \frac{\partial T}{\partial z}\right). \tag{7.20}$$

It is a straightforward matter to solve this equation numerically by computer, and hence predict how the temperature distribution will evolve. An example is shown in Figure 7.17: simulation of the cooling of an amorphous polymer moulding.

To gain a better understanding of what is going on, however, we can exploit the fact that for many polymers the thermal conductivity is relatively independent of temperature, varying by less than 1% per kelvin, and a reasonable approximation to eqn 7.20 is given by

$$\frac{\partial T}{\partial t} = \alpha \frac{\partial^2 T}{\partial z^2} \tag{7.21}$$

where α is the important grouping of properties known as the thermal diffusivity:

$$\alpha = \frac{k}{\rho c_p}.$$

1   For explanation, the reader should refer to any good textbook on heat transfer: for example *Heat Transfer* (7th edn) by J. P. Holman, McGraw-Hill, New York, 1992.

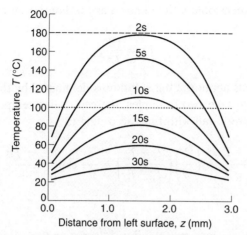

7.17 Computed through-thickness temperature distributions in a 3 mm thickness plate of the amorphous polymer poly(methyl methacrylate), at various times during cooling from the liquid state at 180°C while in contact with a cold metal tool at 20°C. The glass transition of this polymer occurs at 100°C (dotted line). Solid glassy skins are predicted to grow inwards from each surface and to meet at the centre after about 12 s.

The dimensionless parameter determining the limiting mode of heat transfer is the Biot number Bi, defined in terms of k, L and surface heat transfer coefficient h:

$$Bi = \frac{hL}{k}. \tag{7.22}$$

We can see from Table 7.1 that for polymers k is of order 10^{-1} $W\,m^{-1}\,K^{-1}$; most polymer products have L of order 1 mm, and for a polymer melt pressed against a cold metal mould or cooled by flowing cold water, h will be at least of order 10^3 $W\,m^{-2}\,K^{-1}$. It follows from eqn 7.22 that $Bi \gg 1$. This means that the temperature difference at the surface, $T - T_e$, is small in comparison with the through-thickness variations of temperature. We may approximate by considering the limit $Bi \to \infty$, giving the boundary condition

$$T = T_e \text{ at } z = 0 \text{ and } z = 2L.$$

For the polymer in Figure 7.16 cooling from an initial temperature T_i under these conditions, the solution to eqn 7.21 can be obtained analytically as a Fourier series:

$$\frac{T - T_e}{T_i - T_e} = \sum_{m=1}^{\infty} \frac{4}{\pi m} \exp\left(-\frac{m^2 \pi^2}{4} Fo\right) \sin\left(\frac{m \pi z}{2L}\right), \quad m = 1, 3, 5, \ldots,$$

$$\tag{7.23}$$

in terms of a dimensionless time known as the Fourier number:

$$Fo = \frac{\alpha t}{L^2}.$$

When *Fo* exceeds about 0.2 the solution is dominated by the first term in the series and, to take two examples, the centre line temperature T_c and mean temperature $\bar{T}$ fall with time according to

$$\frac{\bar{T} - T_e}{T_i - T_e} = \left(\frac{2}{\pi}\right)\left(\frac{T_c - T_e}{T_i - T_e}\right) = \frac{8}{\pi^2} \exp\left(-\frac{\pi^2 Fo}{4}\right) \tag{7.24}$$

where

$$T_c = T(L), \quad \bar{T} = \int_0^{2L} T \, dz/2L.$$

These results provide a picture of how cooling progresses. In particular they show that any specific measure of temperature, for example $\bar{T}$, decays with time exponentially. The actual time taken to achieve a given degree of cooling—to reach a given value of *Fo*—is proportional to L^2/α. Taking $Fo = 1$ as a typical requirement, it is now clear why thermoplastic polymer products generally have wall thicknesses of only a few millimetres. For most polymers the product ρc_p is $\sim 2 \times 10^6 \ \mathrm{J\,m^{-3}\,K^{-1}}$. Combined with thermal conductivities as given in Table 7.1, this gives values of α of order $10^{-7} \ \mathrm{m^2\,s^{-1}}$, and hence cooling times in the range 10 to 100 s, typical of injection moulding and extrusion processes. Longer cooling times would render the process economics less attractive, by slowing the rate of manufacture. The important point in design is that cooling times increase as L^2. The designer strives therefore to keep wall thicknesses as low as possible.

Figure 7.17 shows how cooling starts at the surface where temperature gradients are highest, and progresses more gradually towards the centre of the component. It follows that solidification starts at the surface with the formation of a solid skin, which then thickens inwards until the whole cross-section is solid. Of course, a crystallizing polymer gives out latent heat as it solidifies, ignored in eqns 7.20 and 7.21; this would produce some distortion of the isochronal lines in Figure 7.17.

Since cooling implies thermal contraction, the non-uniform cooling just described gives rise to thermal stresses. The wall of the component is plate-like, and its geometry forces thermal contraction in the x–y plane to be equal at all depths z. This means that in-plane thermal contraction of the central region, cooling more slowly, is resisted by the already cold and rigid surfaces, producing a final pattern of residual stresses with tension in

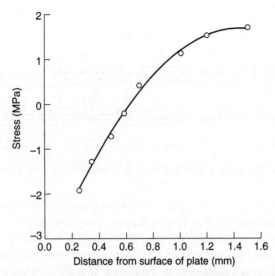

7.18 Measured residual thermal stress in a 3 mm thickness plate of poly(methyl methacrylate), quenched from 110°C into oil at 18°C. The stress is compressive near the surface of the plate, tensile in the interior.

the interior balanced by compression at the surface: see Figure 7.18. Clearly, the presence of this 'built-in' stress will have implications for the mechanical performance of the component in use. It may also pose problems for the manufacturer in terms of product quality. If the residual stress pattern is not symmetrical about the mid-plane—as may be the case if the two surfaces are cooled at different rates—then a bending moment is necessary to hold the component flat. When released from constraint, it will spontaneously warp. A further problem which may arise is that the residual tensile stress in the interior of the part exceeds the stress for cavitation, causing the formation of voids. To prevent these problems, the processor reduces cooling rates, for example by using a warm mould; residual stresses will then have time to relax during cooling. This is one of many instances where the requirements of rapid manufacture and high product quality are in conflict: the production engineer must judge the correct balance, appropriate to circumstances.

7.4 Extrusion

Over 60% of the world's output of plastics is processed by extrusion (7.N.4). The products include tubing, pipes, profiles (for example, window frames), film, sheet, monofilament, parisons for extrusion blow-moulding, and insulated wire for electrical use. These have one thing in common,

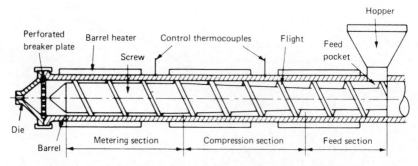

7.19   Illustration of a single-screw plasticating extruder (after Fenner).

which is that the extrusion process manufactures an endless product of constant cross-section that is cut, sawed, chopped, rolled, or otherwise reduced to a portable length. The function of the extruder (see Figure 7.19) is to convert solid feedstock into a homogeneous melt and to pump it through a die at a uniform rate. Following the die, a train of equipment handles the molten extrudate in order to ensure that it cools to precisely the right shape and with the required molecular orientation.

7.4.1   Extruder barrel

The barrel must be extremely strong in order to withstand the high liquid polymer pressures developed: it has characteristics analogous to those of the barrel of a field gun, which is the origin of the name. The size of the extruder is defined by the internal diameter, which is normally in the range 2.5 to 15 cm. The length-to-diameter ratio ranges from about 5 to 34. The shorter machines (L/D below 20) are generally used for processing elastomers, and the longer machines (L/D above 20) for thermoplastics. The most common feed from the hopper (7.N.5) is by gravity (flood feed): the screw continuously extracts the resin it can handle from the hopper. It is necessary frequently to pre-dry the granular feedstock. The point of entry into the barrel (feed throat) is cooled by circulating water. In a flood-fed system, if the pellets become heated in the hopper, they may weld into a mass and block the flow. Barrel temperatures are controlled by electrical heaters monitored by thermocouples. In the barrel, holes for the thermocouples are drilled through the steel of the barrel to just outside the wear-resistant layer, close to the melting polymer. The temperature of the die is also controlled by a thermocouple.

The working of the plastic by the screw generates additional heat which can be of such magnitude, for example, to permit extrusion to continue even if the power to the heaters is switched off. Because of the heat produced by mechanical working, many extruders today are equipped with a barrel cooling system: the barrel must be provided with means to both add and extract heat.

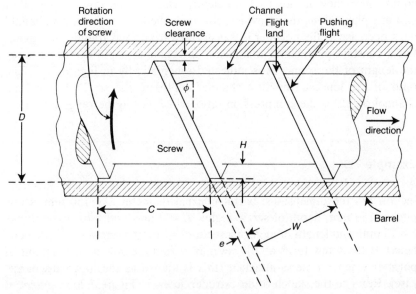

7.20 Screw details.

7.4.2 Extruder screw

The design of the screw has received great attention in recent years. The screw has to

- transport solid feedstock
- compress and melt the solid, and
- homogenize, meter, and generate sufficient pressure to pump the melt against the resistance of the die.

The screw, working on the Archimedean principle (7.N.6), rotates at ~1 revolution per second. The polymer, in granular solid form, as melt, or as a mixture of the two, is contained within the channels (see Figure 7.20) between the flights. It is pushed along the barrel by the forward edge of the flight (pushing flight). The clearance between the barrel and the nearest surface of the flight (the flight land) is very small, of order 10^{-2} mm, and is constant along the screw. The flight land and the inner surface of the barrel are made of hardened steel in order to resist wear. The alloy steel barrel normally has the thin wear-resistant layer centrifugally cast on the inside. The barrel must tolerate an internal pressure of ~100 MPa without elastic deformation above the ~0.15% strain at which the wear resistant layer will crack.

Compressed solid granules have a larger volume than the same mass of liquid, since they do not pack efficiently. Because the rate of mass flow past any plane along the screw is a constant, the area of the screw channel must decrease as the polymer is transformed from solid granules to liquid.[1] For a constant-pitch screw this means the channel depth decreases along the length of the screw, as is indicated schematically in Figure 7.19. The ratio of the longest (the first channel depth) to the smallest (the final channel depth) is the compression ratio, which lies in the range 2 to 4.

Example 7.5

A polymer is extruded with a screw extruder, whose single metering section has the dimensions: barrel internal diameter $D = 150$ mm; screw pitch $= 150$ mm; depth of screw flights $H = 9$ mm; width of screw flights $e = 15$ mm; and length of metering section $= 1$ m (the screw is a single-start helix). If the screw turns at 100 rev min^{-1}, find the volume throughput of polymer when no die is attached (this is known as the 'open discharge' case). Refer to the sketch of an extruder screw in Figure 7.20.

Solution

The extruder screw acts as a pump. In order to understand its action, consider Figure 7.20 and imagine the screw to be stationary while the barrel wall rotates (the effect would be the same). The circumferential speed of the barrel relative to the screw is then

$$\frac{100 \times \pi \times D}{60} = 0.785 \text{ m s}^{-1}.$$

The melt is sheared. Since there can be no net movement of melt normal to the screw flights, the net direction in which the melt is dragged is along the helical channel of rectangular section, in which direction the barrel has an (imagined) velocity component

$$V = 0.785 \cos \phi \text{ m s}^{-1},$$

where ϕ is the helix angle of the screw. From the pitch and circumference, ϕ can be found

$$\phi = \arctan\left(\frac{150}{150\pi}\right) = 17.7°.$$

Hence $V = 0.785 \cos 17.7° = 0.748$ m s^{-1}. The width W of the melt channel, normal to the flights, is given by

$$W = 150 \cos 17.7° - 15 = 128 \text{ mm}.$$

---

1 Note that the polymer velocity also changes along the barrel.

The depth of the melt channel equals the height of the screw flights (neglecting the small clearance):

$$H = 9 \text{ mm.}$$

The volume flow rate Q_0 at which the melt is dragged by shear along the helical screw channel, against no resisting pressure gradient, is then given by (see Problem 7.5)

$$Q_0 = \frac{VHW}{2} = \frac{0.748 \times 0.009 \times 0.128}{2} = 4.31 \times 10^{-4} \text{ m}^3 \text{ s}^{-1}.$$

Comment

Note that in this simplified treatment we have ignored any pressure drop across the metering zone that arises from pressure build-up in the feed and compression zones. This is a reasonable approximation. When the extruder is in use, however, there is actually a pressure **rise** along the metering zone, as the melt approaches the entrance to the die. This cannot be ignored. It causes a back-flow, and consequently reduces the net volume flow rate below the open discharge value Q_0 (see 7.N.7).

The change in channel depth along the screw depends on the manner in which the polymer 'plasticates': whether the polymer crystals melt abruptly (nylon 6.6) or slowly (branched polyethylene); or whether the polymer passes through a glass transition and thus plasticates slowly. Four screws are shown in Figure 7.21:

A a three-zone screw with feed section of constant flight depth, a compression section (it is the geometry and position of the compression section that are designed to match the melting or softening characteristics of the polymer), and a metering section;

B a three-zone screw with a vented section in which the pressure is dropped so that gases may be exhausted from the melt by a vacuum line, or merely by exposure to the atmosphere (a hole is drilled through the barrel wall);

C a 'PVC-type' screw for amorphous polymers which progressively soften through the glass transition; and

D a 'nylon-type' screw for crystalline polymers with a particularly sharp melting point.

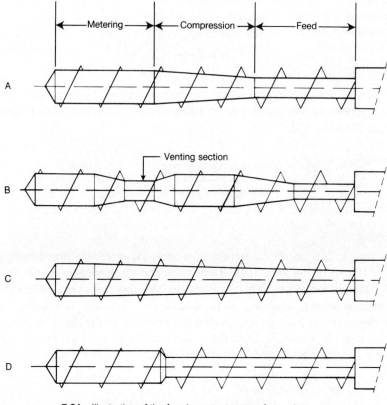

7.21   Illustration of the four important types of extruder screw.

The plasticization process involves:

- development of a melt film on the wall of the barrel;
- scraping off the melt by the pushing flight; and
- pressurizing the melt and delivering it at a metered rate to the die.

Example 7.6

A polymer with viscosity 250 Pa s is to be extruded using a vented screw extruder, with details as follows. Diameter of barrel $D = 75$ mm, screw speed $N = 40$ rev min^{-1}, screw helix angle $\phi = 17.7°$, and screw flight width $e = 7.5$ mm.

The first metering zone has: length = 250 mm, and flight depth $H_1 = 4$ mm. The second metering zone has: length = 500 mm, and flight depth $H_2 = 6$ mm.

The extruder is to produce film 1 m wide with a die whose melt channel is of length 10 mm. Find the film die gap required for satisfactory extrusion with the two metering zones just matched. What would happen if either (a) smaller, or (b) greater die gaps were used? (See Figure 7.21B for a sketch of a vented screw.)

Solution

With a vented screw, it is essential to ensure that the system is balanced: the extruder must run satisfactorily with equal mass throughputs of polymer passing through the first and second metering zones and the die. Consider each in turn.

(1) The first metering zone acts essentially in open discharge (although there will, in practice, be a small pressure drop, as the pressure built up in the first compression zone is relieved). Therefore, its volume throughput Q_1 may be simply calculated using the method of Example 7.5:

$$Q_1 = Q_{O1} = \frac{VH_1W}{2}.$$

The velocity V of the screw relative to the barrel, resolved along the screw channel, is

$$V = \frac{\pi ND \cos \phi}{60} = \frac{\pi \times 40 \times 0.075}{60} \cos 17.7° = 0.1497 \text{ m s}^{-1}.$$

The width W of the screw channel is related to the pitch ($\pi D \tan \phi$) and screw flight width e:

$$W = \pi D \tan \phi \cos \phi - e = \pi \times 0.075 \times \sin 17.7° - 0.0075 = 0.0641 \text{ m}.$$

Hence

$$Q_1 = Q_{O1} = \frac{0.1497 \times 0.004 \times 0.0641}{2} = 1.92 \times 10^{-5} \text{ m}^3 \text{ s}^{-1}.$$

(2) The second metering zone has a volume throughput Q_2 given by the method of Example 7.5:

$$Q_2 = Q_{O2} - Q_{P2}.$$

The open discharge value for this zone is simply

$$Q_{O2} = \frac{VH_2W}{2} = \frac{0.1497 \times 0.006 \times 0.0641}{2} = 2.88 \times 10^{-5} \text{ m}^3 \text{ s}^{-1}.$$

The backflow caused by die entrance pressure P_d is given by (see 7.N.7)

$$Q_{P2} = \frac{WH^3}{12\mu} \frac{P_d}{L_2},$$

where L_2 is the length of screw channel in this zone:

$$L_2 = 0.5/\sin 17.7° = 1.64 \text{ m}.$$

Substituting, we obtain

$$Q_{P2} = \frac{0.0641 \times (0.006)^3}{12 \times 250} \times \frac{P_d}{1.64} = 2.81 \times 10^{-12} \, P_d \text{ m}^3 \text{ s}^{-1}.$$

(3) The die has a volume throughput Q_d given by the expression for flow through a rectangular channel (see Problem 7.10). If W_d, H_d, and L_d are the width, depth and length of the channel in the die, then

$$Q_d = \frac{W_d H_d^3}{12\mu} \frac{P_d}{L_d}.$$

Thus

$$Q_d = \frac{1 \times H_d^3}{12 \times 250} \frac{P_d}{0.01} = \frac{H_d^3 P_d}{30}.$$

(Note: this is an approximation, we have neglected the additional pressure drops at die entrance and exit.)

If the various parts of the extruder are balanced, then there is equal mass flow rate throughout. If density variations are neglected, then volume throughputs in all parts of the extruder must be equal:

$$Q_1 = Q_2 = Q_d.$$

Substituting from (1), (2), and (3),

$$1.92 \times 10^{-5} = 2.88 \times 10^{-5} - 2.81 \times 10^{-12} \, P_d = \frac{H_d^3 P_d}{30}.$$

Of these, the left-hand equation yields the die pressure drop P_d:

$$P_d = \frac{2.88 \times 10^{-5} - 1.92 \times 10^{-5}}{2.81 \times 10^{-12}} = 3.42 \times 10^6 \text{ Pa},$$

ensuring that the two metering zones are matched. The right-hand equation then yields the die gap H_d:

$$H_d = \left(\frac{30 \times 1.92 \times 10^{-5}}{3.42 \times 10^6} \right)^{1/3} = 5.52 \times 10^{-4} \text{ m,}$$

ensuring thus the die is matched to the metering zones.

It is increasingly recognized that a conventional screw consisting only of feed, compression, and metering sections often fails to homogenize the liquid satisfactorily. This is particularly important for:

- polymers with solid constituents (for example TiO_2, mineral, clay, glass, of carbon fibre); and
- compounding techniques which prepare alloy polymers by homogeneously mixing different polymers with a wide divergence in melting points (or glass transitions) and viscosities.

With a standard screw, consisting of feed, compression, and metering sections, the exercisable process parameters are feed rate, rotor speed, and barrel temperature. This is known to be insufficient, and increasing attention is now aimed at screw modification to increase homogeneity by improving mixing.

In single-screw machines, the leading improvements in screw design are based on the barrier principle. The purpose of the barrier, or dam, is to improve homogeneity by subjecting the melt to intensive shear. The melt is forced through a narrow gap ($\sim \frac{1}{2}$ mm) formed between the top of the barrier and the barrel wall, thus causing solid granules and agglomerates to break down. At the same time, the limited clearance over the barrier tends to act as a sieve, preventing solid particles from being conveyed forward before they are broken down or partially melted. The barrier is often combined with further ways of giving better distributive mixing, or improved melting. Barrier devices aimed at improved mixing are normally short elements positioned at the end, or somewhere in the metering section, of the screw.

At the time of writing, the better designs of barrier flight screw give the highest efficiency. The way performance has improved may be summarized by taking examples of 90 mm extruders, 25–30 diameters in length, running with pipe-grade LPE and with a maximum extrudate temperature of 220°C. A conventional (three-zone) screw can produce ~ 2 kg h^{-1} per rev min^{-1}, and with a maximum speed of 75 rev min^{-1} this gives 150 kg h^{-1}. A barrier flight screw developed for LPE could be run at 100 rev min^{-1} giving 267 kg h^{-1} (2.67 kg h^{-1} per rev min^{-1}).

Twin-screw extruders have become of increasing importance for difficult mixing operations. The screws can either turn in the same (co-rotating) or opposite direction (counter-rotating): they are contained within a figure-of-eight-shaped barrel. Co-rotating, intermeshing twin-screw compounders have emerged as the most versatile equipment for controlled mixing. Twin-screw extrusion technology is now firmly established, and considerable operational experience is available. Twin-screw extruders often employ a barrier principle in order to improve melting and mixing.

7.4.3 Die and calibration equipment

The die is bolted to the end of the extruder. Usually, a breaker plate or screen (see Figure 7.19) is placed between the die and the screw tip: this increases direct flow along the axis by inhibiting rotation, and filters the polymer liquid.[1] The die shapes the product but, because of viscoelastic swell, the cross-section of the extrudate expands as it leaves the die. It is then necessary to bring it to the correct size by pulling, as indicated schematically in Figure 7.22. In order to achieve precisely the correct diameter, the extrudate is shaped as it cools, all the while under a tension from the haul-off.

A pipe die is shown in Figure 7.23. The polymer melt flow is separated into an annulus, enters the die land, and exits in tubular shape. It swells on exit and is drawn down to the correct diameter: this is checked by means of a calibrator. The extruded (and still liquid) pipe enters the calibrator through the rubber seals (A in Figure 7.24) and passes stainless steel guides (B). Within the calibrator, thermostatted water cools the pipe. In the calibrator, the wall thickness is gauged by ultrasonics: an alarm is generated if the wall thickness falls above or below set values. Frequently, the pressure of water is below atmospheric, so that with atmospheric pressure in the tube, it is forced against sizing rings in the calibrator.

The draw-down (achieved by the haul-off and sizing rings) brings the

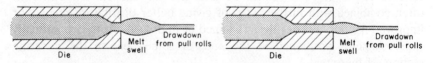

7.22 Illustration of melt swell in short and long die lands. The swell is counteracted by tension from the pull rolls (after Richardson).

---

1 The removal of dirt particles, etc. is particularly important in pipe extrusion: pipe lifetimes under pressure are increased when particular care is taken to purify the polymer.

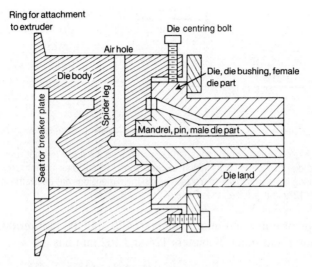

7.23 Pipe or tubing die for inline extrusion (after Richardson).

cross-section down to a size **below** that of the die.[1] The ratio of annular areas is the draw-down:

$$\text{draw-down} = \frac{\text{annular area of die}}{\text{annular area of pipe}}.$$

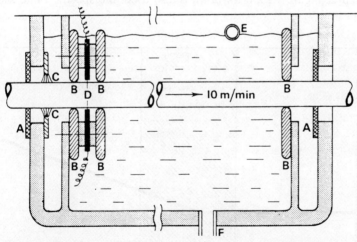

7.24 Calibrator with ultrasonic gauging for pipe and tubing extrusion: A = rubber seals; B = stainless steel guides; C = brushes (to remove bubbles); D = ultrasonic probes; E = water drain; and F = water inlet.

---

1 This is illustrated for rod extrusion in Figure 7.22.

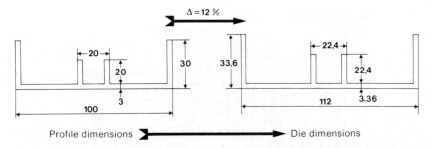

Profile dimensions ➤ ━━━━━━━━━━━━━━━➤ Die dimensions

7.25 On the left a profile obtained from the die shown on the right. The hot liquid profile, as it leaves the die, has dimensions which **exceed** those of the die: tension and calibration equipment bring it, when cooled, to a size in which the dimensions are below those of the die; as indicated, 12% in this case (after General Electric Plastics).

Thus a pipe of outer and inner diameters 120 and 100 mm produced from a die of outer and inner diameters 124 and 102 mm has

$$\text{draw-down} = \frac{(\pi/4)(124^2 - 102^2)}{(\pi/4)(120^2 - 100^2)}$$

$$= 1.13.$$

The draw-down of pipes is normally in the range of 1.10 to 1.20. For tubing (by convention, tubing is of diameter below 12 mm; pipe is above 12 mm), the draw-down is usually large (in the range 2 to 6). For profiles, as indicated in Figure 7.25, draw-down is also used: the specific value shown is 1.12.

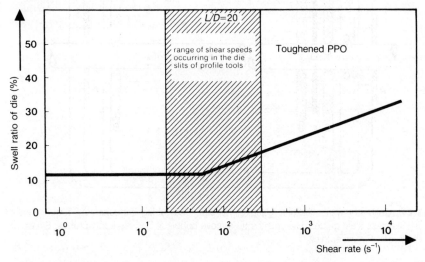

7.26 Dependence of swell ratio of die for a toughened PPO resin on shear rate of the wall (after General Electric Plastics).

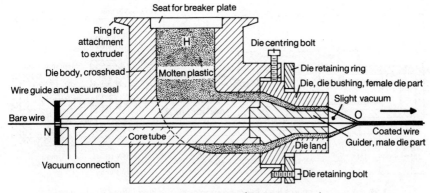

7.27   Wire coating tubing die (after Richardson).

In order to maintain dimensional stability, all features of the extrusion process must be carefully controlled: the degree of parameter interaction is great. For example, the normal swell rate for toughened PPO is in the range 10 to 20%, depending on shear rate and land length. If the shear rate changes then the die swell will change (see Figure 7.26). This means that the rate of rotation of the extruder and the melt temperature must be very carefully controlled. The rate of take-off of the extrudate will obviously affect the draw-down, so this, too, must be carefully controlled. The draw-down will also depend on the temperature of the cooling water in the calibrator, which must not be too low, since this may generate non-uniform cooling of the profile with resulting stresses and deformation within the extrudate.

There are many other extrusion operations of great technical significance. The oldest and most vital of all is wire coating: a die is shown in Figure 7.27. Bare wire is pulled at high speed (~metres/second) into the die at N; it leaves, coated, at O. The melt flow H is turned through a right angle, separated into an annulus, and extruded as a tube. A slight vacuum is drawn as indicated in Figure 7.27, and this facilitates the movement of the hot liquid tube on to the wire.

7.5   Injection moulding

There are several species of injection moulding machine (one is shown in Figure 7.3), but the most important by far is the **reciprocating-screw**. In reciprocating-screw injection moulding the screw plasticizes the solid polymer, thus forming a metered volume of homogeneous melt. It then ceases to rotate and acts as a ram, being thrust forward to inject the melt into the mould. We will confine attention to a representative machine of this type. Reciprocating screw injection moulding is used to process thermosets as

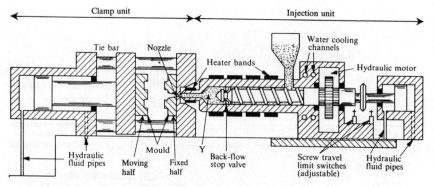

7.28   A reciprocating screw injection moulding machine (after Ogorkiewicz).

well as thermoplastics: the machines for these two purposes differ slightly, as will be described below.

A single-screw injection moulding machine is illustrated in Figure 7.28. The machine is shown with an opened and empty mould. Liquid polymer has been generated at Y by screw rotation, as in an extruder. The nozzle, of small diameter, is sealed by cold, solidified polymer. In the moulding cycle, the next step is the closing of the mould. The screw is then rammed forward in order to inject the polymer liquid. As the screw travels forward, a non-return valve at the tip of the screw prevents liquid from travelling backwards between the flights along the screw. The screw is mechanically prevented from rotating when acting as a ram. The cycle continues with the liquid polymer in the mould cavity cooling and solidifying. Meanwhile the screw starts to rotate again, drawing a new supply of liquid polymer up to position Y ready for the next moulding, and moving backwards to create space for it. The final step in the cycle is opening of the mould and ejection of the moulding. As indicated in Figure 7.28, three hydraulic systems are used:

- to rotate the screw to plasticate, homogenize, and pressurize the polymer;
- to ram the screw forward in order to inject the liquid into the mould and then to hold the appropriate pressure while the moulding cools; and
- to translate the moving half of the mould backwards and forwards so as to open and close the mould (7.N.8), and to hold the mould closed during injection in order to prevent liquid leaking along the parting plane: this leakage is known as flashing.

The mould (see Figures 7.3, 7.28, 7.29) comprises two halves: the **fixed half** attached to the **stationary platen**; and the **moving half** attached to the **moving platen**. The **impression** is formed between the two halves. It is

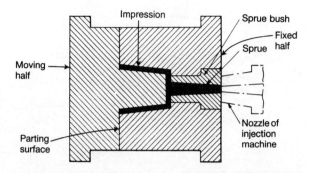

7.29 Feed system for a single-impression mould (after Pye).

filled by liquid flowing from the nozzle and then into the **sprue**. In the filling of multi-impression moulds (see Figure 7.30) the liquid runs from the nozzle into the sprue, and then along channels known as **runners** cut into the parting surfaces. It finally flows through a severe constriction (the gate) and into the mould. By tapering the walls and by ensuring that it is the male portion of the mould that moves, the cooled moulding lifts away after the mould is opened, taking the sprue and runners with it (see Figure 7.31). The mouldings are then broken away from the runners and sprue.

There is considerable contraction of the moulding as it cools, and it is this that causes it to grip the male part. The volume contractions (from temperature of the liquid to 20°C) fall just below 10% for glassy polymers and between 10 and 20% for crystalline polymers (see Table 7.2). If the mould were filled at a pressure of one atmosphere, it would show voids and excessive sink marks after cooling, due to volume contraction. This formidable problem is overcome as follows; see Figure 7.32.

• During stage A–B the liquid polymer in the mould is pressurized. It will be recalled from Section 7.2 that the application of 1000 atmospheres pressure to a polymer liquid yields a contraction of ∼10%. As

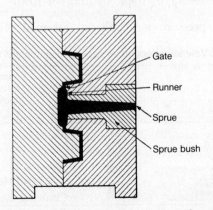

7.30 Feed system for a multi-impression mould (after Pye).

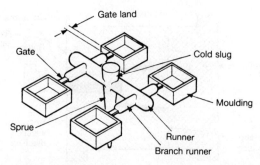

7.31 Feed system of multi-impression mould as portrayed by the assembly when it is removed from the mould (after Pye).

the mould pressure can reach a substantial fraction of 1000 atmospheres (100 MPa), about 10% extra liquid enters the mould than would do so if it were filled at 1 atmosphere pressure.

- During stage B–C the polymer cools under pressure, the mould being continuously packed with more liquid to compensate for thermal contraction. **Over**-packing must be avoided: it leads to polymer being extruded between the faces of the mould halves ('flashing'), and to moulded-in stresses, warping and difficult ejection.

- At point C the gate freezes over, sealing the mould cavity and preventing further filling. As the gate is narrow, it often solidifies (crystallizes or reaches T_g) ahead of much of the rest of the moulding.

- During stage C–D the polymer cools at constant volume under decreasing pressure, until atmospheric pressure is reached at point D.

- During stage D–E the moulding cools freely and undergoes normal thermal contraction. The final fractional shrinkage of the moulding, compared with the volume of the mould cavity is clearly $1 - v_E/v_D$ (typically of order 1%), whereas free thermal shrinkage from the liquid state would be $1 - v_E/v_A$ (typically of order 10%).

Cooling under pressure is clearly an essential feature of injection

Table 7.2   Volume reduction on cooling at atmospheric pressure

Material	Cooled from: (°C)	to: (°C)	Approximate volume contraction (%)
Acrylic	150	20	7
Nylon 6.6	285	20	14
Nylon 6	260	20	13
Polyethylene (density 916 kg m^{-3})	190	20	18
Polystyrene	195	20	7

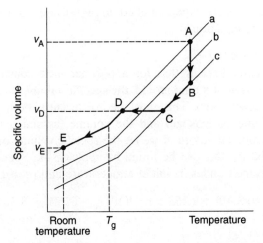

7.32 A schematic diagram showing the *P–V–T* path followed by a typical point within an amorphous polymer injection moulding. Lines a, b, c represent isobars of increasing pressure.

moulding, vital to the production of components with well-controlled dimensions. The achievement of this dimensional accuracy, however, relies upon the ability of the injection moulding machine to withstand huge pressures in the polymer melt. In particular, a large clamping force is required to keep the mould tightly closed when the pressure is at its peak. The ingenious double-toggle clamp is one means by which this is achieved (see 7.N.8). The maximum available clamping force is such an important feature of the machine that it is one of the two main parameters used to specify the capacity of an injection moulding machine. The other important parameter is the 'shot-size' or maximum volume of polymer that can be injected, usually expressed as the mass of an equivalent volume of polystyrene.

Example 7.7

Poly(methyl methacrylate) (PMMA) is to be injection-moulded. The cavity will be sealed by freezing of the gate when the mass of polymer in the cavity is at 165°C and under the packing pressure of 40 MPa. Assuming slow (uniform) cooling, predict the temperature of the melt at which the pressure in the cavity falls to atmospheric, employing the following equation of state which links pressure P (in Pa), specific volume v (in $m^3\,kg^{-1}$), and absolute temperature T for PMMA

$$(10^{-3}P + 2.158 \times 10^5)(10^3 v - 0.734) = 83.1T.$$

(This equation of state when applied to a polymer is known as the Spencer–Gilmore equation.)

Solution (Refer to Figure 7.32)

The essential point here is that for a polymer melt constrained under pressure by the walls of a rigid mould, **the specific volume is constant** after the gate has frozen over. This must be so, because both the mass of polymer (assuming no leakage) and the volume (assuming the mould is rigid) are constant. Therefore, if we let the specific volume of the PMMA in the mould be v_0, this can be found by applying the Spencer–Gilmore equation to the melt under its initial temperature and pressure:

$$(10^{-3} \times 40 \times 10^6 + 2.158 \times 10^5)(10^3 v_0 - 0.734) = 8.31 \times 438.$$

Hence

$$v_0 = 10^{-3}\left(0.734 + \frac{83.1 \times 438}{2.558 \times 10^5}\right) = 8.763 \times 10^{-4} \text{ m}^3 \text{ kg}^{-1}.$$

As the melt cools in the cavity, so long as the pressure exceeds atmospheric, it is pressed against the walls of the mould and the specific volume remains constant. In order to compensate for the falling temperature at constant v_0, there is a corresponding fall in pressure. Eventually, pressure P falls to atmospheric ($=1.013 \times 10^5$ Pa). The temperature T at which this point is reached can be found by applying the Spencer–Gilmore equation again:

$$(10^{-3} \times 1.013 \times 10^5 + 2.158 \times 10^5)(10^3 \times 8.763 \times 10^{-4} - 0.734) = 83.1T.$$

Hence

$$T = \frac{0.1423 \times 2.159}{83.1} \times 10^5 = 369.7 \text{ K } (96.5°C).$$

Comment

On cooling below 96.5°C, the pressure remains at 1 atmosphere, the specific volume falls, and the impression must pull away from the walls of the cavity. If the cooling is not slow and uniform (which is what happens in a real mould in injection moulding), then this ideal treatment has to be modified.

7.5.1 Hot runner moulds

It is obviously inefficient to take a fresh impression of the sprue and runners (see Figure 7.31) for each cycle:

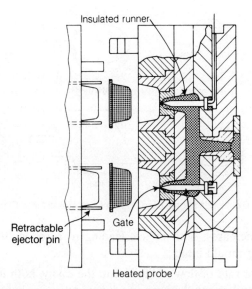

Insulated runner

Retractable
ejector pin

Gate

Heated probe

7.33 An example of a hot-runner mould: an insulated runner mould with heated probes (after Modern Plastics Encyclopedia).

- because it wastes material (with thermoplastics this may be reprocessed but the material is to some extent degraded on each passage through the cycle); and
- because the mouldings must be clipped by hand from the sprue and runners.

This problem can be overcome by establishing, by one means or another, temperature control over the liquid channels throughout the stationary side of the mould. There are, and have been, many variants of this technique, one of which we show in Figure 7.33. The basic idea is to heat the polymer liquid right up to the gate. When the mould cools, the gate freezes over, but the liquid in the heated runners in the fixed half does not freeze: it is insulated and its temperature is controlled by heated probes. When the moving half is withdrawn, the impression moves with it, with a fracture occurring at the gate. The impressions are then pushed off the male portion of the mould by ejector pins and collected: the operator has merely to trim away extraneous material at the position of the fracture.

7.5.2 The gate

The gate and its position (see Figure 7.34) are of considerable importance. The purpose of the gate is:

- to cause work to be done when the liquid is forced through the constriction so that the temperature increases by ($\sim 20°C$) and the

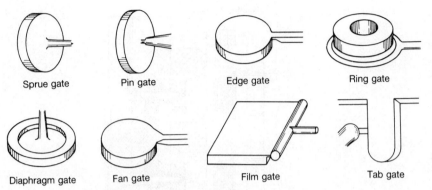

Sprue gate   Pin gate   Edge gate   Ring gate

Diaphragm gate   Fan gate   Film gate   Tab gate

7.34   Eight types of gate used in injection moulding (after ICI plc).

shear rate increases, both effects leading to a reduction in viscosity, thus assisting mould filling;

- to control the rate of flow of liquid into the cavity, both in the primary filling state and, secondly during packing; and
- to enable the mould to be mechanically insulated from the liquid in the barrel by freezing-over during mould cooling. This occurs easily in the gate because of the constriction.

The eight most common gates are illustrated in Figure 7.34. The gate should be positioned so as to generate an even flow of liquid in the impression, thus causing the mould to fill uniformly so that the advancing fronts reach the impression extremities at the same time. An example of correct and incorrect gating is shown in Figure 7.35. The correct gate for this cup-shaped moulding is the sprue gate (b): the cylindrical section fills in a balanced way without a weld line. The incorrect gate is the edge gate (a). A weld line forms, which is a possible source of mechanical weakness and a visual blemish.

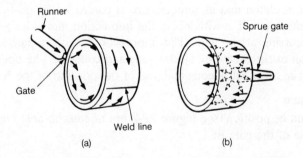

Runner

Sprue gate

Gate

Weld line

(a)   (b)

7.35   For a cup-shaped moulding the correct gate is (b), the sprue gate: an edge gate (a) produces a weld line (after Pye).

The position and dimensions of the gate require close attention at the earliest stages of component and mould design. The component designer and the mould designer should work together during design of the component. If the mould designer has to prepare a design for an already 'designed' component its cost may exceed by a factor of ~100% the cost of the properly designed component (i.e. one in which component and mould considerations are taken into account simultaneously). The gate should be placed in a position:

- where it presents no finishing problems or where it is inexpensive to remove;
- where bending or impact will not take place during service since the gate area will normally have an above average residual stress and be a weak spot; and
- where it will facilitate the expulsion of air by the injected liquid. This normally occurs through minute vents at the parting surface, or around an ejector pin. Unless this is done, adiabatic compression of the air can cause an extreme rise in temperature (which can be high enough to degrade the polymer).

The size of the gate is critical, and it is standard practice to make gates too small initially, so that suitable modifications to shape and size can be made following mould trials. The three dimensional variables of the gate control the moulding operation: the thickness t (see Figure 7.36) controls the time after injection at which the gate freezes, and consequently controls the time available for packing; the flow can be increased by increasing the width w, or by decreasing the land l, without seriously affecting freeze time.

7.5.3 Control of pressure, temperature, and time

The definition of moulding quality is very complex and includes mechanical properties, surface quality, dimensions, and density. In order to obtain reproducible and acceptable quality, it is essential to keep the injection

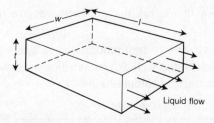

7.36 Dimensions of the gate.

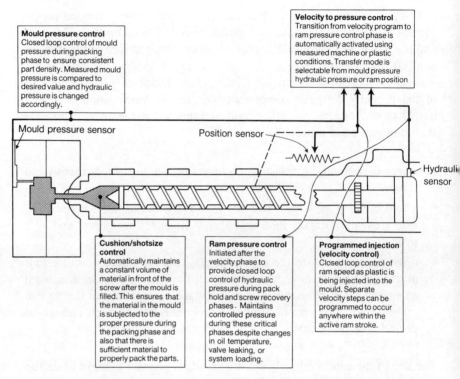

Mould pressure control
Closed loop control of mould pressure during packing phase to ensure consistent part density. Measured mould pressure is compared to desired value and hydraulic pressure is changed accordingly.

Velocity to pressure control
Transition from velocity program to ram pressure control phase is automatically activated using measured machine or plastic conditions. Transfer mode is selectable from mould pressure hydraulic pressure or ram position

Mould pressure sensor

Position sensor

Hydraulic sensor

Cushion/shotsize control
Automatically maintains a constant volume of material in front of the screw after the mould is filled. This ensures that the material in the mould is subjected to the proper pressure during the packing phase and also that there is sufficient material to properly pack the parts.

Ram pressure control
Initiated after the velocity phase to provide closed loop control of hydraulic pressure during pack hold and screw recovery phases. Maintains controlled pressure during these critical phases despite changes in oil temperature, valve leaking, or system loading.

Programmed injection (velocity control)
Closed loop control of ram speed as plastic is being injected into the mould. Separate velocity steps can be programmed to occur anywhere within the active ram stroke.

7.37 Multi-parameter control of injection moulding (after Barber-Colman Company).

moulding process under precise control: the process must not be permitted to run off-course because of, for example, variations in resin quality from batch to batch, changes in hydraulic oil temperature, or sightly malfunctioning hydraulic valves.

Modern machines are directed by microprocessor-based controllers. The input to the control systems comes from:

- thermocouples measuring barrel, nozzle, and mould temperatures;
- a hydraulic pressure sensor measuring oil pressure driving the ram[1] (ram pressure, see Figure 7.37);
- a pressure sensor measuring the polymer pressure in the mould (see Figure 7.37); and
- a ram position sensor (potentiometer) which, during injection, measures ram velocity as well as ram position as shown in Figure 7.37: a clamp position sensor (potentiometer).

---

1 The screw is termed 'screw' or 'ram' depending on the function being described.

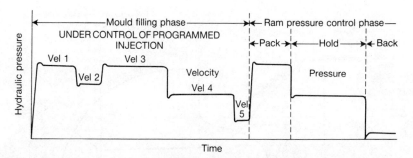

7.38 Dependence on time of the hydraulic pressure driving the ram during mould filling and packing: control sequence (1) and (2a) (after Barber-Colman Company).

Data from these sources are used to optimize the cycle in order to obtain the required quality of moulding with the shortest possible cycle time. During the ensuing production, the optimized process parameters must be repeated as precisely as possible from cycle to cycle.

It has been found expedient to control the dominant moulding action (mould filling and packing) in two alternative modes: (1) and (2a), **or** (1) and (2b).

(1) by moving the ram forward at a sequence of velocities, high velocities at first and low velocities at the end (see Figure 7.38) as the filling stage nears completion: and

(2a) by switching to ram pressure control for the packing state (see Figure 7.38).

The logic of this is elementary: once the mould is filled, the ram essentially stops moving and becomes of little use for packing control.[1] For (1) the data are derived from the ram position sensor: the correcting parameter is the hydraulic ram pressure. For (2a) the data are measurements of hydraulic ram pressure, which is also the correcting parameter.

The pack pressure (see Figure 7.38) is applied for a preset time, which determines the mass of the moulding: the system then transfers to the lower hold pressure until the gate freezes. After a set time at hold pressure, the screw rotates and moves backwards. During this phase (in order to achieve consistent melt viscosity[2] from shot to shot) the back pressure is held at a low but constant value.

---

1 Another reason for controlling velocity during filling is that velocity influences surface finish, shrinkage, and anisotropy.

---

2 As the screw rotates at fixed speed, the back pressure affects the screw torque and hence the work done and so the melt temperature; the melt temperature increases if the back pressure is increased.

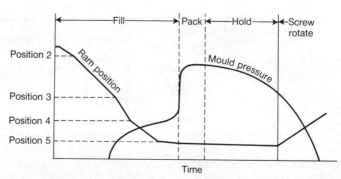

7.39 Dependence of ram position and mould-pressure on time during mould filling and packing: control sequence (1) and (2b) (after Barber-Colman Company).

The ram pressure governs the cavity pressure, but it is fairly clear that due to striction of the screw in the barrel, etc., a measure of cavity pressure would be a more direct control parameter. This provides an alternative mode to (2a). The ram is moved forward at controlled velocities (1) until the filling is complete, and then for packing:

(2b) the microprocessor switches to the cavity pressure sensor (see Figure 7.39).

Control is achieved by comparing cavity pressure with the set value: if a difference exists, the controller will increase or decrease back pressure (see Figure 7.39) on the next cycle. The correcting parameter is the hydraulic ram pressure.

A separate control is exerted over the screw **cushion**. The cushion is the volume of liquid that remains in front of the screw after injection (see Figure 7.37): the cushion affects the pressure transmitted from the system (ram hydraulics and ram) to the cavity during packing. The ram position sensor measures the cushion: if it is low the screw is withdrawn further during the rotation phase of the next cycle.

The mould temperature controls the degree of residual stress in the moulding and, in crystalline polymers, the crystallinity (see Chapter 2). Both parameters affect the mechanical properties, the strength in particular. Typical mould temperatures are given in Table 7.3. The optimum mould temperature is a compromise between the drive to lower cycle time (which implies lowering the mould temperature) and the drive to improve mechanical properties (increasing mould temperature). For example, note that the mould temperatures for nylon 6.6 are between 60 and 90°C—in the region of and above the glass transition (see Figure 4.24). Solid-state stress relaxation goes on at this temperature incomparably faster than at

Table 7.3 The liquid in the barrel is rammed through the nozzle (from a barrel temperature T_B) into the mould, which is maintained at the temperature T_{mould}

Polymer	T_B (°C)	T_{mould} (°C)
Polypropylene	230–275	40–80
Nylon 6.6	270–285	60–90
Poly(methyl methacrylate)	210–250	55–80
Poly(vinyl chloride)	150–180	20–50
Polyethersulfone	310–390	140–160

temperatures, for example, in the region of 20 to 30°C. The molecules in the liquid flow during injection in the presence of large shear and elongational stresses and these must be allowed to stress relax in the mould to acceptable levels. The moulding, when it reaches the preset mould temperature, must be sufficiently stiff and strong that when the mould has been opened it may be pushed off the moving half by the ejector pins and then removed by hand (or by robot).

The mould temperature is controlled by passing temperature-equilibrated fluids through interconnecting drilled holes in the mould. It is the function of the thermostatted fluid to remove heat as quickly as possible from the mould and transfer it to the cooling system. The requirements are:

- to achieve as uniform a temperature distribution as possible in order to arrive at uniformity in the moulded part;
- to remove heat from the mould cavity evenly, so as to avoid warpage; and
- to remove heat rapidly so that the fastest practical moulding cycles can be achieved—see Section 7.3.

Determining the correct pattern of cooling lines is an important step in mould design.

7.5.4 Thermosets

Savings in cost most often accrue from the injection moulding of thermosets (such as polyesters), instead of thermoplastics, because thermoset polymer is usually cheaper than thermoplastic and the cycle times are sometimes shorter (the material cross-links in the mould so that cooling is not necessary before the part is removed). The purpose of the screw is to plasticize and homogenize the precursor material (which may contain short fibres) in preparation for injection into a heated mould in which cure takes place. The basic machine is similar to that described above for thermoplastics, but there are important differences in detail (see Figure 7.40):

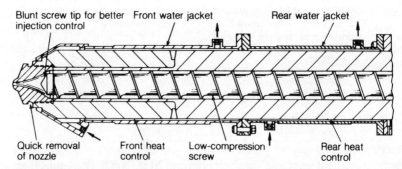

7.40   Characteristics of thermoset screw and barrel (after *Modern Plastics Encyclopedia*).

- The thermoset precursor material must not be heated for long in the barrel, or it will cure prematurely and set solid. Barrels are therefore quite short (L/D in the range 12 to 14).

- There is no non-return valve: the constriction it would generate would heat the precursor material and cause premature cure. The screw is prevented from rotating during injection by a brake, or by other means.

- The compression ratios are low (in the range 0.9 to 1.1), as higher compression ratios would generate too much heat.

- There must be a built-in device for quick release of the nozzle in the event of premature cure, so that the barrel may be cleared quickly.

- Heating is required only at start-up; thereafter water coolers are required in order to remove the heat of preliminary reactions as the precursor material proceeds down the barrel.

- The plasticized precursor material is injected into a heated mould. The mould temperature depends on the polymer system, but is, as a rule, considerably higher (~ 100 K) than the barrel temperature.

- In the heated mould, the liquid precursor quickly cross-links and solidifies: the mould is opened and the part is removed whilst hot.

The quality of moulded thermosets is highly dependent on a precisely repeatable cycle time. This is true also of thermoplastics, but is not so critical as in thermosets. The one phase of the cycle not controlled is part removal from the mould: this is at present done by the machine operator. Whilst he does this, the safety gate protects him and the restart of the next cycle is under his control and is delayed until he completes removal of the part and closes the safety gate. The human variability thus introduced (a change from cycle to cycle of only a few seconds) can affect product repeatability. This is one factor leading to the replacement of the human operator by robots for part removal.

7.5.5 Reaction injection moulding (RIM)

In RIM two (or more) liquid reactants are polymerized in the mould; the reactions must be rapid so that cycle times are short. If short reinforcing fibres or other fillers are introduced into the mould the process is termed RRIM, **reinforced reaction injection moulding**. There is an obvious saving of energy in RIM, since in conventional processes other intermediate steps are necessary; for example, with nylon 6.6, polymerization, pelletizing, and, finally, injection moulding. RIM is used primarily for polyurethanes (mainly elastomeric for automotive application). Nylon 6 (polycaprolactam) and epoxy processes have also been developed. RIM and RRIM are particularly suited to the production of large area parts, such as spoilers, fenders, bumpers, and front and rear facias for automobiles.

The basic principles of the RIM and RRIM processes are shown in Figure 7.41. The mixing chamber is mounted directly on the mould. Micro-mixing of the reactants is achieved through high-pressure impingement of the two liquid jets. It is essential that:

- the large storage reservoirs of liquid reactants be thermally equilibrated (they must not, for example, crystallize as the plant temperature changes);
- the liquid reactants be mixed in the correct proportions before injection; and
- the correct volume of the mixed liquids be injected rapidly.

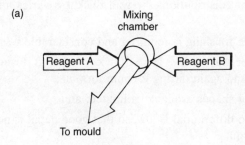

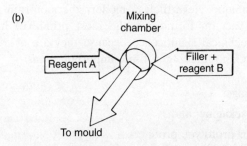

7.41 Basic principles of: (a) RIM, and (b) RRIM.

Pressures within the mould are low since the injected liquid, before it commences to polymerize, has a low viscosity and can easily take up the most complex mould shapes. For a foamed polyurethane RIM system for an automotive product, typical process times are: $\sim$2 s to inject a shot of several kilograms into the mould; $\sim$10 s to gelation (during which time foaming is occurring); $\sim$40 s curing time in the mould: mould opening $\sim$30 s; and demoulding and mould preparation for the next shot $\sim$50 s. This gives a total cycle time of the order of 2 minutes. After demoulding, polyurethane products require a post-cure (typically $\sim$120°C for 1 hour) in order to complete the polymerization reaction. RIM and RRIM are two areas of particularly rapid growth with considerable potential for development.

7.6 Thermoforming

In the 1950s a technique know as vacuum forming was developed for manufacturing large objects such as, for example, a bath. A sheet of thermoplastic is clamped over a negative mould, softened by heating, and then a vacuum drawn within (by pumping the air out through small holes machined into the mould). The heated sheet is consequently sucked down into the mould and takes up the desired shape. Vacuum forming requires low capital and is extremely simple. The disadvantages are:

- poor material distribution: the wall thickness varies and is difficult to control;
- considerable finishing is required and considerable waste ensues;
- the polymer has to be heated twice, once to form the sheet (by extrusion) and again during vacuum forming;
- the range of shapes available is limited; and
- the pressure differential is low, so that good detail is not obtainable by vacuum forming.

It will be noted that injection moulding in all these respects appears to be a superior technique. Nevertheless, modern thermoforming techniques (of which simple vacuum forming is the most primitive) have become of considerable technical significance because they are at a decided advantage over injection moulding in three important areas:

- large formings;
- thin-wall packaging; and
- short-run or prototype products.

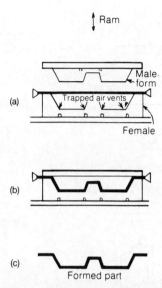

7.42 Thermoforming with matching moulds: (a) the heated sheet in position over the negative (female) mould; (b) the male is rammed down forcing the sheet into the female mould; and (c) the form is cooled and removed from the mould (after Modern Plastics Encyclopedia).

The technical developments that have brought this about can be grouped under three headings.

1. **The tool** can be negative (as for the bath) or positive. A negative mould gives an accurate outer shape; a positive tool gives an accurate inner shape. Thermoforming is now performed also with matched moulds (see Figure 7.42). Excellent reproduction of mould detail and dimensional accuracy can be obtained. The moulds are cheap and can be made of wood, metal, plaster, or epoxy.

2. **Pressure**: vacuum forming gives a pressure differential, at most, of 1 atmosphere. The use of positive pressures (up to 10 atmospheres) with or without vacuum on the other side, has led to the achievement of much superior forms. Because of the pressures involved, the machines become somewhat massive as the area of the forming increases.

3. **Pre-stretch**: the purpose of pre-stretching is to give a better material distribution in the wall and to permit deeper formings to be produced. There are several variants of this. Figure 7.43 shows the pressure-bubble vacuum-snapback technique.

(A) The heated plastic sheet is clamped and sealed across the pressure-box.
(B) Controlled air pressure blows the bubble, giving the pre-stretch of 35 to 40%.

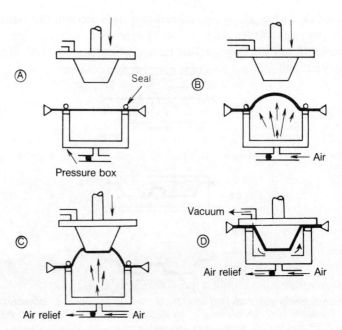

7.43 Thermoforming: the four stages in the pressure-bubble vacuum-snapback technique (after *Modern Plastics Encyclopedia*).

(C) The plug is moved down into the bubble; air pressure in the bubble keeps contact between the bubble and the plug.

(D) The plug is lowered to its final position, a vacuum is pulled between plug and sheet, and the pressure in the pressure-box causes the sheet to take up the shape of the plug. The speed at which the plug is lowered is an important parameter and is used to optimize material distribution and molecular orientation in the wall.

In this technique good reproduction of detail is achieved (a leathergrain pattern, for example) by a combination of vacuum on one side of the sheet, and pressure on the other.

The polymers used in thermoforming must be of fairly high relative molecular mass since the heated sheet must be form stable. The significant property is **melt elasticity**. The elastic effects are produced by entropy–elastic forces between physical cross-links (molecular entanglements). The polymers used include polystyrene, ABS, acrylics, polycarbonate, PVC, polypropylene, and linear polyethylene.

Sandwich construction is used to yield a surface with specific properties. For example in refrigerator panels there is a problem with environmental stress cracking in the presence of milk, fats, and oils. One solution is a surface of expensive, glossy and oil-resistant high-acrylonitrile ABS, with a cheaper low acrylonitrile ABS centre. The sandwich is coextruded (7.N.9). Coextrusion is not the only method for upgrading performance: laminates, heavily-filled and foamed polymers are also thermoformed.

7.7   Blow moulding

Blow moulding is a widely used technique for producing hollow containers (usually bottles) in vast numbers, extremely cheaply. There are three major variants:

- extrusion-blow moulding;
- injection-blow moulding; and
- stretch-blow moulding.

They have in common the formation of the precursor, a simple hollow tube known as a **parison**. One end of the parison is closed (for instance, simply nipped) so that it can be inflated in the heated, softened state. If inflates until it touches the walls of the cooled mould. The parison at once takes up the shape of the mould and cools. The mould is then opened and the bottle removed.

   In **extrusion-blow moulding** (see Figure 7.44(A)), the extruder extrudes a parison into the open mould. (The extruder ceases to extrude whilst the rest of the cycle takes place.) The mould closes and the parison is inflated (B). When the parison has taken up the shape of the mould (C) the mould is cooled and finally the mould opens, and the bottle is removed (D). The extruder then extrudes another parison (A) and the cycle is repeated.

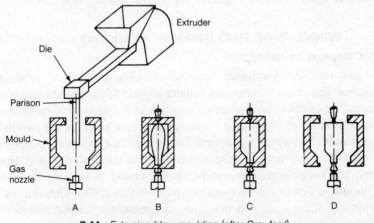

7.44   Extrusion-blow moulding (after Crawford).

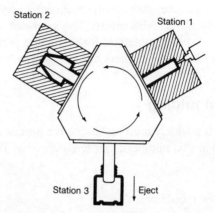

Station 2

Station 1

Station 3 | Eject

7.45 Three-station rotary machine for blow moulding (after *Modern Plastics Encyclopedia*).

In **injection-blow moulding** the parison is injection moulded on to a steel rod (see Figure 7.45, station 1). The rod and parison are then rotated to station 2 (the blow mould) where air is blown into the parison to cause it to take up the shape of the mould. The rod and bottle are next rotated to station 3, at which the mould opens and the bottle is ejected.

In **stretch-blow moulding**, highly sophisticated equipment causes the blown bottle to have **biaxial** orientation in the walls: the orientation is frozen-in by rapid cooling. In the other two techniques, extension and injection-blow moulding, some orientation is produced, but in stretch-blow moulding large, controlled, biaxial stresses are produced by a combination of blowing and stretching (the latter by means of an axial mechanical pull). The advantages of biaxial orientation are higher transparency and gloss, improved permeation, rigidity, impact resistance, and higher stability against internal pressure. The polymers used are PET (in particular), PVC, PP, and, to a lesser extent PS, ABS, and polycarbonate (see 2.N.9).

7.8   Compression and transfer moulding

7.8.1 Compression moulding

This was the first technique used for forming plastics in production quantities. The mould comprises a matched pair of male and female dies. A measured quantity of partially cross-linked polymer (this technique is not now used for thermoplastic polymers) is placed between the two halves of the mould. The upper die is then lowered and the polymer compressed; it is simultaneously heated by heat transfer from the heated mould. Before setting, the hot polymer completely fills the mould. It is then left to cure. The mould is finally opened and the part removed. The metering of the appropriate amount of polymer, and also some preheating, is being increasingly performed by screw machinery. This can increase by up to 400%

the output per mould cavity. The advantages of compression moulding over other processing techniques are:

- the polymer flows over shorter distances, thus reducing frozen-in stresses;
- polymer is not forced through small gates which can lead to points of weakness in a moulding;
- low mould maintenance costs;
- fairly low initial mould costs; and
- mould design is not complicated by sprue and runner layout.

A major disadvantage is that the mould temperature must be kept relatively low so that reaction times are lengthy: if the polymer surface in contact with the hot mould cures too quickly (before the rest of the charge is heated and therefore liquified) a poor moulding can result and the mould cavity pressure may then be inhomogenous. These problems are reduced by preheated screw feed and also in the related technique of transfer moulding.

7.8.2 Transfer moulding

In this process a measured charge of polymer (again partially cross-linked) is heated in a pot (see Figure 7.46(a)) from which it is rammed directly into the heated mould (b). Forcing the polymer through the entry gate results in a useful homogenization and a rise in temperature. The molten polymer then fills the mould under pressure from the ram. The polymer is held at temperature whilst the cross-linking reaction proceeds to completion. The mould is then opened and the now rigid part is removed. This is an important technique for the production of precision shapes, such as electronic and computer cases.

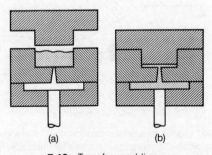

(a)   (b)

7.46 Transfer moulding.

Notes for Chapter 7

7.N.1

Melt fracture in dies is a complex and ill-understood phenomenon. Experimental evidence links it to the presence of excessive stress on the melt, and at least in some cases the critical stress is believed to occur at the **entry** to the die capillary. We can gain some understanding of this by analogy with the fracture of solids. We assume melt fracture occurs when the melt is elongated under too high a stress. Since we are concerned with a melt, the existence of a critical stress implies the existence of a critical strain-rate, for a given viscosity. Consider the rate of elongation that takes place within the die. Figure 7.47 shows an element of melt passing through a cylindrical die with diameter D which reduces to a value D_e at exit. To simplify, we assume that the polymer moves as if it were a plug of melt, with all shear confined to a narrow region very close to the wall; this is a reasonable assumption under some circumstances (see Section 7.2.3.3). In the contraction section, the melt is squeezed by the die as the diameter decreases, giving a rate of radial strain

$$\frac{\mathrm{d}\varepsilon_r}{\mathrm{d}t} = -\frac{1}{D}\frac{\mathrm{d}D}{\mathrm{d}t} = -\frac{1}{D}\frac{\mathrm{d}D}{\mathrm{d}z}\frac{\mathrm{d}z}{\mathrm{d}t}.$$

The last term in this equation is simply the velocity, which for a volume flow-rate Q must be $4Q/\pi D^2$, and the gradient of D is clearly $2\tan\alpha$, where α is the half-angle of the conical die channel. The rate of radial strain becomes

$$\frac{\mathrm{d}\varepsilon_r}{\mathrm{d}t} = -\frac{8Q}{\pi D^3}\tan\alpha.$$

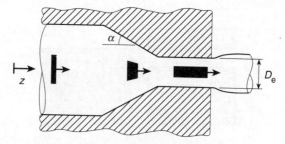

7.47 An element of melt passes through a cylindrical die. The die channel narrows, causing the element to be squeezed and to elongate. If the rate of elongation is too high, melt fracture will occur.

The melt is incompressible to a good approximation; hence the axial and radial strain-rates are related:

$$\frac{\mathrm{d}\varepsilon_z}{\mathrm{d}t} + 2\frac{\mathrm{d}\varepsilon_r}{\mathrm{d}t} = 0,$$

and we see by combining these equations that the maximum elongation rate is encountered at the narrowest point of the conical region, the 'throat' of the die:

$$\left(\frac{\mathrm{d}\varepsilon_z}{\mathrm{d}t}\right)_{\text{max}} = \frac{16Q}{\pi D_e^3}\tan\alpha.$$

To avoid melt fracture, the die must be designed to prevent an excessive strain-rate being reached at the desired throughput rate Q. It follows that the two die parameters D_e and α play important roles in preventing melt fracture. Since the limiting parameter is actually the stress, the critical elongation-rate increases with decreasing viscosity, and an effective means of escaping melt fracture is therefore to increase the melt temperature.

7.N.2

According to the WLF theory, temperature affects viscosity through the dependence of volume on temperature. The dominant parameter is volume: any change in volume leads to a change in viscosity. Thus increasing pressure, which lowers volume, increases viscosity: at 1000 atmospheres the viscosity increase is **roughly** equivalent to a drop in temperature of 50°C. Increasing temperature, which increases volume, decreases the viscosity. The basic concept is that molecules can flow in the liquid only if the neighbouring molecules cooperate. The degree of cooperation required depends on the volume. A useful image is the way in which people in an elevator find it increasingly difficult to get out when the elevator becomes more and more crowded.

7.N.3

The sensitivity of viscosity to molecular length will be no surprise to any reader who cooks spaghetti: long strands are more difficult to stir than a similar quantity of the chopped variety. The reason in both cases—polymer molecules and spaghetti—is the increasing difficulty of unravelling a bundle of string-like objects as their lengths increase. A long molecule mixed with others can be thought of as trapped within a tube. As it slides past its neighbours we might expect it to be resisted by a viscous force proportional to its length. Figure 7.48 confirms this—up to a point. The viscosity of polystyrene is proportional to RMM provided this does not

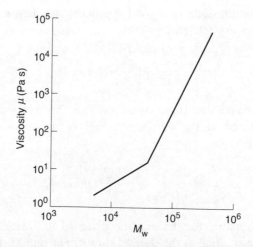

7.48 Graph showing the melt viscosity of polystyrene at 217°C, plotted versus weight-average RMM. Two regimes are apparent. Below the critical RMM M_c the gradient of the log–log plot is 1; above M_c it is 3.4. For polystyrene, $M_c \approx 39\,000$.

exceed a critical value M_c. Beyond M_c, the viscosity rises more rapidly. Why? The reason is that very long molecules are 'entangled' with one another. We use the word in a topological sense: each molecule is hooked around a number of its neighbours so that even if it is pulled taut it cannot be straightened along its whole length, but only zigzag fashion **between** entanglements—see Figure 7.49. Bulk flow of the melt then necessitates pulling molecules through these tangles. The result is a great increase in resistance to flow, and the viscosity rises more rapidly with RMM as shown in Figure 7.48. In fact the value of RMM at which the transition occurs is a

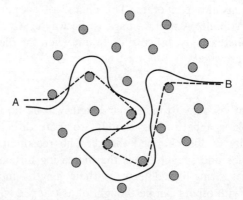

7.49 Illustration of the topological constraint provided by 'entanglements' in a liquid polymer. The circles represent chains with direction normal to the page; they provide a forest of obstacles to the movement of molecule AB. Even when pulled taut, AB is forced to follow the tortuous path shown as the dashed line.

good measure of the point where an entanglement **network** is formed, each molecule being entangled with at least two others. If the RMM between entanglements is M_e, the critical RMM must clearly be

$$M_c = 2M_e$$

For the majority of polymers, M_e lies in the range 2000–20 000.

7.N.4

The figure of 60% ignores the extrusion step common to essentially all plastics in which pellets are formed. Pelletization is achieved by extruding the polymerized material obtained from the reactor into continuous strands, about 3 mm in thickness, and then chopping them into pellets. From the viewpoint of energy economy it is not efficient to melt to make pellets and then to re-melt the pellets to fabricate. The RIM process (see Section 7.5.5) avoids this inefficient step.

7.N.5

The design of the hopper is of considerable importance. Any interruption in the flow of granules will impair the product. Feeding aids such as hopper crammers and feed metering have been used to advantage to increase output and overcome problems of bridging (blocking) in the feed hopper.

7.N.6

As the screw rotates, the polymer (solid, liquid, or mixed solid/liquid) is restrained from rotating by friction at the barrel surface, which generates a propulsion of the polymer along the barrel. This is in the manner of a rotating bolt with a nut restrained from rotating by a spanner: the nut will move down the screw. Even if the polymer is incompletely restrained from rotating, there is still a nett propulsion along the barrel.

7.N.7

When a screw extruder is extruding polymer through a die, there is normally a large pressure drop across the die. The screw is pumping against a pressure gradient. The pressure rise along the screw from atmospheric up to the entrance to the die must equal the subsequent pressure drop back to atmospheric at the die exit. As an example, consider the extruder described in Example 7.5, extruding a polymer of viscosity $\mu = 300$ Pa s through a die with a pressure drop of 10 MPa across it. Ignoring all but the metering section of the screw (of length 1.00 m), the length of screw channel over which the pressure rises is

$$L = 1.00/\sin 17.7° = 3.29 \text{ m.}$$

Thus the (positive) pressure gradient is given by

$$\frac{dP}{dz} = \frac{10^7}{3.29} = 3.04 \times 10^6 \text{ Pa m}^{-1}.$$

This pressure gradient produces a back-flow Q_p which, alone, would be given in the Newtonian approximation by the usual expression for pressure-driven flow through a rectangular duct (see Problem 7.10):

$$Q_p = \frac{WH^3}{12\mu} \frac{dP}{dz} = \frac{0.128 \times (0.009)^3}{12 \times 300} \times 3.04 \times 10^6 = 7.88 \times 10^{-5} \text{ m}^3 \text{ s}^{-1}.$$

In the Newtonian approximation the combined effects of the two opposing flows Q_0 and Q_p are additive (this is usually assumed in practice as an approximation even in non-Newtonian cases), thus the new throughput is now

$$Q = Q_0 - Q_p - 4.31 \times 10^{-4} - 7.88 \times 10^{-5} = 3.52 \times 10^{-4} \text{ m}^3 \text{ s}^{-1}.$$

In reality, the pressure gradient cases a further slight reduction in Q from 'leakage' flow through the narrow gap between the flight tips and the barrel wall (screw clearance, see Figure 7.20). When a screw becomes excessively worn, this can give a serious reduction in throughput. What if the flow is non-Newtonian, and the viscosity is not known? Then an approximation for the prevailing shear-rate can be taken to be that of the drag flow (see Example 7.2), i.e.

$$\dot{\gamma} \approx \frac{V}{H} \left(= \frac{0.748}{0.009} = 83.1 \text{ s}^{-1} \text{ in this case} \right)$$

and a corresponding viscosity can be employed to evaluate Q_p.

7.N.8

Hydraulic clamps are one of three basic clamping mechanisms in use today, the other two being mechanical (or toggle clamps) and hydromechanical clamps. A typical **double-toggle clamp** is shown in Figure 7.50. In the mould-open position (a), the actuating piston has pulled the crosshead back, so placing the links in a crumpled configuration. As the actuating piston moves to the right, the links commence to straighten. Initially, the mechanism works at low mechanical advantage and high velocity. As the moving platen approaches the mould-closed position (b), the linkages become almost fully extended. Here the mechanism works at high mechanical advantage and low velocity, which prevents the mould halves from slamming together and applies the greatest possible force for mould

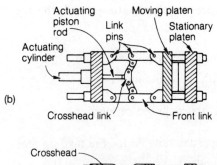

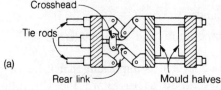

7.50 Typical double-toggle clamp: (a) open; and (b) closed (after *Modern Plastics Encyclopedia*).

closing. When the mould commences to open, the initial velocity is low, which avoids damaging the freshly moulded impression. Toggle-clamps are popular in the smaller machines (up to 400 tonnes clamp locking force) but above this value the weight of toggle mechanism becomes a drawback and hydraulic clamps are favoured (see Figure 7.28). Above 1000 tonnes locking force, hydromechanical clamps utilizing a combination of mechanical and hydraulic functions are used. In this range the fully hydraulic system has an increasing disadvantage in that for each cycle very large volumes of oil have to be moved.

7.N.9

Coextrusion is widely used in film making. For example, when packaging wine or fruit juices it is vital to retain fragrance and to inhibit the entry of undesirable gases (oxygen in particular). The best barrier film for this function is ethylene–vinyl alcohol copolymer (EVAL):

$$\text{+CH}_2\text{—CH}_2\text{)}_{n'}\text{(CH}_2\text{—CH)}_{n''} \qquad \text{(XXXII)}$$
$$\vert$$
$$\text{OH}$$

However, EVAL as a monolithic film is quite unsatisfactory: what is required, ideally, is a multi-layer, composite film. One excellent composite, for example, is a five-layer film ABCBA in which A is branched polyethylene (35 μm), B is a bonding polymer (5 μm), and C is EVAL (8 to 12 μm). Another barrier component widely used is nylon; for example, for a wide range of food packaging a typical combination is ABCBD, in which A is branched polyethylene, B is bonding polymer, C is nylon 6, and D is ethylene–vinyl acetate copolymer. In general, the required properties of the inner layer are:

- good heat sealing properties; and
- compatibility with the contents.

For the outer layer, the required properties are:

- mechanical properties suitable for the take-up reel;
- good strength; and
- higher melt temperature than the inner layer.

With a nylon barrier layer it is important that it be encapsulated (by the outer and inner layers) so as to retard the absorption of water. For thermoforming applications, these multilayers films are best produced by film blowing (see Figures 7.1 and 7.2). For an ABCBD film four extruders handling the four resins are required to operate in parallel, pumping liquid polymer to the die. The films are formed and brought together in the liquid state where they bond tightly together prior to the film blowing.

Problems for Chapter 7

7.1   At very low strain rates, polymer melts flow approximately as incompressible Newtonian fluids. Show that under these conditions the elongational viscosity λ and shear viscosity μ are related by

$$\lambda = 3\mu.$$

(Hint: separate tensile stress into hydrostatic and pure shear components.)

7.2   A film of polymer melt is stretched at a rate of elongation $\dot{\varepsilon}_t$, while constrained so that its width remains constant. Show that, under the conditions referred to in Problem 7.1, the true stress σ_t in the direction of stretching is given by

$$\sigma_t = \tfrac{4}{3}\lambda\dot{\varepsilon}_t.$$

7.3   A thread of molten polymer is extruded at a volume flow rate Q and under a tension F. Assuming isothermal conditions and Newtonian flow with constant elongational viscosity λ, show that the velocity v of the thread increases with distance z from the die as follows:

$$v = v_0 \exp\left(\frac{Fz}{Q\lambda}\right),$$

where v_0 is the velocity at the die exit (ignore die swell). Sketch a graph of v versus z as predicted, and compare this with how you would expect v to vary with z in a real extrusion process. Explain the differences.

7.4   A tubular film of linear polyethylene was extruded with straight sides (achieved by applying internal air pressure). The volume flow rate was 7×10^{-6} m³ s⁻¹, and the haul-off tension was found to be 15 N. The following measurements were made of the axial velocity v at various distances z from the die.

z (mm)	300	325	350
v (mm s⁻¹)	230	343	535

Find the apparent elongational viscosity at position $z = 325$ mm. (Hint: refer to Problem 7.2.)

7.5   Two flat, rectangular plates with lateral dimension $w \times l$ lie parallel, one above the other, a distance h apart. The gap between them is filled with molten polymer. One plate is fixed; the other moves parallel to it in the direction of dimension l at velocity v, causing the fluid to be sheared.

(1) Show that the fluid is displaced past the stationary plate with volume flow rate Q given by

$$Q = \frac{vhw}{2}.$$

(2) Find the equal and opposite shear forces which must be applied to the plates in order to sustain the deformation, assuming the polymer to be:
 (i) Newtonian with viscosity μ; and
 (ii) non-Newtonian, obeying the power law (eqn 7.18) with apparent viscosity μ_{a0} at shear stress τ_0, and with power-law exponent n.

7.6   A capillary rheometer is used to measure viscosity. A pressure transducer records the pressure drop ΔP across the die, and the die capillary has a diameter of 2 mm. Experiments are carried out at three volume flow rates Q, for dies of each of three lengths l. Table

Table 7.4 ΔP values for various values of l and Q in Problem 7.6

	$Q = 2$ $(10^{-7}$ m³ s⁻¹$)$	$Q = 4$ $(10^{-7}$ m³ s⁻¹$)$	$Q = 8$ $(10^{-7}$ m³ s⁻¹$)$
$l = 5$ mm	1.502	2.100	2.935
$l = 10$ mm	2.337	3.266	4.565
$l = 20$ mm	4.006	5.599	7.826

7.4 gives values of ΔP (in MPa) measured for polypropylene at 190°C. For each flow rate Q find the apparent viscosity, assuming Newtonian behaviour.

7.7 Use the data in Problem 7.6 to find the power-law exponent n for polypropylene at 190°C. Hence deduce the wall shear rate and corresponding apparent viscosity for each value of Q. (Equations for power-law flow through a capillary are given in Example 7.4.).

7.8 Find the percentage change in viscosity per degree change in temperature for polystyrene ($T_g = 97$°C) at the following temperatures:

(1) 170°C—assume that viscosity is governed by the WLF equation (see 7.N.2 and Problem 5.27, with $\mu_T = a_T \mu_{T_g}$).

(2) 230°C—assume that viscosity is governed by the Arrhenius equation with activation energy 104 kJ mol^{-1}.

7.9 Die swell is important in the design of gates for injection moulds: too little swell can lead to 'jetting' of the melt stream across the mould cavity instead of a smooth filling of the mould. The primary variables may be taken to be the gate dimensions t, w, and l (see Figure 7.36), melt temperature T, injection volume flow rate Q, and polymer molecular mass RMM. Construct a table to summarize the effects on die swell of increasing each variable alone, and add brief notes of explanation. (Recall that a measure of the duration of viscoelastic 'memory' is the relaxation time discussed in Chapter 4.)

7.10 A Newtonian fluid, of viscosity μ, flows through a slit-like channel of narrow rectangular section, driven by a pressure drop per unit length $\Delta P / \Delta L$. The channel is of depth h and width (transverse to the flow) w, where $w \gg h$.

(1) Find how the shear stress τ varies with distance y from the wall of the channel.

(2) Derive the velocity profile through the depth of the channel.

(3) Show that the volume flow rate Q is given by

$$Q = \frac{wh^3}{12\mu}\left(\frac{\Delta P}{\Delta L}\right).$$

(4) Show that the wall shear rate $\dot{\gamma}_w$ is given by

$$\dot{\gamma}_w = \frac{6Q}{wh^2}.$$

7.11 A power-law fluid obeying eqn 7.18 flows through the slit-like channel described in Problem 7.10.

(1) Find how the shear stress τ varies with distance y from the wall of the channel.

(2) Derive the velocity profile through the depth of the channel.

(3) Show that the wall shear rate $\dot{\gamma}_w$ is related to the volume flow rate Q by

$$\dot{\gamma}_w = \frac{2(2n+1)}{n} \frac{Q}{wh^2}.$$

7.12   A tube of branched polyethylene (MFI = 2) is extruded at 170°C through an annular die of diameter 50 mm and die gap 2 mm. Within the die the average linear velocity of the melt is 2 m min^{-1}. Calculate the pressure gradient in the die, using viscosity data given in Figure 7.14. Use the approximate procedure of assuming a constant viscosity appropriate to the maximum shear stress in the die gap. (Hint: a thin-walled tube such as this can be treated as if it were a flat sheet.)

If the die swell ratios in thickness and circumferential directions are 1.50 and 1.23, respectively, find the final dimensions and velocity of the extrudate.

7.13   Another polymer melt is extruded through the tube die described in Problem 7.12. This melt may be assumed to obey the power law, with viscosities 4500 and 1000 Pa s at shear rates 10 and 100 s^{-1}, respectively. A pressure drop per unit length of 10^8 Pa m^{-1} is applied to the die. Find the volume flow rate at which the tube is extruded. (A useful equation for power-law flow through a slit-like channel is given in Problem 7.11.)

7.14   A polymer melt of density ρ and specific heat capacity c_p is extruded through a die, across which the viscous resistance of the melt causes a pressure drop ΔP. Show that under adiabatic conditions (no heat loss to the wall of the die) the temperature rises by ΔT, where

$$\Delta T = \Delta P / \rho c_p.$$

7.15   An extruder is fitted with a screw with a single metering zone, as follows.

Flight depth	6.3 mm
Flight width	12 mm
Barrel diameter	120 mm
Length of metering zone	400 mm
Helix angle ϕ	17.7°

ABS is to be extruded at 190°C, with a screw speed 60 rev min^{-1}. Using viscosity data from the graph below, construct a graph of volume throughput Q versus the pressure rise ΔP along the metering zone for ΔP between 0 and 10 MPa. (Hint: use the approximation of constant viscosity at a value corresponding to an approximate shear rate in the channel of the screw.)

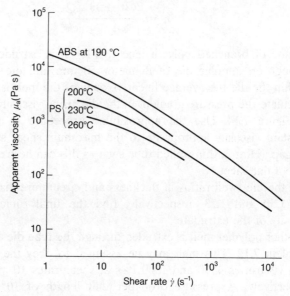

7.16 Two dies are under consideration for producing ABS tube with the extruder of Problem 7.15. Details are as in Table 7.5. (For the purpose of this question neglect pressure drops at die inlet and exit.) For each die construct a graph of volume throughput Q versus pressure drop ΔP across it, for Q between 0 and 2×10^{-4} m^3 s^{-1}. Use viscosity data for ABS at 190°C given above, with the approximation of viscosity being constant at the value appropriate to the maximum shear rate.

Superpose the Q versus ΔP graphs with that for the extruder screw (Problem 7.15) in order to find the operating point (Q and ΔP) for each screw–die combination. (At the operating point, Q and ΔP are equal for screw and die.)

Table 7.5   Information for Problem 7.16

	Diameter of die opening (mm)	Die gap (mm)	Length of channel in die (mm)
Die A	150	2	20
Die B	150	4	20

7.17 Polystyrene film is to be extruded using a 120 mm diameter vented screw with the following details.

First metering zone: length = 360 mm; flight depth = 5.7 mm; and flight width = 12 mm.

Second metering zone: length = 600 mm; flight depth = 8.25 mm; and flight width = 12 mm.

Take the screw helix angle ϕ to be 17.7°, the screw speed to be 80 rev min^{-1}, and the temperatures at the first and second stages to be 200°C and 230°C, respectively. Viscosity data for polystyrene are given in the graph above. Take the density of polystyrene within the screw to be approximately constant at 981 kg m^{-3}.

Use the above information to find expressions for the volume throughput Q as functions of pressure rise ΔP for each metering zone. Hence find the mass throughput (in kg h^{-1}) at which they will perform together satisfactorily (i.e. at which they will be matched), and the die pressure drop required to achieve this condition.

7.18   Rheometry experiments on a certain sample of polyethylene (PE) at 210°C show that, in steady shearing flow, shear stress τ is related to shear rate $\dot{\gamma}$ as follows:

$$\tau = 1.4 \times 10^4 \dot{\gamma}^{1/2} \text{ Pa.}$$

(1) A PE filament is to be extruded at 210°C, using a cylindrical die with diameter 2 mm over a length 15 mm. Find an expression for the volume flow-rate Q in terms of the pressure drop ΔP along the die channel.

(2) The die is to be used with an extruder whose operating characteristic (Q versus ΔP) for this PE is given by

$$Q = 3 \times 10^{-5} - 2.8 \times 10^{-12} \Delta P \text{ m}^3 \text{ s}^{-1}.$$

Predict the output volume flow-rate of the extruder–die combination (ignoring pressure drops at die entry and exit), and hence find the speed at which the filament must be wound up to achieve a final diameter of 1 mm.

7.19   To avoid melt fracture in the polyethylene extrusion described in Problem 7.18, the maximum elongational strain-rate in the die must not exceed 10^2 s^{-1}. The die is therefore provided with a conically tapering entrance channel of half-angle α. Estimate the maximum allowable value of α, and comment on any approximations used (see 7.N.1).

7.20   A thin-walled thermoplastic pipe is to be manufactured using a single-screw extruder and die.

(1) Draw labelled sketches of the important internal details of extruder and die.

(2) The molten polymer is a power-law fluid, where shear stress τ and shear rate $\dot{\gamma}$ are related by $\tau = k\dot{\gamma}^n$ and k and n are constants (see note below). Show that the relations between volume flow-rate Q and pressure drop ΔP for the die and extruder screw may be written

$$Q = A\,\Delta P^{1/n} \qquad Q = B - C\Delta P^{1/n}$$

respectively, where A, B, and C are independent of ΔP. Relate A, B, and C to dimensions on your sketches in (1). Note: use the expression for wall shear-rate given in Problem 7.11.

(3) In an attempt to increase productivity, the screw speed is increased by 10%, and the melt temperature is increased, giving a 5% decrease in k. Determine the percentage change in output rate, stating any assumptions you make.

7.21 A thread of Newtonian polymer melt is extruded at a volume flow-rate Q and hauled off under a tension F as it cools. The velocity at the die exit (after die-swell) is v_0 and the absolute temperature falls with distance z from the die according to

$$T = \frac{T^*}{1 + az}$$

where a and T^* are constants. The viscosity follows the Arrhenius equation (4.N.6.4) with activation energy ΔH. Derive an expression for the velocity profile $v(z)$.

A nylon monofilament is manufactured as described above, with $T^* = 550$ K and $a = 0.2$ m^{-1}. It is to have a final diameter of 100 μm and to be produced at the rate 5 ms^{-1}. Under optimum conditions the nylon crystallizes at 420 K under a true stress of 10^4 Pa. Find the required value of v_0 and hence the required die diameter if the die swell is 1.5. (Take the shear viscosity of nylon to be 100 Pa s at 550 K and the activation energy to be 40 kJ mol^{-1}.)

7.22 As the tube of Problem 7.12 is extruded, it is hauled off under tension, and inflated in order to maintain the diameter constant. Jets of cold air induce crystallization (the 'frost-line') a distance 1 m downstream of the die. By this point the tube has been drawn down to a wall thickness of 50 μm. Find the required tension in the tube and the internal air pressure required. (Note: ignore natural cooling between the die and frost line, and assume Newtonian behaviour with an elongational viscosity of 24 kPa s.) Refer to Problems 7.2 and 7.3.

7.23 In a new design for a system of thermoplastic shelving, the shelves are to be reinforced lengthwise by identical ribs, with cross-section as shown below. In use the upper surface will be visible. It is proposed to extrude the shelves in ABS (thermal diffusivity = 1.2×10^{-7} m^2 s^{-1}). To avoid distortion, the through-thickness mean temperature of any part of the moulding must be no more than 60°C after water-cooling. The temperature of polymer melt and water bath are respectively: $T_{melt} = 190$°C, $T_{water} = 15$°C. For an economically viable production rate, the cooling time must not exceed 10 s.

(1) Find the maximum value of the unknown thickness d consistent with the limit on cooling time.

(2) Suggest why the ribs are to have only 65% of the thickness of the plate (see Section 8.3).

(3) When the first batch of shelves is produced it is found that they are warped (i.e. bent) after cooling, with the ribs on the convex side. Suggest a reason for this and propose a remedy.

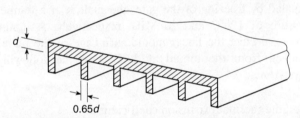

0.65d

7.24 Each half of a standard audio tape cassette has the form sketched below. It is essentially a flat plate of thickness 2 mm, with three openings as shown. A certain manufacturer produces them by injection moulding in transparent polystyrene (with thermal diffusivity 1.2×10^{-7} m^2 s). The gate is at position A, and a four-cavity mould is used with the layout sketched. To avoid distortion during ejection it is found necessary to wait until the mean temperature of the mouldings has fallen to 50°C. Initial mouldings are produced with the machine settings: $T_{mould} = 20$°C, $T_{melt} = 180$°C, cycle time $t_{cycle} = 8$ s. Unfortunately, this leads to an unacceptable reject rate: many mouldings are found to have cracks as indicated at position B. From experience, the moulder judges that any of the alternative modifications (i) to (iii) could remedy the situation, keeping other parameters constant.

(i) raise T_{melt} to 230°C and adjust t_{cycle} accordingly;

(ii) raise T_{mould} to 40°C and adjust t_{cycle} accordingly;

(iii) move the gate to position C (this could mean changing to a two-cavity mould, and a new mould would cost £8000).

(1) Explain the most likely cause of the cracking.
(2) For each proposal (i) to (iii), suggest why it might reduce the chance of cracking at B, and calculate the cost of implementing it, if a batch of 10^6 mouldings remains to be produced. Hence suggest the best remedy. (Take factory cost to be £25 per hour per injection moulding machine.)

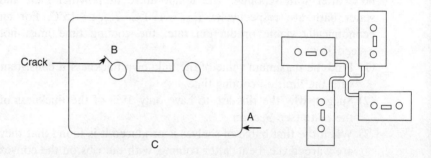

7.25 A plate of polystyrene is to be injection-moulded. When the cavity is sealed by freezing of the gate, the melt is at a temperature and pressure of 120°C and 20 MPa, respectively. Assuming uniform cooling, predict the linear mould shrinkage of the plate when it is removed from the mould at 20°C. Use the following data for polystyrene.

Volume thermal expansion coefficient at temperatures below T_g	$= 2.0 \times 10^{-4} \text{ K}^{-1}$
Volume thermal expansion coefficient at temperatures above T_g	$= 6.0 \times 10^{-4} \text{ K}^{-1}$
T_g	$= 97°\text{C}$
Compressibility at 120°C	$= 0.55 \text{ GPa}^{-1}$

7.26 A certain grade of polyethylene in the molten state obeys the following equation of state linking pressure P (in Pa), specific volume v (in $\text{m}^3 \text{ kg}^{-1}$), and absolute temperature T:

$$(10^{-3}P + 3.2 \times 10^5)(10^3 v - 0.822) = 2.9 \times 10^2 T.$$

In the solid state, at 20°C the density is 970 kg m^{-3}.
(1) Find the free linear shrinkage of a block of polyethylene on cooling uniformly from 130°C to 20°C.
(2) If it is injection moulded, find the packing pressure required in a moulding which is sealed with the melt at 130°C, to ensure

that the linear mould shrinkage on cooling to 20°C does not exceed 2%. (Assume uniform, i.e. slow, cooling.)

7.27 Flat circular plates, each with a concentric circular hole, are to be produced by injection moulding with a central diaphragm gate (see Figure 7.34). Final dimensions at 20°C must be: outer diameter = 200.0 mm; inner diameter = 40.0 mm; and thickness = 3.00 mm. The material will be a grade of polycarbonate with properties given below, and the mould will be steel at 80°C. With a particular setting of the machine controls and when the gate is just solidified, the mean temperature of the molten polymer in the mould cavity has fallen to 187°C and the packing pressure is 30 MPa. By assuming uniform cooling, estimate the volume shrinkage to be expected and hence the required dimensions of the mould cavity at 20°C (assume here that the polycarbonate is isotropic). Suggest a minimum speci-fication for a suitable injection moulding machine in terms of clamping force and shot size expressed in grams of polystyrene. (The polycarbonate is amorphous with the following properties: T_g is 150°C; volume coefficients of thermal expansion are 6.5×10^{-4} K^{-1} and 2.1×10^{-4} K^{-1} above and below T_g respectively; compressibility is 0.30 GPa^{-1} at 187°C; for polystyrene the density is 1060 kg m^{-3}; for steel the coefficient of linear thermal expansion is 11×10^{-6} K^{-1}.)

7.28 Describe and explain briefly the states of molecular alignment expected in the moulded plates described in Problem 7.27: (i) at the mid-plane of the plate, and (ii) at the surface of the plate. Sketch the distorted shapes of the plates that would be expected due to anisotropic thermal contraction when (i) or (ii) dominates. How might the distortion be prevented?

7.29 The diagram below shows a wall air vent to be manufactured by injection moulding from a particular grade of thermoplastic PVC (see note below). It has the form of a flat plate with nine through-thickness holes. Assuming the mean temperature in the mould cavity lags behind the temperature in the gate by 20 K during cooling, estimate the minimum packing pressure required to ensure that linear mould shrinkage, measured at 20°C, is less than 0.5%. What is the minimum capacity of injection moulding machine that should be used, in terms of clamping force, if the mould has a single cavity? (Note: for the purpose of this problem, the PVC can be taken to be amorphous with a glass transition temperature T_g of 75°C, independent of pressure; coefficients of volumetric thermal expansion above and below T_g are 6.3×10^{-4} K^{-1} and 1.8×10^{-4} K^{-1} respectively; bulk modulus of the molten PVC is

3.0 GPa.) See Problem 8.15 for a continuation of this Problem.

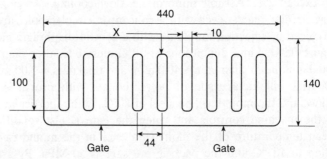

All dimensions in mm

7.30 It is likely that you are writing with a pen which has at least one component that is an injection-moulded plastic. Examine carefully either this or another convenient plastic moulding.

(1) Sketch the component, indicating on the sketch the position of: (i) the gate; (ii) the parting line where the mould halves met; and (iii) any weld lines which are visible.

(2) List advantages and disadvantages of locating the gate of the moulding in alternative positions.

7.31 Any thick block of polymer, quench-cooled from the melt, usually contains residual thermal stress which is tensile in the interior of the block and compressive at the surface (see Figure 7.18). Injection moulded components, however, especially near the gate, are sometimes found to contain residual stress which is compressive in the interior and tensile at the surface.

(1) Explain these two observations.

(2) Write notes on their implications for load-bearing applications of thermoplastics.

8 *Design*

8.1 Introduction

In order to achieve a successful design with polymers, the engineer must pay proper attention to their unique physical and chemical properties, and to their special processing characteristics, whilst keeping economic factors in mind. When mistakes are made, they are usually due to:

- inappropriate selection of materials;
- insufficient awareness of differences in behaviour between polymers and metals;
- designing components purely for function without adequate attention to manufacturing processes and mould design; and
- . application of traditional concepts used in designing with metals, rather than approaching the task afresh with polymers in mind.

This chapter will discuss the points to be noted in selecting polymeric materials, and the factors to be taken into account in making a detailed design.

8.2 Materials selection

8.2.1 The selection procedure

The first step in selecting a material for a particular application is to define the properties that are essential to the performance of the component. In some cases, the requirements limit the choice to one or two materials. For example, expansion joints in some chemical installations require a combination of flexibility with resistance to extremely corrosive environments: the bellows shown in Figure 8.1 are made from PTFE, despite its high price and difficulties in processing, because no materials other than fluorocarbon polymers meet the requirements. On the other hand, several less expensive polymers have sufficient resistance to dilute sulfuric acid to make a satisfactory battery case; toughened polypropylene is chosen for its combination of chemical resistance, stiffness, toughness, processability, and moderate price.

To take another example from the car industry: transparency is the first requirement in rear light lenses, which are made from PMMA. However, PMMA does not have a high enough softening temperature or fracture resistance to be suitable for headlamps; the material being studied as a

8.1 Bellows units machined from PTFE for use as expansion joints in the chemical industry (after Henry Crossley (Packings) Ltd.).

replacement for glass is polycarbonate. For other automobile components, stiffness and strength over the expected range of operating conditions are the basis for choice, although the designer must also bear in mind that some components are likely to come into accidental contact with automotive fluids, including acid, fuel, and oils, all of which can cause fracture problems.

Table 8.1 sets out the main factors to be taken into account in selecting polymeric materials. The relative importance of each factor will of course vary with the application. The initial shortlist of possible materials can be drawn up by eliminating polymers that do not meet key requirements. Chemical resistance and transparency have already been cited as examples of essential properties in particular applications.

The automobile industry provides other good examples of polymer selection. Body panels must have a sufficiently high softening temperature to withstand the conditions in the paint ovens.[1] For volume car production it is also essential to choose a material that can be processed at a rate

---

1 It is best not to paint, if the obtainable moulded colours are aesthetically acceptable, as painting lowers moulded part toughness (cracks can develop in a brittle paint film and propagate into the plastic).

Table 8.1 List of factors to be considered in selecting polymeric materials

Mechanical	Type and magnitude of normal service stresses Loading pattern and time under load Fatigue resistance Allowable deflections Overloads and abuse; impact resistance
Thermal	Normal range of operating temperatures Maximum and minimum service temperatures
Environmental	Solvent and vapour attack Reactions with acids, alkalis, water, etc. Water absorption effects Ultraviolet light exposure and weathering; oxidation Erosion by sand, rain, etc. Attack by fungi, bacteria, or insects
Electrical	Resistivity Dielectric loss Antistatic properties Tracking resistance
Hazards	Flammability Toxicity of additives or degradation products
Appearance	Transparency Surface finish Colour matching and colour retention
General	Tolerances and dimensional stability Weight factors Space limitations Expected service life Acceptance codes and specifications Environmental acceptability Leaching of additives Permeability to vapours and gases Wear resistance
Manufacturing	Choice of process Method of assembly Finishing and decoration Quality control and inspection
Economics	Materials costs Cost of capital plant: moulds, and processing machines Speed of production Number of mouldings/units required Operating costs of component, including maintenance and fuel consumption

sufficient to meet the demands of the assembly plant; whereas in making models for the specialist market, a high production rate is less important. Several types of polymeric material have been used for body panels, notably glass-fibre reinforced polyster resins, sheet moulding compounds (SMC), and reaction injection moulded (RIM) polyurethane; further developments are to be expected in this area.

In the engine compartment, the key properties tend to be resistance to automotive fluids, and mechanical properties over a wide range of operating temperatures. For example, glass-fibre filled nylons are chosen for radiator header tanks because of their high heat-distortion temperatures. Both nylons and polypropylenes have been used for cooling fans: polypropylene has the lower softening point, but is cheaper. Low-temperature properties are also important in some components: through-out the coldest winter, bumpers must resist impact, plasticized PVC trim must remain flexible, and rubber engine mountings must continue to provide vibration isolation.

Once a shortlist has been prepared by adopting **negative criteria** in order to eliminate unsuitable materials, the design engineer can review the more positive side of polymer properties. Manufacturers' handbooks and various encyclopaedias provide tables of data that form a suitable basis for materials selection. Table 8.2 gives figures for specific grades of the major engineering plastics.

It is important to recognize that properties vary widely within each family of plastics as they do in metals, **and that the terms 'polypropylene' and 'ABS' are no more specific than the term 'steel'.** In addition to variations arising from the chemical structure of the polymer itself, the range of products is enlarged by the use of additives, including antistatic and flame-retardant compounds, impact modifiers, mineral fillers, and fibre reinforcements. Furthermore, manufacturers are increasingly turning to alloying and blending of polymers in order to develop novel materials. Because of space limitations, it is not possible to show more than one or two variants.

8.2.2 The 'big four' commodity thermoplastics

The four major thermoplastics—**polyethylene, polypropylene, PVC, and polystyrene**—together represent over 85% by volume of world plastics consumption. Because of their lower prices, these commodity materials dominate the market, and in any materials selection procedure there are good economic reasons for considering them first before turning to the more expensive engineering plastics.

Polyethylenes are tough, ductile, and easily moulded, but their moduli, yield stresses, and melting points are relatively low. Polypropylenes are stiffer, have higher yield stresses and melting points, and also mould well, but in comparison with polyethylenes are more prone to fracture, especially at low temperatures. This is a consequence of the rise of yield stress of PP below its T_g. Rubber toughening substantially improves low-temperature fracture resistance, at some sacrifice in modulus. Short glass fibres increase the stiffness of PP, but increase costs and greatly reduce ductility. The stiffening effect of the fibres enables them to be used at temperatures approaching the melting point—a principle that applies equally well to other semi-crystalline plastics: without the reinforcement, creep is a major problem in the temperature range between T_g and T_m. As

Table 8.2 Data on representative grades of thermoplastics. Families of polymers are listed in descending order of consumption. Strain rate and other conditions are specified in ASTM standards

	ASTM test	HDPE	PP	30% GF PP	uPVC	ABS	PMMA	50% RH PA6	dry PA6.6	PA6.6 50% RH	33% GF/T§ PA6.6 50% RH	PC	POM	MPPO	PBT	30% GF PBT	40% GF PPS†	PSF	PES
Relative price		1.0	1.0	1.6	0.7	1.8	1.8	2.8	2.9	2.9	3.0	3.5	2.5	3.0	3.1	3.1	6.9	12.0	12.0
Young's modulus (GPa)	D790	0.8	1.5	6.5	3.0	2.1	3.2	1.2	2.8	1.2	5.5	2.3	2.8	2.5	2.2	7.5	11.7	2.7	2.6
Yield/fracture* stress (MPa)	D638	28	33	86*	55	41	72*	41	83	59	110*	65	69	55	50	120*	134*	70	82
Elongation at Break (%)	D638	300	50	4	30	20	4	290	60	300	4	100	40	50	300	3	1	50	40
Notched Izod impact strength J m⁻¹	D256	>1300	150	125	70	350	40	75	53	112	140	700	75	200	55	65	75	86	84
Heat distortion temperature (°C)	D648	<23	<23	148	55	110	95	<23	90	<23	232	140	136	130	55	215	>260	174	203
Limiting oxygen index‡	D2863	18	18	18	45	19	17	23	28	31	28	25	16	24	20	32	46	30	36
Coefficient of linear expansion (10⁻⁶ K⁻¹)	D696	90	110	25	60	96	68	70	81	81	18	70	122	60	70	25	22	54	55
Specific gravity	D792	0.96	0.90	1.12	1.40	1.04	1.18	1.13	1.14	1.14	1.32	1.20	1.42	1.06	1.31	1.62	1.60	1.24	1.37

† see 8.N.5
‡ see 8.N.6
§ GF/T means 'glass-filled and rubber toughened'.

both PE and PP have glass transitions that are below room temperature, they tend not only to creep under applied stresses, but also to warp in response to moulded-in stresses and physical ageing processes, including recrystallization (see Chapter 4). Their aliphatic chain structure makes them relatively susceptible to oxidation, especially in sunlight, and adequate stabilization by antioxidants is therefore important. Polyethylene is used for gas and water pipes, automobile fuel tanks, and cable insulation. Polypropylene is used for air ducting, parcel shelves, air cleaners, and many other car applications, and in many domestic appliances (washing machine tanks are a good example).

The PVC family consists of two contrasting types of material: the rigid, unplasticized polymer (uPVC) is quite different from plasticized PVC, which is really a rubber. Relatively little uPVC is injection-moulded, mainly because it degrades rapidly above 200°C, giving off HCl. The main processing method is extrusion: large amounts of PVC are used in pressure pipes for water supply and irrigation, in drain and sewer pipes, and in conduit. Other major applications include extruded window frames, wall cladding, and sheet for use as shuttering when pouring concrete. Because the basic polymer is notch-sensitive, most uPVC grades contain at least a small amount of a rubber toughening agent. The relatively high density of PVC offsets its low price per tonne to some extent.

The polystyrene family has strong claims for consideration when rigid, dimensionally-stable mouldings are required. Polystyrene itself is a transparent, brittle polymer with a limited range of applications, mainly in packaging and disposable cups. Rubber modified or 'high-impact' polystyrene (HIPS) grades are preferred for durable items including radio, TV, and stereo cabinets and tape cassettes. Increased toughness, yield stress, and solvent crazing resistance are achieved by copolymerizing the styrene with acrylonitrile to produce ABS (acrylonitrile–butadiene–styrene), which has the added advantage of giving a much glossier surface on injection mouldings; the drawback is a substantially higher price. Applications of ABS include telephones, radio, TV, and business machine cabinets, and automobile components. As in the case of PE and PP, addition of antioxidants is necessary, because the rubbery polybutadiene phase is especially susceptible to oxidation on exposure to sunlight.

8.2.3 The engineering thermoplastics

The more expensive plastics compete by offering a superior combination of properties to those obtainable from the 'big four' commodity polymers. For transparency and resistance to uv degradation PMMA is outstanding, with polycarbonate as a competitor where there are serious fracture problems. Transparent toughened grades of PMMA have been developed to meet this competition. Polycarbonate offers a unique combination of toughness, stiffness, high softening temperature, and processability that makes it the leading contender in many applications; its main drawback is it susceptibility to solvent crazing, which can be reduced by alloying with PBT. Applications of polycarbonate include street lighting, safety helmets, electrical terminals, coil formers, and housings; the PC–PBT alloy is used for car bumpers. Polysulfone and polyethersulfone are somewhat similar to polycarbonate (8.N.1), in that they are ductile, high-softening, non-crystalline plastics; although they do not offer the same degree of toughness, they are able to operate at higher temperatures. They are also used in load-bearing and insulating parts in the electrical industry, in carburettor parts, bearing cases, radomes, and heat-resistant air ducts.

The other non-crystalline plastic shown in Table 8.2 is modified poly(phenylene oxide) (MPPO), a rubber-toughened alloy made by melt blending PPO with HIPS (8.N.2). This blend can in one sense be regarded as a high-performance member of the polystyrene family, offering toughness, stiffness, and mouldability, together with a higher softening point than ABS (which is one its main competitors). Modified PPO is used for business machine housings, car radiator grilles, radio, television, and camera parts and other components of consumer durables.

The group of semicrystalline engineering thermoplastics, comprising the nylons (PA 6 and PA 6.6), polyacetals (POM), and the thermoplastic polyesters (PBT and PET), form another convenient set, since they have a number of characteristics in common, and compete for similar markets. They are particularly suitable for bearings and gears because of their abrasion resistance, general toughness and fatigue resistance, and resistance to oils, gasoline, and organic liquids. With the aid of glass-fibre reinforcement, they can operate at relatively high temperatures. Toughened grades of all three families of polymer are available, as brittle fracture can be a problem in the standard, untoughened grades when notches or stress concentrators are present. As Table 8.2 shows, there are significant differences in properties within the group, most notably in the effect of moisture: whereas nylons swell and soften as a result of absorbing water, polyacetals and polyesters are relatively little affected. Nylons, polyacetals, and thermoplastic polyesters are used for an extremely wide range of engineering applications, from drill housings to timing chain sprockets.

8.2.4 Thermosets and composites

The principal thermosetting moulding materials are the phenol–formaldehyde (PF), urea–formaldehyde (UF), melamine–formalehyde (MF), polyester, and epoxy resins (8.N.3, 6.N.1, 6.N.2). The prices of the cheapest resins are just below those of the polypropylenes, whilst some of the epoxy resins used in aerospace applications are over ten times as expensive. The formaldehyde-based resins all evolve gases during cure, which can cause porosity. Both PF and UF resins are used mainly in electrical fittings. Some MF resin is also used for plugs and switches, but the main applications are in tableware. Polyesters and epoxy resins are the preferred materials for making fibre composites. However, phenolic-based composites are currently being specified for trains because of problems of flammability with polyesters.

Polyester resins (6.N.1) cost about 20% more than PP, and can be considered as commodity materials. They are used in a variety of composites—mostly with glass reinforcement—which are characterized by high stiffness, strength, and toughness, resistance to chemical attack, and dimensional stability; maximum operating temperatures are about 100°C. The term GRP usually refers to glass-reinforced polyesters, although it can have wider connotations. Major applications of GRP include building and construction, ship and boat hulls, chemical plant, and various forms of transport.

For more demanding applications, where greater strength and stiffness are required, and the range of operating temperatures is higher, the preferred materials are epoxy resins (6.N.2) containing carbon or aramid fibres. In order to achieve the stiffness demanded by the aerospace industry, it is necessary to use very highly cross-linked resins, which have relatively low G_{IC} values (in the region of 100 J m^{-2}). Active efforts are being made to improve the balance of stiffness and toughness in high-performance composites, both by modifying the epoxy resin system, and by using other matrixes: one contender is poly(ether ether ketone) (PEEK), which is thermoplastic (8.N.4). New grades of polyimides containing fluorine have been developed for continuous use at up to 250°C.

8.2.5 Rubbers

The susceptibility of NR and SBR to attack by oxygen and ozone is due largely to the presence of carbon–carbon double bonds on the main chain. Ethylene–propylene copolymer rubbers (EPR) and butyl rubbers,[1] which have relatively few double bonds, are much more resistant to oxidation. Both are slightly more expensive than NR. Good all-round mechanical properties make EPR suitable for a wide range of general applications, whereas butyl rubbers show high mechanical hysteresis at room temperature, and are therefore more limited in their applications: hysteresis causes overheating in cyclic loading. The outstanding characteristic of butyl rubbers is their low permeability to gases, which has established them as indispensable materials for tyre inner tubes and liners.

Tyres usually contain at least three other rubbers: the tread is made

---

1   Butyl rubber is basically polyisobutylene (q.v.) containing 1–3% of isoprene as comonomer.

from a blend of polybutadiene (BR), which provides wear resistance, and SBR, which improves grip in wet weather; the remainder of the tyre is built from various combinations of natural rubber, BR, and SBR, in order to achieve optimum fatigue and cut-growth resistance, bonding to metals and fabrics, and low heat build-up.

Ease of processing is an important advantage in all branches of polymer technology, and one that has been important in developing applications for plasticized PVC. Unlike the more conventional rubbers listed above, plasticized PVC is a thermoplastic elastomer, which can be extruded, calendered, or melt-coated on to fabrics and other substrates. Without the support of a fabric or other base, it suffers from a relatively low tensile strength and tear resistance. On the other hand, it is a good electrical insulator, and is inexpensive: there is little difference in price between plasticized and unplasticized grades of PVC. Applications include cable insulation and interior trim for cars.

Another family of polymers that has found extensive applications in cars, partly because of easy processing characteristics, are the polyurethanes (PU), which are usually used in foam form. The RIM process (see Section 7.5.5) is based largely upon PU. Depending upon their chemical structure, polyurethanes vary in stiffness from very rubbery to rigid. All grades offer good resistance to tearing, abrasion, oils, and oxidation.

8.3 Designing for manufacture

In selecting a polymeric material, it is vitally important to decide at the same time on the method of manufacture of the component, and to take account of the effects of processing on properties. Whereas metallic components are formed in a series of operations, including forging, pressing, machining, heat-treatment, and polishing, **polymeric components are usually formed in a single operation**, and offer little opportunity for subsequent treatment. A design may appear attractive on the drawing board, but prove unsuccessful because of insufficient attention to the production process. Before turning our attention to the traditional concerns of the designer in the areas of stiffness and strength, it is necessary to describe the way in which product performance is affected by details in the method of manufacture.

Polymers vary widely in their processing characteristics, and it is important to ensure that the design developed can be produced economically and efficiently from the chosen material. The most obvious differences in processing behaviour are between thermoplastics and long-fibre composites, but there are also significant variations within each class of materials. To take an extreme example, PTFE has such a high melt viscosity that it cannot be injection-moulded, but has to be formed from powder by sintering. Differences in melt viscosity, crystallinity, and resistance to warping under the influence of moulded-in stresses are all factors to be considered in developing a design. Economics are of course a major consideration. When there is a requirement for a large number of a given component, injection moulding is usually the preferred process. Thermoforming is competitive when the reduced capital costs of machine and mould outweigh the additional cost of extruding sheet prior to forming, but it is suitable only for relatively simple shapes. Another factor to be considered in developing a design is how the component is to be joined to the remainder of the product: welding and fastening processes are suitable for thermoplastics, whereas thermosetting resins cannot be welded, but have to be joined by adhesives or mechanical fastening.

8.3.1 Injection moulding (see Section 7.5)

The design of the mould is a major part of the whole design process. A complicated mould costs as much as a large family house, and once made can be altered only to a minor extent. If there are serious problems, they can usually be solved only by throwing the mould away and making a new one. Two basic decisions concern the siting of (i) the parting line between the two halves of the mould, and (ii) the gate or gates. The method of extracting the moulding at the end of the cycle also needs some thought. Removal is aided by allowing a draft angle (i.e. taper) of between $1/8°$ and $1°$, depending upon the stiffness of the material, and by the use of ejector pins or plates. A moulding containing undercuts and through-thickness holes will require a number of retractable 'cores', which will obviously add to costs. External and internal threads can also be made in this way, using core pins that unscrew. Alternatively, well-rounded undercut parts, including external threads, can be snapped or stripped from the mould, provided that the strains reached in the process do not exceed allowable limits. It is unwise to attempt this procedure with glass-reinforced materials, because of their low failure strains.

The way in which the mould fills determines the direction, degree, and type of molecular orientation in the moulding, especially near the surface. As illustrated in Figure 8.2, the material in the melt front is stretched as it advances, and is then cooled rapidly when in contact with the mould wall, with the result that molecular orientation is retained. Fresh material then flows between the frozen surface layers to create a new melt front. This cycle continues until the mould is full. Relaxation takes place rapidly in the melt if it has a low viscosity, and orientation is therefore highest when the melt temperature is relatively low. A compromise may be necessary between product quality and production economics, as low melt temperatures reduce cycle times. Avoidance of thermal degradation is another reason for reducing processing temperatures, especially in the high-softening glassy plastics.

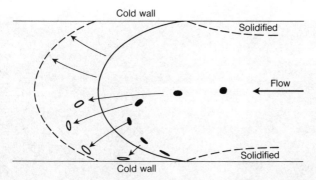

8.2 Schematic diagram showing flow into a mould. A spherical region behind the melt front becomes stretched into an ellipsoidal shape as it moves forward. Its orientation is then frozen-in owing to rapid cooling in contact with the cold mould wall.

When the material contains short glass fibres, or other reinforcements, their orientation will also be determined by the flow pattern. Figure 8.3 shows a section through an injection-moulded part made from glass-reinforced polypropylene: near the surface, the fibres are oriented preferentially in the flow direction, whilst in the central region they are aligned in the direction transverse to flow. A full explanation of these effects is beyond the scope of this book.

In recent years, designers have increasingly used computers to simulate the flow of polymer melts in moulds, in order to optimize mould filling patterns. These computer aided design (CAD) techniques allow the user to experiment with different positions and designs of gates and runners, and to determine where melt fronts will meet to form weld lines. Weld lines are potential sites for crack initiation, and should either be eliminated or located in regions of low stress concentration. They are inevitable when there is more than one gate, or when the moulding contains through-thickness holes, but they can also be formed by constricting melt flow locally, so that the advancing fronts on either side come together. After deciding upon the gating, and determining the basic mould flow pattern, the next step in the CAD analysis is to divide the mould into basic elements (discs, sheets, and cylinders) related to the flow paths, and to calculate pressures, temperatures, and shear rates in the flowing melt, using experimental data on the rheological properties of the polymer. The procedure can be repeated many times until an optimum mould design has been achieved.

An example of computer-aided prediction of mould-filling is shown in Figure 8.4. The component is a radiator fan cowl for a motor vehicle. It is to be moulded from 10% talc-filled polypropylene. As can be seen, the shape is three-dimensional and an analytical solution for the flow pattern is impossible. The first step in the computer simulation is to approximate the shape of the real component by the mesh of triangular elements shown

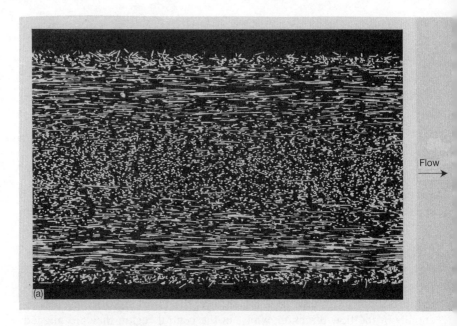

Flow
→

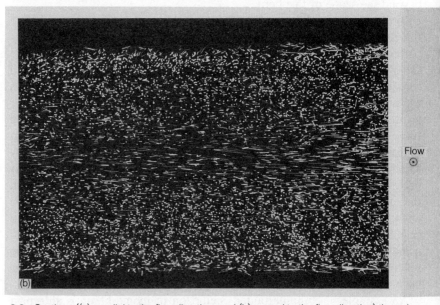

Flow
⊙

8.3 Sections ((a) parallel to the flow direction; and (b) normal to the flow direction) through a glass-reinforced polypropylene injection moulding, showing the short fibres near the surface oriented parallel to the flow direction, whilst those in the central region tend to be transverse to flow (after M. W. Darlington).

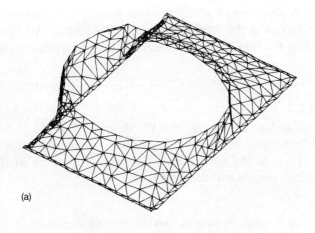

(a)

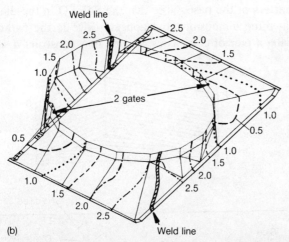

(b)

8.4 Computer-aided simulation of injection moulding of a polypropylene radiator fan cowl. (a) Mesh of triangular elements used in the numerical analysis of melt flow. (b) Positions of melt fronts predicted for various times (in seconds) from the start of mould-filling. The mould is full within approximately 3 seconds. (After C. R. Fenn and M. Sharrock, Chloride Lorival Ltd., and Landrover UK Ltd.)

in Figure 8.4(a). The flow of molten polypropylene through this mesh is calculated by numerical analysis, allowing for the accompanying cooling of the melt by the walls of the mould. Typical results are shown in Figure 8.4(b), where the positions of the melt fronts are marked for different times from the start of filling. In this particular case, two gates are assumed; two weld lines are predicted, one on either side of the opening in the cowl. Note that the siting of the gates is well-judged: the completion of filling is predicted to occur at about the same time on both sides of the moulding. A simulation such as this also predicts temperature and pressure distributions in the moulding. Based on the results, the design can be modified and the simulation repeated until an optimum design is reached.

Example 8.1

A rectangular box measuring 20×12 cm and 10 cm deep is injection-moulded from a gate at the centre of the base. Assuming that at any given time the speed of the flow front is the same everywhere in the mould, determine the positions of the weld lines.

Procedure

The box can be represented in two dimensions as shown in Figure 8.5. Because of symmetry, it is sufficient to show two adjacent sides, plus the attached quarters of the base. Note that the point O in the diagram is the same gate in three dimensions, but appears twice in the two-dimensional drawing. Using a pair of compasses, we can now construct a set of circles

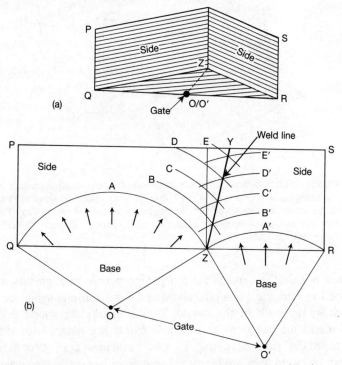

8.5 Construction for determining position of weld line in rectangular box: (a) the box is cut in half along the diagonal PQRS; (b) a further cut is made along the cross-diagonal OZ, and the box is laid flat.

of radii $OA = O'A'$, $OB = O'B'$, etc., showing the position of the flow front at succeeding stages in the mould-filling process. Where the circular arcs meet, a weld line occurs. There will be a similar weld line in each corner region of the box.

Comment

This is a highly simplified version of the mould-filling calculation performed in computer-aided design. In the diagram, the weld line bisects the angle OZO'.

The smooth flow of the melt in a mould is interrupted by **sharp corners and abrupt changes in wall thickness**. Both types of feature have other undesirable effects, and should be avoided where possible. Sharp internal corners act as stress concentrators, and are one of the most frequent sites for crack initiation in moulded parts. Stress concentration factors can be calculated using standard formulae given in books on stress analysis. They should preferably be limited to a value of 1.5, and when a sharp corner cannot be avoided, the radius should be not less than 0.5 mm.

Changes in wall thickness not only interrupt flow, but also give rise to problems of differential shrinkage, including sink marks, voids, warping, and moulded-in stress. Sink marks are particularly prominent opposite ribs in mouldings made from semi-crystalline polymers, as illustrated in Figure 8.6. During moulding, the skin region cools very rapidly while the core is still molten and under pressure. Subsequent cooling and crystallization in the core causes volume contractions of the order of 10% (see Table 7.2).

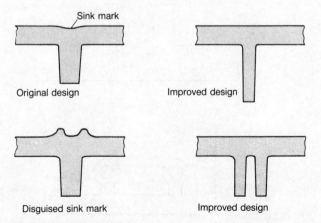

8.6 Sink marks form where the wall thickness changes, and especially opposite ribs and bosses. The effect is minimized by limiting rib thicknesses, if necessary by using double ribs. Where sink is inevitable, styling features can be added in order to disguise it.

Stresses arising from differential contraction are sometimes sufficient to initiate cracks over a long period. Ribs, flanges, and bosses, which increase the stiffness and strength of the part without increasing overall section thicknesses, require careful design in order to avoid these problems. In order to reduce sink marks, ribs should be between 1/2 and 2/3 the thickness of the wall that they are reinforcing. For this reason, a double rib is preferable to a single, wide rib (see Figure 8.6). Unless lateral support is added, the height of a rib should be less than three times the wall thickness. In the automobile industry, where these limits have to be exceeded in some cases, styling features are sometimes added to disguise sink marks from ribs, as shown in Figure 8.6.

Threaded metal inserts may be moulded into the components simply by placing them into the mould, preferably after preheating. This technique is known as **insert moulding**. This type of attachment point is especially useful when the screws have to be removed repeatedly during service. In order to avoid stress concentrations, the inserts should have no sharp corners, and any knurling should be rounded. It is usually advisable to provide strengthening to the moulding at the point of attachment, in the form of a boss with supporting ribs; Figure 8.7 shows some recommended designs. Because plastics have higher coefficients of expansion than

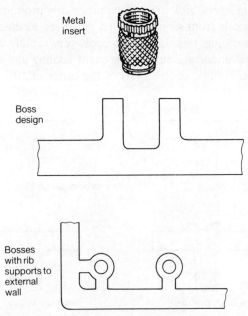

Metal insert

Boss design

Bosses with rib supports to external wall

8.7 Points of attachment using screws. The metal insert is threaded internally, and knurled externally to provide grip. It is supported by a boss, which in turn is supported by ribs of various designs.

metals, they shrink on to the insert on cooling from the melt, locking it into place.

8.3.2 Joining and fastening

When two surfaces made from the same thermoplastic have to be joined, welding is the preferred method. There are several ways in which welding can be accomplished, but they all depend upon melting the surfaces to be joined, and bringing them together under pressure. During this operation, melted material flows outwards from the weld, forming a 'bead' on the side of the component, as illustrated in Figure 8.8.

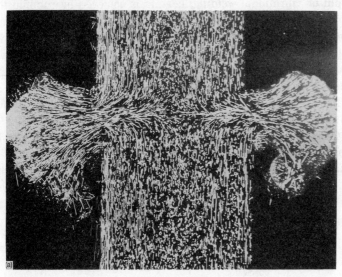

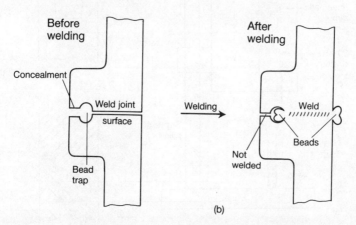

8.8 (a) Section through a hot-plate welded joint in glass-reinforced polypropylene, showing how the melt flows outwards to form a bead. (b) In some designs, the weld bead is hidden.

The simplest type of welding is the **hot-plate or hot-tool process**, in which the surfaces are held in contact with a heated metal plate for about 15 seconds, and then pressed together. This method is used to join polyethylene gas pipes. Some forethought is necessary in designing parts that have to be welded together. The surfaces are preferably planar, and provision must be made for jigging to bring them together. A common requirement is that the weld bead be hidden from view, which can be achieved as illustrated in Figure 8.8. As the bead can act as a stress concentrator, the siting of the weld can be important.

In **spin or vibrational welding**, heating is achieved by rubbing the two surfaces together at a sufficiently high speed **to melt the polymer by friction**. Spin welding is suitable for circular objects, whereas vibrational welding can be achieved by either circular or linear motion. Total cycle times are of the order of 10 seconds. The moulding must be designed to allow the application of pressure to the joint, which usually means including a flange, and the walls of the moulding must be thick enough to prevent excessive flexure during welding.

Another process in which mechanical energy is transformed into heat is **ultrasonic welding**, which is illustrated in Figure 8.9. A piezoelectric crystal transmits longitudinal vibrations at a frequency of about 20 kHz through

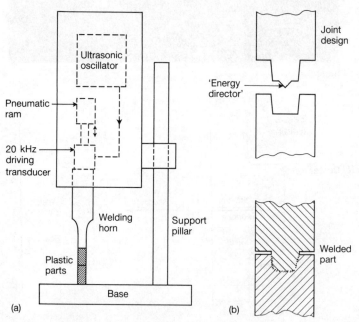

8.9 (a) An ultrasonic welding machine; and (b) a section through a moulded part intended for ultrasonic welding, showing the 'energy director'.

the workpiece via a tuned metal horn. The surface to be welded is shaped to form an 'energy director' in which the ultrasonic energy is converted into heat. Bonding is accomplished within 1 second. This method is rather more restricted in its capabilities than either hot-plate or frictional welding, because standing waves are set up in the horn, and it is difficult to develop uniform heating over a wide area.

The ultrasonic welding machine can also be used for insertion of metal parts and for a type of riveting. Sonic energy transmitted through a small metal part melts the polymer and allows the metal part to be inserted within 1 second. Apart from the heat generated at the metal–plastic interface, the moulding remains cool. This method offers cost advantages over the use of moulded-in metal inserts, and is used, for example, in mounting the hinge joints on to spectacle frames. Whichever method is chosen, supporting bosses should be included in the design. Figure 8.10 illustrates the use of the machine for ultrasonic staking, in which a moulded peg of plastic is heated and formed into a retaining head. The base of the peg and the contour of the flange should be suitably rounded, in order to avoid stress concentrations. The energy concentrator may be designed into the top of the peg or be produced by shaping the tip of the welding horn.

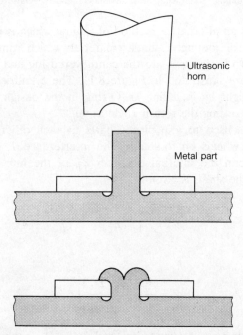

Ultrasonic horn

Metal part

8.10 Use of the ultrasonic machine for 'staking', in which the top of a plastic peg is melted to form a retaining head, in a similar manner to riveting.

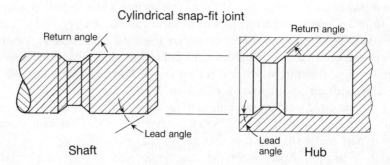

Cylindrical snap-fit joint

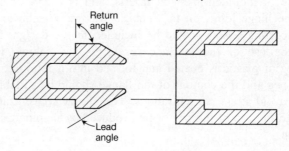

Cantilevered lug snap-fit joint

8.11   Features of snap-fitting using cylindrical and cantilevered lug joints.

Snap-fitting is an alternative method of joining which is not only cheap but also useful for joining dissimilar materials which cannot be welded together. The two basic types are the cantilevered lug and the cylindrical snap-fit, which are illustrated in Figure 8.11. The cylindrical type can be made pressure-tight by including an O-ring in the design. A permanent joint is made by setting the return angle at 90°.

Press-fitting is used to join plastic parts to each other or to metals, especially to fit wheels on to shafts. The interference I, defined as the difference between the internal diameter d_h of the hub and the outer diameter d_s of the shaft, is given by

$$I = d_s - d_h = \frac{d_s}{Z} \left(\frac{Z + \nu_h}{E_h} + \frac{1 - \nu_s}{E_s} \right) \sigma_d, \qquad (8.1)$$

where

$$Z = \frac{1 + (d_h/D_h)^2}{1 - (d_h/D_h)^2},$$

and the subscripts h and s refer to hub and shaft; D_h is the outer diameter of the hub; σ_d is the design stress of the polymer; ν is Poisson's ratio; and

E is Young's modulus. An incorrect interference can result on the one hand in cracking of the part, and on the other in loosening of the joint. Stress relaxation in the polymer and differences in thermal expansion coefficient must both be taken into account. Knurled metal surfaces enable the polymer to creep into depressions and increase joint strength.

8.3.3 Thermosetting polymers

Manufacture of components from thermosetting resins or from rubbers involves not only shaping the material, but also carrying out an exothermic chemical reaction. Proper control of temperature is required in order to obtain an acceptable rate of reaction without overheating and causing unwanted reactions which lead to thermal runaway. It may therefore be necessary to restrict thicknesses. The choice of curing formulation is also important, and where solid curing agents are used, as in epoxy resins, the state of subdivision of the crystals can be critical. Significant shrinkage occurs during curing of resins, and must be allowed for in mould design. Constrained shrinkage can cause void formation and internal stress.

Differential shrinkage causes surface irregularities in fibre composites, where the fibres restrict contraction locally. Special 'low-profile' additives have been developed for use in polyester resins to combat this problem in applications that demand high-quality surfaces (notably in the car industry). The other major problem in the manufacture of components from composites is to minimize defects, including broken, kinked, misaligned, or incompletely wetted fibres, resin-rich areas, voids, or dust contamination. This is mainly a question of production engineering, with an emphasis upon quality control. Fibre orientation in laminates can be predetermined with some precision by means of tape-laying or filament-winding machines, but some care is needed to avoid gaps on the one hand and overlap on the other.

8.4 Designing for stiffness

8.4.1 Plastics

Standard tests for the stiffness of plastics are based on either tensile or flexural measurements. The tensile test has the advantages that the stress is uniform in the gauge length, and that the corresponding strains can be measured directly. On the other hand, the three-point bending test illustrated in Figure 8.12 can be carried out with very simple apparatus. Young's modulus is calculated by applying the standard equations for a beam undergoing small elastic deflections:

$$E = \frac{L^3 P}{4bd^3 \Delta}.$$

$$(8.2)$$

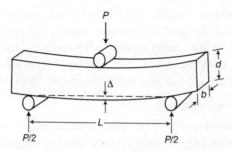

8.12 The three-point bending test.

The maximum stress occurs at the midspan, in the outer fibres, and is given by

$$\sigma_{max} = \frac{3PL}{2bd^2}. \tag{8.3}$$

The maximum shear stress occurs in the neutral plane at the centre of the bar:

$$\tau_{max} = \frac{3P}{4bd}. \tag{8.4}$$

These equations hold for small deflections, when the polymer is linearly viscoelastic.

Values of modulus determined in tension or flexure at one or more temperatures are provided in tables of data supplied by manufacturers. Whilst these single-point data are useful for materials selection, they are obviously inadequate for detailed design of load-bearing components. **Here the engineer must look for information about time-dependence, which is usually obtained from tensile creep measurements**. Most manufacturers' handbooks contain sets of creep curves obtained over a range of applied stresses, for strains up to about 0.03, as shown in Figure 8.13; some handbooks also include creep curves for one or more elevated temperatures.

For ease of reference, the creep data are usually replotted in one or more different ways, as illustrated in Figures 8.14(a) and (b). Isochronous stress–strain curves (Figure 8.14(a)), which are discussed in Chapter 4, are included in most discussions of creep characteristics. From isochronous curves of this type, the engineer can determine the secant modulus of the polymer at any given strain or applied stress and time under load. This **creep modulus** $E(\sigma, t)$ is simply the reciprocal of the apparent creep compliance at the appropriate stress and time: $\sigma / \varepsilon(\sigma, t)$. Another convenient way to present creep data is in the form of isometric curves, as shown in Figure 8.14(b), which are helpful in designing plastic components to a

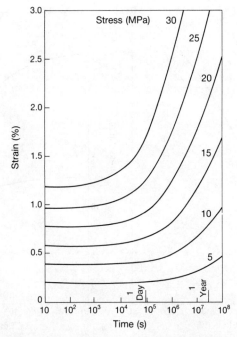

8.13 Creep curves for PES at 150°C (after ICI plc).

given **strain limit**. The precise value of the limiting allowable strain varies with the material, from 0.01 for a brittle material such as glass-fibre reinforced nylon, to 0.03 for a ductile polymer such as unreinforced polypropylene. Moduli are, of course, dependent upon temperature, and in some polymers upon moisture content. These effects are illustrated in Figure 8.15, using data for PA 6.6, one of the most moisture-sensitive polymers.

The problems of exact design for a viscoelastic polymer with non-linear properties are severe. For example, in Figure 8.14(a) the stress–strain curve is linear[1] only at the smallest strains (below ~ 0.2%). Most plastic parts are designed to operate at strains well above 0.2%, and in this case exact stress analysis is impossible. In practice, a 'safe' approximate procedure known as the **pseudo-elastic** design method is used. The salient features of the method, which is very straightforward to apply, are as follows:

- For the component and material under consideration, initial decisions are taken on:

---

1 See Figure 4.2: linear viscoelastic design is outlined in Section 4.4.2.

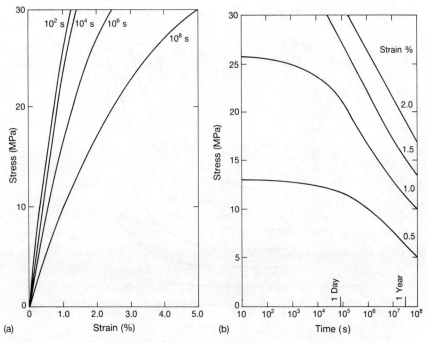

8.14 Creep data for PES at 150°C replotted: (a) as isochronous curves; and (b) as isometric curves (after ICI plc).

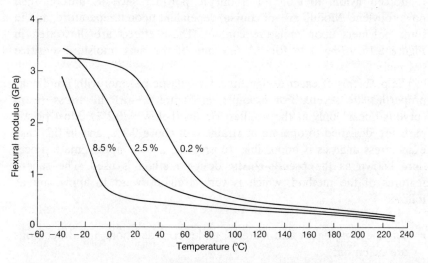

8.15 Effects of temperature on the modulus at short times of PA 6.6 at three different moisture contents: dry as moulded, 50%, and 100% relative humidity, which yield water uptakes (in weight %) of 0.2, 2.5, and 8.5 (after E. I. du Pont de Nemours).

(i) maximum temperature;
(ii) duration of loading; and
(iii) maximum allowable deflections **or** maximum local strain.

- **Linear elastic** stress analysis is then used to find stress and strain distributions and deflections, and the component is dimensioned so as to ensure that these all lie within allowable limits: some iteration may be needed at this stage. The usual methods are deployed:
  (i) standard analytical solutions for simple shapes; and
  (ii) computer based numerical solutions (finite element analysis) for complex shapes.

- Where Young's modulus is needed in the above stress analysis, the tensile creep modulus is inserted, **using the value for the appropriate temperature and duration of loading, and for the maximum tensile strain** in the component (as determined by the linear elastic analysis).

- In designs that are **strain-limited**, the maximum strain for the purposes of specifying creep modulus is simply the maximum allowable strain for the material in question. In designs that are **deflection-limited**, the maximum strain is not known in advance. In these cases, a short iteration is used: an initial estimate is made, and the stress analysis is repeated a few times until consistent results are obtained.

- Where Poisson's ratio is needed, a constant value is assumed. For plastics, ν lies between 0.35 and 0.42; for rubbers ν lies fractionally below 0.5.

The reader will note that the pseudo-elastic method is conservative. The stress analysis uses a modulus that is really appropriate only to the most highly strained regions of the design, and applies it to the whole component. Elsewhere, strains are lower, **and the creep modulus is greater than that used in the analysis**. This results in a small but unavoidable element of over-design: a component designed in this way will be somewhat thicker and more complex (e.g. because of ribbing) than strictly it needs to be to meet the specification.

Example 8.2

You are designing a plastic chair with the aid of a finite-element program, and have chosen a toughened polypropylene for the application. Creep data for the material at 23°C are available in graphical form; for space reasons, they are given here by the equation:

$$\varepsilon = 0.022(3.5 + t^{0.16})\sigma + 1.5 \times 10^{-6}[\exp(0.9\sigma)]t^{1/3},$$

where ε is the **percentage** extension after t seconds under applied stress σ MPa. The maximum allowable strain in the chair after 5 hours under load at 23°C is 2.0%. What value should you assign to the modulus of the PP in the design?

Solution

This is an example of strain-limited design. We apply the pseudo-elastic design method, specifying the duration of loading as 5 hours (which is 18000 seconds). In order to determine the modulus, we need the isochronous stress-strain curve for $t = 18000$ s. Substituting in the equation,

$$\varepsilon = 0.022(3.5 + 18\,000^{0.16})\sigma + 1.5 \times 10^{-6}[\exp(0.9\sigma)](18\,000)^{1/3}$$

$$= 0.183\sigma + 3.93 \times 10^{-5} \exp(0.9\sigma).$$

On plotting the stress-strain curve, we find that the stress at 2.0% strain is 9.65 MPa. The required modulus E is therefore

$$E = \frac{9.65 \times 10^6}{2.0 \times 10^{-2}} = 0.483 \times 10^9 \text{ Pa} = 0.483 \text{ GPa}.$$

Comment

At very small strains, the term in $\exp(0.9\sigma)$ becomes negligible, and the isochronous stress–strain relation is linear:

$$\varepsilon = 0.183\sigma.$$

The modulus then takes its maximum value for the case $t = 18\,000$ s. Taking account of the units of stress and strain as before, we have

$$E = \frac{10^6}{0.183 \times 10^{-2}} = 0.546 \times 10^9 \text{ Pa} = 0.546 \text{ GPa}.$$

The pseudo-elastic design method may be used for components submitted to intermittent loading, provided that the intervals during which the material is unloaded are sufficient to allow virtually complete recovery. Some manufacturers provide recovery data that enable the validity of this assumption to be tested. Alternatively, the Boltzmann superposition principle may be used to determine whether the assumption gives a satisfactory approximation (see Chapter 4). If not, or if the load is varying in a more complex manner, a more complete analysis of deflection behaviour based upon the Boltzmann principle may be necessary. Linearity can be assumed for strains up to about 0.005.

Example 8.3

The toughened polypropylene of Example 8.2 is to be used in constructing a cylindrical water tank. The tank will have a diameter of 3 m and a height of 2.5 m. It will be full for 4 months, and then left empty for 8 months before being refilled. Calculate the wall thickness required to ensure that the residual strain in the wall does not exceed 0.05% at the end of the first year. Neglect constraints imposed by the base of the tank.

Solution

Provided that the strains are sufficiently small, the Boltzmann superposition principle can be applied to this problem. We can treat the creep as the resultant of two responses: one due to a stress σ applied over the year $(3.15 \times 10^7$ s); and the other due to a stress $-\sigma$ applied over the final 8 months $(2.10 \times 10^7$ s). Noting the restriction to small stresses and strains, we shall neglect the term in $\exp(0.9\sigma)$ in the creep equation, and use the simplified equation

$$\varepsilon = 0.022(3.5 + t^{0.16})\sigma.$$

The validity of this approximation can be checked later. Inserting the specified values of t, we have for the strain developed after one year

$$\varepsilon = 0.022\left[3.5 + (3.15 \times 10^7)^{0.16}\right]\sigma - 0.022\left[3.5 + (2.10 \times 10^7)^{0.16}\right]\sigma$$

$$= 0.022\left[(3.15 \times 10^7)^{0.16} - (2.10 \times 10^7)^{0.16}\right]\sigma = 0.0219\sigma.$$

Setting $\varepsilon = 0.05\%$, we have $\sigma = 0.05/0.0219 = 2.28$ MPa. This stress is permissible at the base of the tank, where the water depth is 2.5 m. A cubic metre of water has a mass of 1000 kg. The pressure at the bottom of a full tank is therefore given by

$$P = 2.5 \times 1000 \times 9.81 = 24\,525 \text{ Pa}.$$

The hoop stress in the wall of a thin-walled cylinder of diameter D and wall thickness W is given by

$$\sigma = \frac{PD}{2W}.$$

In the present problem,

$$W = \frac{PD}{2\sigma} = \frac{24\,525 \times 3}{2 \times 2.28 \times 10^6} = 0.016 \text{ m} = 16 \text{ mm}.$$

The maximum strain developed in the cylinder, after 1.05×10^7 s (4 months) under a stress of 2.28 MPa, is 0.84%. The non-linear term in the creep equation contributes an additional 0.003%, and can therefore be neglected. On this basis, use of the Boltzmann superposition principle is justified.

In design problems involving stress relaxation at constant strain, it is normal practice to use the creep modulus to calculate values of stress relaxation. This incurs an error, as discussed in Chapter 4, but this is usually less than 5%.

The pseudo-elastic method is not intended to provide very accurate predictions, but is good enough for most applications, in view of the uncertainties arising from a variety of sources. The engineering formulae apply to homogeneous, isotropic materials, and the creep data are obtained from specimens subjected to uniform tensile stress at constant temperature. In practice, most mouldings show high molecular orientation in the surface layers, with a relatively unoriented core, and operate at fluctuating strains and temperatures. Under these circumstances, the conservatism of the pseudo-elastic method provides a valuable margin of safety.

8.4.2 Fibre composites

Components are made from long-fibre composites by laminating together undirectional layers of fibre-reinforced resin. If all of the layers have the same fibre orientation, the material is weak in the transverse direction; angle-ply laminates are therefore preferred for most applications. Choosing a stacking sequence that is symmetrical about the mid-plane (e.g. $0/+45/-45/90/-45/+45/0$) ensures that in-plane stresses do not produce out-of-plane twisting or bending, and that bending does not cause a change in the dimensions of the mid-plane. In asymmetric laminates, twisting occurs not only in response to applied stresses but also as a result of resin contraction during curing and cooling. Nevertheless, there are designs in which asymmetry is desirable (e.g. turbine fan blades).

In designing for stiffness in angle-ply laminates, the anisotropic moduli of each layer must be taken into account. These can be measured experimentally or calculated using the equations given in Chapter 6. The problem is simplified by considering only in-plane stiffness constants, which are represented by four independent quantities: Young's modulus parallel to the fibres E_1; Young's modulus transverse to the fibres E_2; the shear modulus measured with the shear stress parallel to the fibres G_{12}; and Poisson's ratio with free contraction transverse to the fibres ν_{12} (see Chapter 6). Elastic constants at any arbitrary angle to the fibre direction are functions of these four constants, and are calculated by matrix multiplication. A Mohr's circle method for performing this transformation is described in Chapter 6.

Laminate analysis is based upon the principle of strain compatibility, which means that the in-plane strains in neighbouring layers must be equal. The stresses required in each layer to produce a given strain are first calculated from the previously computed elastic constants of the layer for the appropriate angle to the fibre direction, and then integrated over all layers. The results define the elastic constants of the whole laminate. Simple computer programs take the drudgery out of these calculations.

Example 8.4

The stresses and strains in a unidirectional carbon fibre-epoxy laminate are related by the equations (stresses in GPa)

$$\sigma_1 = 205.9\varepsilon_1 + 2.27\varepsilon_2$$

and

$$\sigma_2 = 2.27\varepsilon_1 + 7.57\varepsilon_2,$$

where 1 refers to the fibre direction, and 2 to the normal direction in the plane of the lamina. Obtain corresponding equations for a cross-plied laminate in which 75% of the fibres lie parallel to the A direction and 25% are parallel to the B direction, at 90° to A.

Solution

For 75% of the plies, A corresponds to the 1-direction. In these plies

$$\sigma_A = 205.9\varepsilon_A + 2.27\varepsilon_B.$$

In the other 25% of plies, A represents the 2-direction, giving

$$\sigma_A = 7.57\varepsilon_A + 2.27\varepsilon_B.$$

The average stress in the laminate is the arithmetical average of the two contributions

$$\sigma_A = (205.9 \times \tfrac{3}{4} + 7.57 \times \tfrac{1}{4})\varepsilon_A + (2.27 \times \tfrac{3}{4} + 2.27 \times \tfrac{1}{4})\varepsilon_B$$

and

$$\sigma_A = 156.3\varepsilon_A + 2.27\varepsilon_B.$$

Note that this calculation is based on the principle that the strains are the same in both sets of plies, and that stresses are added. Applying the same procedure to the B direction,

$$\sigma_B = (205.9 \times \tfrac{1}{4} + 7.57 \times \tfrac{3}{4})\varepsilon_B + (2.27 \times \tfrac{1}{4} + 2.27 \times \tfrac{3}{4})\varepsilon_A$$

and

$$\sigma_B = 2.27\varepsilon_A + 57.2\varepsilon_B.$$

Example 8.5

Calculate Young's moduli in the principal directions for the unidirectional and cross-plied laminates discussed in the foregoing example.

Solution

The data were given in the form of a stiffness matrix, which can be written Q, where

$$\begin{bmatrix} \sigma_X \\ \sigma_Y \end{bmatrix} = \begin{bmatrix} Q_{XX} & Q_{XY} \\ Q_{YX} & Q_{YY} \end{bmatrix} \begin{bmatrix} \varepsilon_X \\ \varepsilon_Y \end{bmatrix},$$

or in abbreviated form, using the summation convention,

$$\sigma_i = Q_{ij}\varepsilon_j.$$

In order to calculate moduli, we must invert the matrix, and express the data in the form

$$\varepsilon_i = [Q_{ij}]^{-1}\sigma_j.$$

We can then specify uniaxial stress and obtain strains resulting from this single stress to define the modulus

$$\varepsilon_i = \frac{\sigma_i}{E_i}.$$

For the unidirectional laminate, we first need the determinant, given by $(205.9 \times 7.57 - 2.27^2) = 1553.5$. The inverted matrix is then

$$[Q_{ij}]^{-1} = \begin{bmatrix} \dfrac{7.57}{1553.5} & \dfrac{-2.27}{1553.5} \\ \dfrac{-2.27}{1553.5} & \dfrac{205.9}{1553.5} \end{bmatrix} = 10^{-3} \begin{bmatrix} 4.87 & -1.46 \\ -1.46 & 132.5 \end{bmatrix}.$$

This gives us

$$E_1 = (4.87 \times 10^{-3})^{-1} = 205.3 \text{ GPa}$$

and

$$E_2 = (132.5 \times 10^{-3})^{-1} = 7.54 \text{ GPa}.$$

For the cross-plied laminate, we have

$$Q_{ij} = \begin{bmatrix} 156.3 & 2.27 \\ 2.27 & 57.2 \end{bmatrix}.$$

The determinant is $(156.3 \times 57.2 - 2.27^2) = 8935.2$. Therefore

$$[Q_{ij}]^{-1} = \begin{bmatrix} \dfrac{57.2}{8935.2} & \dfrac{-2.27}{8935.2} \\[2ex] \dfrac{-2.27}{8935.2} & \dfrac{156.3}{8935.2} \end{bmatrix} = 10^{-3} \begin{bmatrix} 6.40 & -0.25 \\ -0.25 & 17.5 \end{bmatrix}.$$

This gives us

$$E_A = (6.4 \times 10^{-3})^{-1} = 156.3 \text{ GPa}$$

and

$$E_B = (17.5 \times 10^{-3})^{-1} = 57.1 \text{ GPa}.$$

Comment
The same principles are used to calculate the properties of angle-ply laminates. In these materials, it is also necessary to rotate the matrix in order to determine the properties at angles to the principal directions. For further details, see more advanced texts, or one of the computer packages available for carrying out these calculations.

The use of ribs, corrugations, box sections, and sandwich structures to increase the flexural rigidity and strength of the structure is an important element in design with both plastics and composites. Thick wall sections are undesirable, not only because of the increased amount of material needed, and the greater weight of the product, but also because of problems associated with cooling of the moulding. Thick sections lengthen

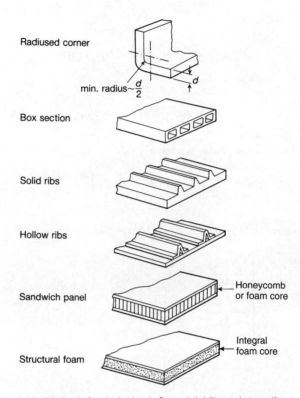

Radiused corner

min. radius~$\frac{d}{2}$ d

Box section

Solid ribs

Hollow ribs

Sandwich panel   Honeycomb or foam core

Structural foam   Integral foam core

8.16   Methods for designing-in flexural rigidity and strength.

mould cycle times and also give rise to differential shrinkage and consequent warping. Figure 8.16 illustrates some standard approaches.

8.4.3 Rubbers

The ability of rubbers to undergo large reversible deformations makes them ideal materials for springs, mountings, and bearings of various types. The stiffness characteristics of these components depend upon the fact that the shear modulus of a typical rubber is about 1 MPa, whereas the bulk modulus is about 2 GPa. In other words, it is very much easier to change the shape of a piece of rubber than to change its volume.

If a block of rubber is sandwiched between wide sheets of steel to which it is firmly bonded (as illustrated in Figure 8.17) the sandwich has a low resistance to shear but a high resistance to compression. Under compression, the rubber bulges outwards at the edges, but is constrained in the centre by the metal and surrounding rubber. In the limiting case, where the area of the sandwich is infinite, in-plane strains are zero, and compression can occur only as a result of a decrease in the volume of the rubber. By contrast, the steel sheets do not restrict the response of the rubber to a

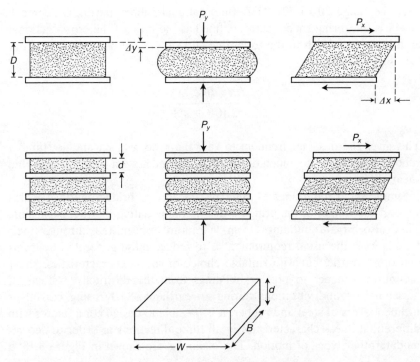

8.17 Effects of bonded steel sheets on the response of rubber blocks to compression and shear (after P. B. Lindley).

shear stress. Whether there is one layer of rubber, or many, the shear stiffness of the sandwich is given by

$$\kappa_S = \frac{P_x}{\Delta x} = \frac{GA}{D},$$ (8.5)

where G is the shear modulus of the rubber, A is the cross-sectional area of the sandwich, and D is the total thickness of rubber. A similar equation can be written for the compression stiffness:

$$\kappa_C = \frac{P_y}{\Delta y} = \frac{E_c A}{D},$$ (8.6)

but in this case the effective compression modulus E_c of the composite sandwich is dependent upon the geometry, and is given by

$$E_c = E(1 + 2kS^2),$$ (8.7)

where E is the Young's modulus of the rubber; k is a numerical constant which decreases from 0.93 to 0.53 with increasing hardness of the rubber over the range 30 to 70 IRHD (international rubber hardness degrees); and S is the spring shape factor, defined as the ratio of the cross-sectional area of the sandwich to the force-free surface area of one layer of rubber (see Figure 8.17):

$$S = \frac{BW}{2d(B + W)}.$$ (8.8)

This gives the designer freedom to vary the ratio $\kappa_C : \kappa_S$, and so 'tune' a rubber suspension to meet different requirements for compression and shear stiffness.

Simple shear mountings of this type are used as bridge bearings, which support the weight of the bridge whilst accommodating thermal expansion. They also isolate buildings from vibrations transmitted through their foundations: the usual requirement is to reduce noise generated by trains and other traffic, but with suitable choice of spring characteristics, shear mountings can act to protect buildings from the destructive effects of horizontal ground vibrations during an earthquake. By using curved or inclined layers of steel and rubber, it is possible to design components with different stiffness characteristics in all three directions in order to accommodate other types of motion. Two examples are shown in Figure 8.18: a

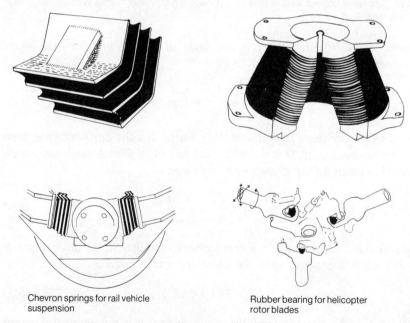

Chevron springs for rail vehicle suspension

Rubber bearing for helicopter rotor blades

8.18 Designs that make use of the low shear modulus and high bulk modulus of rubber (after P. B. Lindley and A. D. W. Leaver).

chevron mounting for use on underground trains; and a helicopter rotor bearing which can support the compressive forces generated by the revolutions of the rotor blades whilst accommodating other motions of the blades.

Example 8.6

Design a square-section bridge bearing ($W = B$, Figure 8.17) to accommodate a horizontal movement Δx of up to 50 mm (due to thermal expansion), and to support a maximum load of 120 tonnes, with a maximum deflection $\Delta y = 3$ mm under vehicle loading $F = 25$ tonnes. The maximum permitted shear strain γ in the rubber is 0.5. For stability, the width W of the bearing should be 5 times the total thickness D of rubber. Use a natural rubber of 60 IRHD, for which $k = 0.57$, and $E = 4.45$ MPa.

Solution

Let n be the number of rubber layers of thickness d. Vertical deflection per layer $\Delta y = \Delta Y/n$

$$D = nd = \Delta x/\gamma = 0.05/0.5 = 0.1 \text{ m}$$

$$W = 5D = 0.5 \text{ m}$$

$$\kappa_c = F/\Delta y = nF/\Delta Y = (25\,000 \times 9.81/0.003)n = 81.75n \text{ MN m}^{-1}$$

$$E_c = \kappa_c D/A = n(81.75 \times 10^6 \times 0.1/0.5^2) = 32.7n \text{ MPa}$$

For one layer of thickness d (using eqn 8.8), with $d = 0.1/n$ m,

$$S = 0.5^2/(4 \times 0.5d) = 0.125n/0.1 = 1.25n.$$

From eqn 8.7,

$$32.7n = 4.45(1 + 2 \times 0.57 \times 1.25^2 \times n^2).$$

Solving for n yields

$$n = 3.98, \qquad d = 0.0251 \text{ m}$$

The bearings should consist of four rubber layers, each measuring 500 × 500 × 25 mm.

Comment

The value of $nd = D = 100$ mm permits the required horizontal movement (50 mm); the value of $d = 25$ mm (slightly less than the design computation of 25.1 mm) ensures that the vertical movement is less than 3 mm, as required.

8.5   Designing for strength

The strength of a component is limited by the yield strength and fracture resistance of the material from which it is made. Designing to avoid general yielding in a polymeric component is relatively straightforward, provided that allowance is made for the time and temperature dependences of the yield stress, which are discussed in Chapter 5. **It is much more difficult to ensure that the component will not fail as a result of brittle crack propagation**.

The fracture resistance of a component depends not only upon the bulk properties of the material from which it is made, but also upon the size, orientation, location, and type of cracks or other defects present. Defects may be introduced during manufacture, or formed at a later stage in the life of the component. As discussed in Chapter 5, their severity can be expressed in terms of an equivalent (sharp) crack length. In a well-made component, this length is usually below 50 μm, but larger defects are sometimes unavoidable: for example, weld lines behave like surface cracks which in severe cases run as much as 300 μm into the sub-surface region of the moulding. Voids and aggregations of pigment particles are other examples of defects introduced during manufacture. Voids can usually be eliminated from moulded or extruded thermoplastics, **but present a real problem in manufacturing components from long-fibre composites and in fluorocarbons formed by sintering**.

Even if the component contains no significant defects when it is made, cracks may initiate and grow subsequently for a variety of reasons. Not only external loading, but also thermal gradients, internal stresses arising from processing, and differential absorption of water or other liquids, can cause cracking. Cyclic loading due to mechanical stressing or heating and cooling is more damaging than static loading. A common feature of these problems is that fracture occurs at stresses below the yield stress; subcritical crack growth preempts yielding.

Both organic and inorganic liquids can aggravate the problem of subcritical crack growth under external or internal stressing. Avoidance of fracture problems is essentially a matter of listing the liquids and vapours that the polymer is likely to come into contact with, and selecting materials accordingly. Manufacturers issue comprehensive tables showing how their products perform in organic liquids, acids, and alkalis. In general, the less polar polymers such as polystyrene are more susceptible to physical attack in the form of solvent crazing, whereas the more polar polymers such as nylons are prone to chemical attack by acids.

Embrittlement of the surface as a result of photo-oxidation is another major cause of fracture problems in polymers. The photons that make up the ultraviolet component of sunlight have enough energy to split chemical bonds and thus set off a chain reaction in which the surface layer becomes slowly oxidized. Over a period of about a year, the material may become embrittled to a depth of over 100 μm. This brittle surface layer readily forms cracks at low strains, **which can propagate into and through the unaffected material below.** As discussed in Chapter 5, whether the crack will grow depends upon its length, the stresses acting on it, and the fracture toughness of the material ahead of it. Not all polymers are equally affected by this 'outdoor ageing' problem: PVC and acrylic polymers (including PMMA) are relatively resistant. On the other hand, PE, PP, BR, SBR, and natural rubber are readily attacked: unprotected PE film becomes completely brittle after a month or two in the sun.

The obvious solution to the ageing problem is to choose intrinsically resistant materials for outdoor applications. For example, window frames are made from extruded sections of rigid PVC toughened with an acrylate rubber (in preference to BR or other rubbers containing the easily oxidized carbon–carbon double bond). Further protection is afforded by adding:

- pigments which absorb or reflect light (PE for agricultural use is loaded with carbon black);
- antioxidants which combine with the free radicals responsible for oxidation; and
- chemical compounds that absorb ultraviolet light (cf. suntan oil).

Manufacturers' handbooks give data on tensile and flexural yield or fracture stress, tensile elongation at break, notched Charpy or Izod impact strength, and fatigue life. Whilst these data are useful for materials selection, they are not always sufficient for quantitative design. To date, fracture mechanics has been used to only a limited extent in design with plastics, and data are not provided in data books. The lack of approved standards for testing of polymers has been an obstacle to the wider use of fracture mechanics. However, this problem is now being rectified.

For the reasons discussed above, the designer should not rely on short-term strength data. Recognition of the problem of sub-critical crack growth has led some manufacturers to include in their handbooks graphs showing long-term strength under static loading: an example is given in Figure 8.19.

Dynamic fatigue is more widely recognized as a cause of fracture, and most handbooks provide fatigue data in the form of S–N curves, as illustrated in Figure 8.20. These curves usually flatten beyond 10^6 cycles, in which case it is possible to define a fatigue endurance limit, i.e. the minimum stress required to cause failure within 10^7 cycles. Both the static

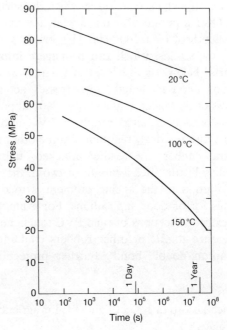

8.19 Time to failure of PES under static loading at three temperatures (after ICI plc).

and the dynamic tests reflect the response of the material to the small defects that are present in the unnotched specimens, and in a sense are measuring the distribution of intrinsic flaw sizes as much as the fracture resistance of the polymer. Consequently, when the component is to be used in a critical application, there is a good case for mounting a full-scale fracture mechanics study.

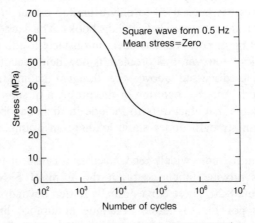

8.20 Fatigue endurance S–N curve for PES at 20°C (after ICI plc).

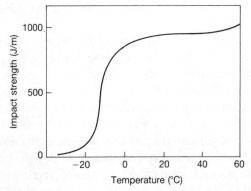

8.21 Notched Izod impact strength of polycarbonate over a range of temperatures (after General Electric Plastics).

Standard Charpy or Izod impact data cannot be used for design calculations. They are provided simply as a basis for materials selection. A low impact strength is an indication that the material may be prone to brittle fracture in service. However, it is important to note that the notches used in standard tests are not as sharp as natural cracks, and that a high notched Izod impact strength does not guarantee that the material will never suffer brittle fracture in service. Impact data are most useful when they span a range of temperatures, as illustrated in Figure 8.21, because they then indicate possible problems when the material is to be used at low temperatures. Both Charpy and Izod impact tests are useful in studying outdoor ageing, because the effects of any embrittlement at the surface of

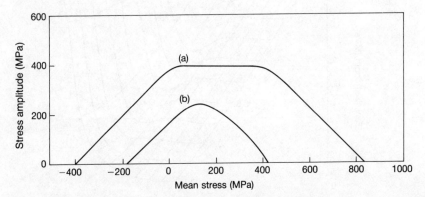

8.22 Modified Goodman diagram showing fatigue stress amplitude required to cause failure in 10^7 cycles as a function of mean stress for two carbon-fibre reinforced epoxy resin materials: (a) unidirectionally reinforced, stressed parallel to fibres; and (b) cross-plied laminate, with 45% of plies parallel to stress, and 55% transverse. Fibre volume fraction 60% (after R. Tetlow).

the notch root are brought out most clearly in a bar subjected to rapid three-point bending.

The fracture resistance of fibre-reinforced plastics is less sensitive to defects, because the fibres act as separate load-bearing elements, even when a crack has formed in the matrix. Furthermore, the toughening effect of fibres is retained at low temperatures. On the other hand, the strength properties are strongly dependent upon fibre orientation, and are highly anisotropic in areas where the fibres are aligned. Provided that the fibre concentration and alignment are properly defined, data on tensile strength and fatigue endurance are directly applicable in design. Long-fibre composites show exceptionally good fatigue resistance, retaining a high proportion of their short-term strength over 10^7 cycles. Figure 8.22 presents data for a carbon-fibre reinforced epoxy resin, showing the stress amplitude needed to cause failure in 10^7 cycles as a function of mean stress.

Strength calculations on angle-ply laminates are based upon the elastic analysis described in Section 8.4.2. The strains produced in the laminate by a given set of applied stresses are first calculated, using the computed laminate stiffness constants. Stresses corresponding to these strains are then calculated for each layer of the laminate. These stresses are then expressed in terms of stresses parallel to and normal to the fibres, and the combination of stresses is compared with strength criteria for unidirectional material, which should preferably be obtained by experiment.

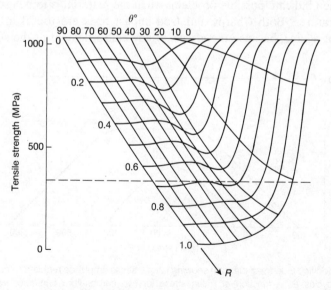

8.23 Strength chart for a $[0, \pm\theta°]$ carbon–epoxy laminate containing 60 vol.% fibres, showing the maximum allowable stress in 0° direction as a function of θ and R, the fraction of angled plies (after R. Tetlow).

Results for a three-direction carbon fibre-expoxy composite $(0, \pm \theta)$ under tensile stress in the $\theta = 0°$ direction are shown in Figure 8.23. The composite contains 60 vol.% fibre; R is the fraction of plies which are angled at $\pm \theta$ and $1 - R$ is the fraction at $\theta = 0$.

Example 8.7

A series of three-layer carbon fibre-epoxy composites is to be designed according to Figure 8.23, for values of $\theta = 90°, 60°, 45°, 30°$, and $20°$. What is the fraction of plies in the $\theta = 0°$ direction required to give a strength of 350 MPa?

Solution

The dotted line is drawn in Figure 8.23 for strength 350 MPa. We require $1 - R$. Reading the value of R from the figure yields the following table.

Table 8.3   Information for Example 8.7

θ	R	$1 - R$
90°	0.67	0.33
60°	0.70	0.30
45°	0.70	0.30
30°	0.69	0.31
20°	1.0	zero

Comment

These results emphasize the point that angled plies make little contribution to the strength of the laminate if the ply angle θ is greater than about 25° (between $\theta = 90°$ and $\theta = 30°$ we require 33% to 31% of plies to be parallel to $\theta = 0$ in order to give the required strength). Note that for $\theta = \pm 20°$, even with $R = 1.0$ (i.e. no plies at $\theta = 0$) the strength exceeds the design stress, 350 MPa.

One problem area for fibre-reinforced thermosetting resins is interlaminar splitting, which occurs particularly easily when the void content is high. A spanner dropped on to a laminated sheet could initiate undetected damage, which might result in failure by buckling when the sheet is loaded in compression. Measurements of interlaminar shear strength (ILSS) are made by subjecting short bar specimens to three-point bending; the ILSS is calculated using eqn 8.4. The aircraft industry also uses *ad hoc* tests in which a panel is subjected to compression after receiving an impact from a round-ended striker. The development of new, tougher composites, including thermoplastics-based materials, is aimed partly at solving this problem.

Example 8.8

(1) A unidirectional carbon fibre-epoxy laminate 3 mm thick is cut into a 10 mm wide strip and subjected to three-point bending with a span of 20 mm. Failure occurs by interlaminar splitting at a load of 2400 N. Calculate the interlaminar shear strength of the composite.
(2) When the span is increased to 40 mm, failure occurs by tensile fracture from the surface at a load of 1575 N. Calculate the tensile strength of the unidirectional composite, and find the span at which the transition from shear to tensile failure takes place.

Solution

(1) From eqn 8.4,

$$\tau_{max} = \frac{3P}{4bd} = \frac{3 \times 2400}{4 \times 10 \times 10^{-3} \times 3 \times 10^{-3}}$$

$$= 60 \times 10^6 \text{ Pa} = 60 \text{ MPa}.$$

(2) From eqn 8.3,

$$\sigma_{max} = \frac{3 \times 1575 \times 40 \times 10^{-3}}{2 \times 10 \times 10^{-3} \times (3 \times 10^{-3})^2}$$

$$= 1.05 \times 10^9 \text{ Pa} = 1050 \text{ MPa}.$$

Comparing eqns 8.3 and 8.4, we find

$$\frac{\sigma_{max}}{\tau_{max}} = \frac{2L_{crit}}{d},$$

when $L = L_{crit}$ (the critical span at which the transition from shear to tensile failure occurs). Rearranging,

$$L_{crit} = \frac{\sigma_{max} \times d}{2\tau_{max}} = \frac{1050 \times 3 \times 10^{-3}}{2 \times 60}$$

$$= 26.25 \times 10^{-3} \text{ m} = 26.25 \text{ mm}.$$

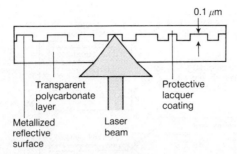

8.24 Mode of operation of a compact disc. The laser beam enters from below, traverses the disc, is reflected (or **not** reflected if it hits a pit) from the metallized surface and returns back through the CD carrying the digital information (after Bayer AG).

8.6 Case histories

8.6.1 Optical memory devices (compact discs)

In the CD, digital information is stored on a transparent polycarbonate disc in the form of microscopic indentations in the surface. The pitted surface is metallized for laser beam reflection and then coated with a protective lacquer. The laser beam enters from below (see Figure 8.24. When the beam hits a pit there is no reflection; when it hits a flat element between the pits it is reflected. The no-reflection/reflection sequence delivers a series of zero/one signals to a photodetector which converts the digital information to analogue music signals.

The pits lie along a helical track (see Figure 8.25. The pits are of depth 0.1 μm; the track pitch is 1.6 μm and the pit length is 1 to 3.3 μm. The

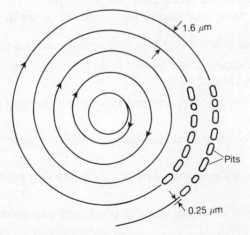

8.25 The information on a compact disc is laid on a helix which starts close to the centre and moves outwards; the laser beam is focused to a spot of diameter 1 μm, smaller than the pitch of the helix (after Bayer AG).

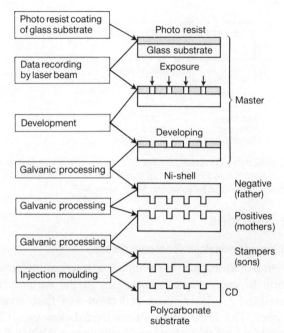

8.26 Steps in the mastering and replication of a compact disc (after Bayer AG).

laser beam reaches a sharp focus (a spot of diameter 1 μm) on the metallized track. When the beam passes through the disc outer surface on the inward and outward path it is of diameter 1 mm and is not therefore affected by scratches, dust or finger marks. The information-bearing surface is protected by the metal layer and the lacquer.

The steps in the preparation of the mould are shown in Figure 8.26:

- a master disc is prepared (glass with a photosensitive layer);
- the layer is exposed to a pulsed laser beam that transfers digitalized signals from the file;
- the master disc is then developed, leaving laser induced pits in the layer **(positive)**;
- galvanic processing then produces a nickel **negative** (known as the father);
- from the father a second galvanic processing step produces a **positive** mother;
- from the mother a third galvanic processing step produces a **negative** son; and
- the final **positive** CD is produced by placing one of the sons in a mould and injection moulding with polycarbonate.

The design requirements governing the choice of polymer are:

- high transparency;
- minimum water absorption (the CD is asymmetric, with one side metallized; water-induced expansion of the plastic will not be isotropic and must therefore be minimized);
- sufficient impact resistance with a reasonable heat deflection temperature (in order to guarantee longevity in a hot climate);
- excellent moulding characteristics in order to produce high dimensional stability, accurate surface reproduction and extremely low moulded-in stress (and therefore low optical birefringence); and
- a highly pure polymer to guarantee error-free data retrieval.

These requirements called for a glassy polymer with a reasonably high T_g. Polycarbonate was found to be the ideal system. Development work by Bayer AG showed that standard grades were not satisfactory. A specific resin was developed with a low RMM (to decrease viscosity and to reduce moulded-in stress). The RMM was not reduced enough to lower the toughness to an inappropriate level.

The standard audio CD is of diameter 120 mm and has a storage capacity of 600 megabytes; it could store the entire text of the *Encyclopaedia Britannica*, equivalent to 1200 floppy disks. The development of polycarbonate as the material for larger optical memory discs (133 mm, 200 mm, and 300 mm) is in progress.

8.6.2 Motorcycle drive sprocket

The drive sprocket is injection-moulded in rubber-toughened PA6.6, which is resistant to repeated impact over a wide range of temperatures, and which shows excellent abrasion resistance. The plastic sprocket absorbs the shock of gear changes better than metal, causes less wear on the chain, and provides a quieter ride. The polymer is resistant to lubricants and other liquids. Reduced friction improves power transmission and, coupled with a lower weight, results in better performance.

8.6.3 Cross-country ski bindings

The ski bindings are injection moulded in polyacetal, which was chosen for its rigidity, light weight, resilience, toughness, retention of properties at low temperatures and in strong sunlight, and ease of accepting moulded-in inserts. A major requirement was resistance to stress cracking in the presence of the strong solvents used to remove ski wax. Reduced weight

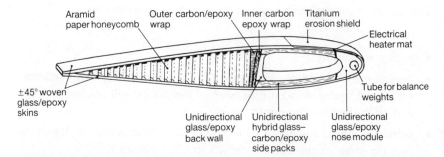

8.27 Section through composite rotor blade for the Westland Sea King helicopter (after R. Sanders and P. S. Grainger).

and elimination of snow and ice build-up on the bindings enhance the performance of the competitive skier.

8.6.4   Helicopter rotor blade

The blade has a true composite construction, as shown in Figure 8.27, with areas of unidirectional glass and carbon fibre, woven glass cloth trailing edge skins laid at ±45° over a polyaramid paper honeycomb core, and a titanium erosion shield. The polymeric matrix for the fibres is epoxy resin. The metal blades that it replaces were retired after 9400 hours in service, whereas the composite blades do not need replacement during the 40 000 hour life of the airframe. Safety is improved, and overall running costs reduced.

8.6.5   Cautionary tale no. 1

A manufacturer of kitchen fans found that ABS louvres were fracturing at the centre. He complained about the material (according to ancient custom), and demanded an improvement. Subsequent investigation showed that the louvres were double end-gated, and therefore had a weld line in the middle of the bar. This might have been satisfactory, as the loads imposed during service were small, were it not for the fact that the designer had also arranged that the louvre bars were flexed in order to clip them into position. The mould was redesigned with a single end-gate, and the designer was relegated to disposable picnicware.

8.6.6   Cautionary tale no. 2

An inventor took his new PMMA kitchen bowl from the dishwasher, which had just completed its duties. The bowl shattered when he poured a bottle of gin into it. Subsequent investigation showed that the PMMA absorbed about 2% of water under hot, wet conditions in the dishwasher, and that only the surface layers dried out during the dry heating stage. The surface was therefore under tensile stress due to differential swelling when the ethanol exhibited its powers as a solvent crazing agent.

Notes for Chapter 8

8.N.1

The terms 'polycarbonate', 'polyethersulfone', and 'polysulfone' generally refer to the specific chemical structures shown below, although they could legitimately be applied to any polymer based respectively on the carbonate, ether–sulfone, or sulfone linkages. All three polymers are made by condensation polymerization, and contain the benzene ring (aromatic structure), which is particularly stable against thermal or oxidative degradation, and which also contributes to chain stiffness when (as in the molecules shown below) the substituent groups forming the adjacent parts of the chain are attached in the **para** positions (i.e. on opposite sides of the ring).

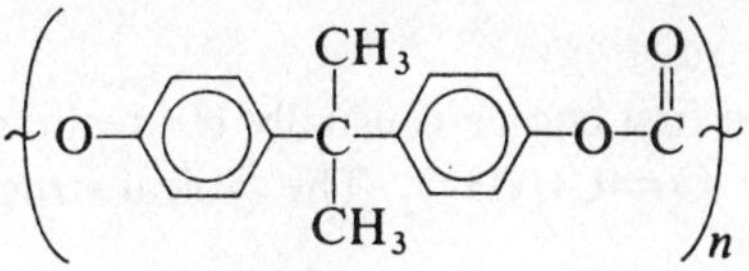

Polycarbonate (PC)   (XXXIII)

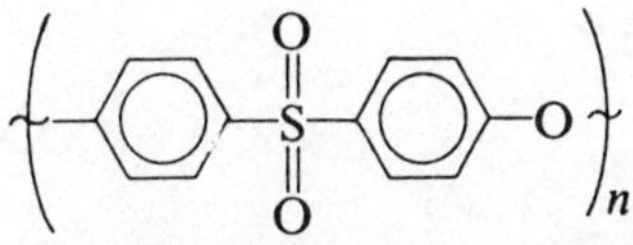

Polyethersulfone (PES)   (XXXIV)

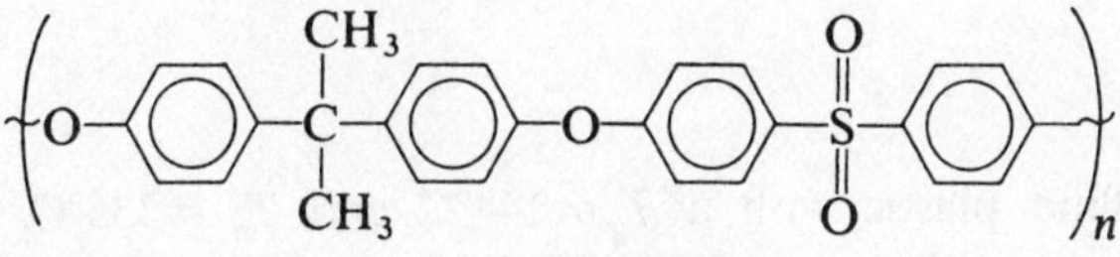

Polysulfone (PSF)

(XXXV)

8.N.2

Poly(2,6-dimethyl-1,4-phenylene oxide)—which, thankfully, is referred to as PPO—has the unusual property of being completely miscible with another homopolymer, in this case polystyrene. This enables it to be toughened by alloying with HIPS. The structure of PPO is

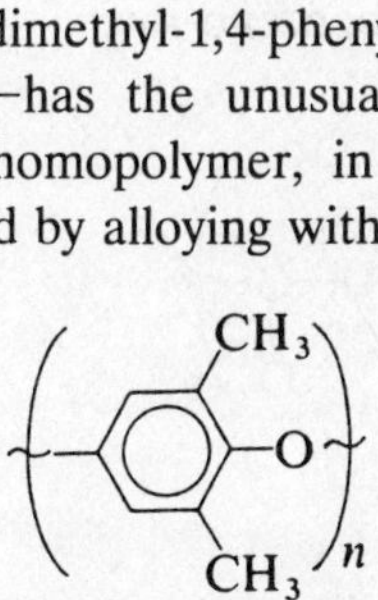

Poly(phenylene oxide) (PPO)   (XXXVI)

The T_g of the alloy is intermediate between that of PS ($\sim 100°C$) and the T_g of PPO ($\sim 220°C$). In addition to improving impact resistance, the HIPS also reduces the melt viscosity.

8.N.3

Dr Leo Baekeland's work on the condensation reaction between phenol (C_6H_5OH) and formaldehyde (HCHO), which was carried out in 1905, resulted in the the first true synthetic polymer (Bakelite). The resin is highly cross-linked, and suffers from the disadvantage that only dark colours can be produced. The typical unit is

$$\sim CH_2 \underset{\overset{|}{CH_2}}{\overset{\overset{OH}{|}}{\bigcirc}} CH_2 \sim \qquad \text{(XXXVII)}$$

Where light colours are required, the first choice is usually UF resin, made by reacting formaldehyde with urea: $O{=}C(NH_2)_2$. The typical structural unit is

$$\begin{array}{c} \sim N{-}CH_2 \sim \\ | \\ C{=}O \\ | \\ \sim N{-}CH_2 \sim . \end{array} \qquad \text{(XXXVIII)}$$

8.N.4

PEEK is a semi-crystalline plastic with a $T_g \approx 144°C$ and $T_m \approx 335°C$. Processing temperatures are in the range 360–400°C. Unreinforced PEEK is very ductile, reaching strains in excess of 1.0 in the tensile test. Like other high-softening plastics, it is based on the benzene ring

$$\left(\bigcirc {-}O{-}\bigcirc{-}O{-}\bigcirc{-}\underset{\overset{\|}{O}}{C} \right)_n \qquad \text{(XXXIX)}$$

8.N.5

Poly(phenylene sulfide) (PPS) is a highly crystalline plastic with a $T_m \approx$

288°C. It has no known solvents below 200°C. Injection moulding temperatures are 320–360°C. Recommended mould temperatures are ~130°C in order to maximize crystallinity. The chemical structure is

(XL)

8.N.6

The **limiting oxygen index** (LOI) is the minimum concentration of oxygen, expressed as a percentage, that will support combustion of the polymer. It is measured by passing a mixture of O_2 and N_2 over a burning specimen, and reducing the O_2 level until a critical level is reached.

Problems for Chapter 8

Note: In answering Problems 8.1–8.6 it will be helpful to consult the commercial literature available from manufacturers, and standard reference books which tabulate the properties and applications of plastics.

8.1 With the aid of Table 8.1, list the factors to be considered in selecting a polymer for each of the following applications, stating whether the factor is of primary, secondary, or only minor importance:
(1) pipe for (underground) mains gas supply;
(2) bottle crate (stackable);
(3) industrial safety helmet;
(4) domestic kettle;
(5) rigid suitcase shell;
(6) socket for 100 W electric light bulb;
(7) incubator for premature baby unit; and
(8) bottles for carbonated drinks.
Suggest one or more candidate materials for each application.

8.2 What circumstances might cause a manufacturer to change the material used for a particular application from:
(1) HDPE to PP?
(2) PP to HDPE?
(3) PA6.6 to POM?
(4) PMMA to PC?
(5) PC to PES?

8.3 List the factors of primary, secondary, and minor importance to be taken into account in selecting a polymeric material for car bumpers and fenders. Note that the bumper is required to survive an impact at up to $2\frac{1}{2}$ miles per hour without damage, and in order to achieve fuel economy should have the minimum possible weight consistent with performance. With the aid of your list, select up to three candidate polymers.

8.4 Three applications of polypropylene in the modern family car are given in Section 8.2.2. By referring to the literature, compile a more complete list of current applications of polypropylene polymers and copolymers in automobiles, and suggest further possible applications. Where PP is used in competition with metals or other plastics for the same type of component, summarize the respective advantages and disadvantages.

8.5 Taking account of the properties given in Table 8.2 and in 8.N.5, together with any other relevant information available to you, suggest some applications for glass-filled PPS.

8.6 Select a polymeric material from which to make a daisywheel for a computer printer. The daisywheel will have an overall diameter of 70 mm, and consist of 56 individual arms, each 22 mm length, radiating from a central hub. Printing speeds will be approximately 30 characters per second. In what ways can the correct choice of material and production process help to meet the special requirements for stiffness, strength, and low inertia inherent in this application?

8.7 You are required to make a rectangular tray measuring 45×30 cm, with a depth of 12 cm in ABS. An acceptable tray can be produced by thermoforming 3 mm extruded sheet, which is available as 200×120 cm sheets at £2300 per tonne. It would be necessary to use blanks measuring 55×40 cm, which would be trimmed to 47×32 cm after forming. For thermoforming, the mould would cost £4000, cycle time would be 90 seconds, and total factory costs would be £12 per hour. A somewhat better tray, weighing 495 grams, could be made by injection moulding. The mould cost is £16 000, cycle time is 30 seconds, and factory costs are £25 per hour. A scrap rate of 5% can be assumed. Moulding granules cost £1600 per tonne. Scrap from either process can be sold at £500 per tonne. Calculate the minimum number of trays you would have to make to justify injection moulding.

8.8 A designer of garden wheelbarrows is considering a new design where the frame of the wheelbarrow will be manufactured from a thermoplastic. After considering many materials she narrows the choice to: 30% (by weight) glass-fibre reinforced polypropylene (GF-PP), or 30% (by weight) glass-fibre reinforced thermoplastic polyester

(GF-PET). The designs employed for the two materials differ only in their wall thicknesses, which have been determined by stress analysis of the frame. A batch of 250 000 frames is required, to be produced by injection moulding within a period of 12 months. Factory running costs are charged at £40 per hour per injection moulding machine, and the cost of producing a mould is £15 000. The factory operates three 8-hour shifts per day, 7 days per week, 48 weeks per year, and a machine usage rate 90% and a scrap rate of 1% can be assumed. The following table gives other relevant data.

	GF-PP	GF-PET
Cost of material (£/kg)	0.92	1.78
Wall thickness (mm)	8.0	4.1
Mass of moulding (kg)	4.5	3.3
Cycle time (s)	150	55

(1) List the important requirements of a material for the wheelbarrow frame.
(2) Comment on the designer's choice of GF-PP and GF-PET, suggesting why similar related materials were rejected.
(3) Suggest why the cycle time of GF-PET may be so much less than that of GF-PP.
(4) Find the total cost per frame for each of GF-PP and GF-PET.
(5) Taking all factors into account, which material would you recommend?

8.9 You have to design a 150 mm heavy-duty pulley wheel, with a shaft diameter of 20 mm, to be injection moulded in nylon. Make a set of outline drawings, showing at least three contrasting designs for the area between hub and rim, which may be plain or stiffened, solid or spoked. The stiffening may be provided by radial ribs or by corrugations. Indicate how the design affects the pattern of flow in the mould, and consequently the quality of the moulding. Which of your designs will present the fewest problems in manufacture?

8.10 Your company is involved in the design and manufacture of unreinforced and glass-reinforced plastic parts for the automobile industry. Write a case for the management, to show the benefits to be gained from investment in a computer-aided design system.

8.11 Blow moulding is a standard method for manufacturing automobile fuel tanks in high-density polyethylene. The tanks are often of very irregular shapes in order to utilize available space. What other method or methods could be used to make HDPE fuel tanks, and why is blow-moulding preferred?

8.12   A box having the dimensions given in Problem 8.7 is injection-moulded from a gate at the centre of the base. Assuming that at any given time the speed of the flow front is the same at all points in the mould, determine the positions of the weld lines.

8.13   Refer to Example 8.1. Because of the differences in flow path, the material in the skin of the box moulding is oriented differently on either side of the weld line YZ. Using a geometrical construction, or otherwise, determine the angle between the two orientation directions: (1) near Z; and (2) near Y, assuming that the surface orientation is parallel to the flow direction.

8.14   Apply the principles outlined in Problem 8.13 to the box described in Problem 8.12, and determine the maximum and minimum angles between orientation directions in the surface material on either side of the weld lines.

8.15   Continue Problem 7.29. Sketch one of the PVC mouldings described in the question, indicating where you would expect weld-lines to form. If, during installation, it is necessary to bend the vents, such that the maximum bending stress occurs at position X, suggest what problem might occur and how you could overcome it by repositioning the gates. If the mould has two cavities, sketch a possible runner layout for the gates in their new locations.

8.16   A 12.5 mm diameter steel shaft is to be pressed into a POM hub of outer diameter $D_h = 25$ mm. The design stress of the POM may be taken as half the yield stress (see Table 8.2). Taking $\nu_s = 0.33$, $\nu_h = 0.4$, $E_s = 210$ GPa, and $E_h = 2.8$ GPa, calculate the allowable interference. What is the allowable interference if the shaft is also made of POM?

8.17   The creep of a rubber-toughened polypropylene at 20°C can be fitted to the equation

$$\varepsilon = 0.015(4 + t^{0.15})\sigma + 1.5 \times 10^{-6}t^{1/3}\exp(0.9\sigma),$$

where ε is the percentage extension after time t seconds under applied stress σ MPa. Plot a series of creep curves showing ε as a function of $\log t$ in the range $\varepsilon = 0$ to 3% and $t = 10$ to 10^8 s, at stress intervals of 2 MPa. Use these curves to construct: (a) isochronous curves of σ against ε at $t = 10^2$, 10^4, 10^6, and 10^8 s; and (b) isometric curves of σ against $\log t$ at $\varepsilon = 0.5$, 1.0, 1.5, and 2.0%.

8.18   A chair seat to be injection-moulded in toughened polypropylene will consist essentially of a square plate of side 400 mm. It will be simply supported along two opposite edges, and must be able to carry a load of 85 kg, uniformly distributed, for up to 5 hours without deflecting by more than 5 mm. The strain must nowhere exceed 3%. Using the creep data given in Problem 8.17, find the thickness of polypropylene required. Use the following expressions for the central deflection Δ and maximum bending moment M_{max} in a simply-supported plate subjected to a uniformly-distributed force W:

$$\Delta = \frac{5}{384} \frac{Wl^3}{EI} (1 - v^2)$$

and

$$M_{max} = \frac{Wl}{8},$$

where E, I, v, and l have their usual meanings. Take $v = 0.4$ for PP.

8.19   A company plans to manufacture the chair seat described in Problem 8.18 by injection moulding. In order to ensure economical and satisfactory moulding, it has decided to limit the thickness of the plate to 4.5 mm, and to use ribs on the underside of the plate to provide the necessary stiffness and strength. Design a set of ribs for the chair seat, specifying their number, width, depth, and spacing. To what extent would your design be affected if the company also decided to cover the chairs with cushioning?

8.20   The PP water tank described in Example 8.3, with a wall thickness of 16 mm, is filled with water to a depth of 2.5 m for 3 months, left empty for the following month, then filled to a depth of 2.0 m for 8 months. Calculate the change in diameter at the base of the tank at the end of each of these periods. The constraint due to the base of the tank can be neglected for calculation purposes.

8.21   At 45°C, the creep of an acetal polymer is given by

$$\varepsilon = 0.018(1.3 + t^{0.1})\sigma + 10^{-5} t^{1/3} \sinh(0.38\sigma),$$

where ε is percentage strain, stress σ is in MPa, and t is in seconds. Plot an isochronous curve of modulus $E(t)$ against log ε in the range $\varepsilon = 10^{-4}$ to 3×10^{-2} at $t = 100$ hours, and determine the strain level at which $E(t)$ falls from its zero-strain value by 1%.

8.22   Pipes to carry natural gas at moderate pressures are now made from extruded medium density polyethylene (MDPE).

     (1) Treating it as a pressurised thin-walled tube, determine the minimum wall thickness d for an MDPE pipe to meet the specification: diameter = 104 mm; design pressure rating = 400

kPa; design lifetime = 50 years. Use data for MDPE from the Note below.

(2) An actual pipe to this specification, but with $d = 8$ mm, was cut into short lengths. When these were sliced axially to create semicircular halves, they were found to bend inwards spontaneously to a final diameter of 96 mm.

  (i) Determine the level of thermal residual strain that this observation indicates, assuming it to vary linearly across the pipe thickness.

  (ii) With respect to this residual strain, comment on: its likely effect on the pipe lifetime; its cause; how it might be reduced.

Note: creep data are available from tensile tests on injection-moulded bars of MDPE. Creep strain ε is given as follows in terms of stress σ (in MPa) and creep time t (in hours):

$$\varepsilon = 1.35 \times 10^{-3}\sigma^{1.3}t^{0.11}.$$

Poisson's ratio $\nu = 0.4$. The design strain limit for MDPE is 0.02 (apply this to the maximum principal strain).

8.23 Planks of softwood timber (density 390 $\mathrm{kg\,m^{-3}}$) with approximate cross-section 250×20 mm are widely used in the construction industry. A thermoplastic substitute is proposed. To be viable it must meet the following criteria over a design lifetime of 10^7 s at 20°C:

(1) it must have a longitudinal flexural stiffness (EI) of at least 2000 $\mathrm{N\,m^2}$;

(2) it must withstand a longitudinal bending moment of 760 N m; and

(3) it must withstand a surface pressure of 20 kPa.

In addition, the external cross-sectional shape must be rectangular with width 250 mm and depth up to 40 mm, and the weight should be as low as possible. It is proposed to manufacture the new product as an extruded hollow box section, reinforced if necessary by longitudinal webs (see the 'box section' in Figure 8.16). The chosen material is short glass-fibre reinforced polypropylene (GF-PP) with properties given below.

(1) Show that it is possible to meet all the criteria above, and find the required number of webs.

(2) Comment on the approximations used in obtaining your answer to (1).

(3) List the advantages and disadvantages of the new product as an alternative to wood.

Note: GF-PP has the following properties at 20°C. Density = 1120 $kg\,m^{-3}$. Tensile creep strain (measured on end-gated injection moulded bars) can be expressed as

$$\varepsilon = 1.78 \times 10^{-5}\sigma^{1.6}t^{0.07},$$

where stress σ is expressed in MPa and time t in seconds. The recommended design strain limit is 7.5×10^{-3}. To ensure economic manufacturing rates during extrusion, the maximum wall thickness should not exceed 4 mm.

8.24 Part of a meter valve mechanism will consist of a rectangular cantilever beam spring injection-moulded in acetal. The free end will be held at a constant deflection in order to maintain a moment of at least 0.5 N m over a period of 3 years at a temperature of 45°C. The depth of the beam has already been fixed for reasons of space at 5 mm. Taking the design strain limit as 1%, calculate the minimum width of the beam, using data given in Problem 8.21 for the acetal polymer.

8.25 An electrical fitting, injection-moulded in PES, will operate at 150°C. It will be bolted to a steel support, and the total clamping force must not fall below 10 kN after 10^6 seconds. Washers with inside and outside diameters of 10 mm and 20 mm, respectively, will be used. The maximum allowable strain for PES at 150°C is 0.7%. Calculate the number of bolts needed, using data given in Figures 8.13 and 8.14.

8.26 An engine of mass 4000 kg vibrates in the horizontal plane. Design a set of rubber pads to support the weight of the engine and also to isolate its vibrations from the foundations. The pads should be square, and must not compress by more than 1 mm under the weight of the engine. In order to isolate vibration, the resonant frequency of the engine and mounting combination must not be greater than 4 Hz. Base your calculations on a rubber having Young's modulus $E_0 = 2.2$ MPa, shear modulus $G = 0.64$ MPa, and $k = 0.73$.

8.27 A $+45/-45/-45/+45$ sheet of cross-plied carbon fibre-epoxy laminate is made using identical plies of the same material as specified in Examples 8.4 and 8.5. A stress σ_A (GPa) applied in the 0° direction (i.e. at 45° to both sets of fibres) produces in-plane strains ε_A and ε_B. The stress–strain relationship is

$$\sigma_A = 59.3\varepsilon_A + 49.7\varepsilon_B.$$

(1) Calculate Young's modulus in the A direction.

(2) Calculate Young's modulus in the 0° and 90° directions for a 0/+45/−45/90/−45/+45/0 laminate made from the same material.

8.28   A certain grade of polysulfone (PSF) containing liquid additives is found to craze if held at a static stress of 12 MPa for periods of more than 4 weeks. Internal stress due to differential thermal contraction in injection-moulded parts also causes occasional crazing. By treating an injection moulding as a two component system, with an outer skin that is cooled rapidly to 20°C when in contact with the cold mould, and a core which then cools from 180°C within the volume defined by the frozen skin, calculate the ratio of skin thickness to section thickness that is needed to initiate crazing in a long rectangular bar. Use data on PSF from Table 8.2, taking $\nu = 0.35$. Assume that crazing occurs at the same stress under uniaxial and equal biaxial tension (see Problem 5.8). Suggest measures for overcoming the internal crazing problem.

8.29   What fracture problems might arise as a result of:

(1) painting plastic parts for automobile applications;

(2) using nylon parts in outdoor fittings for a telephone system to be installed in termite-infested regions of the world (note: termites secrete formic acid);

(3) introducing a new cleaning fluid for PMMA aircraft windows; and

(4) making the radius too small at the base of the teeth of a POM gear wheel?

8.30   A new diving helmet is being designed to operate at maximum depths of 250 m. The windows will be circular PMMA discs 200 mm in diameter. For the chosen PMMA, the fatigue endurance limit is 20 MPa, and Poisson's ratio $\nu = 0.4$. Allowing a safety factor of 2.5, calculate the minimum thickness d of PMMA needed, using the following formula for the maximum stress σ_{max} in a disc of radius R clamped at the circumference and under pressure P:

$$\sigma_{max} = 3PR^2(1 + \nu)/8d^2.$$

Under high bending stresses, semicircular cracks (half of penny-shaped crack of radius a) initiate at the surface of the PMMA. For this geometry, $Y = 2/\pi$. Taking K_{IC} for PMMA as 1 MPa m$^{0.5}$, calculate the minimum crack size that could initiate catastrophic fracture in a window of the thickness that you recommend, when the helmet is at its maximum operating depth. Assume $\rho = 1030 \text{ kg m}^{-3}$ for seawater. Visual inspection can be relied upon to detect cracks if $a > 2$ mm. At what depth do cracks of this size initiate catastrophic fracture in the window that you have designed?

8.31 Epoxy resin containing 60 vol.% continuous carbon fibres has been selected for the rotor blades of an axial-flow air compressor. Each blade will be 150 mm long and 30 mm wide. In use, they will be loaded in bending as cantilever beams under a uniformly distributed load of up to 3 N mm^{-1}. Treating the blades as rectangular in cross-section, and allowing a safety factor of 6 on strength, calculate the minimum allowable thickness for the blades. The data in Figure 8.23 should be used for the calculation.

Further reading

Note: *The table at the end of this section indicates which items in the further reading list are relevant to particular chapters in the book*

1. Agarwal, B. D. and Broutman, L. J. (1980). *Analysis and Performance of Fibre Composites*. John Wiley, New York. A relatively advanced text that gives a clear and comprehensive account of the difficult subject of stiffness and strength in aligned fibre composites. (Relevant to Chapters 6 and 8.)
2. Bassett, D. C. (1981). *Principles of Polymer Morphology*. Cambridge University Press, Cambridge. A highly readable account. (Relevant to Chapter 2.)
3. Billmeyer, F. W. (1984). *Textbook of Polymer Science* (3rd edn). John Wiley, New York. A popular account of polymer science, wide ranging and comprehensive with emphasis on the chemistry. May be valuable to a mechanical engineer as a reference text. (Relevant to Chapter 1.)
4. Blythe, A. R. (1979). *Electrical Properties of Polymers*. Cambridge University Press, Cambridge. An introduction to electrical properties of polymers: a topic growing in importance as polymers begin to be used more widely in electrical devices.
5. Bovey, F. A. and Winslow, F. H. (1979). *Macromolecules—An Introduction to Polymer Science*. Academic Press, New York. A comprehensive survey covering the synthesis, characterisation and physical chemistry of polymers. (Relevant to Chapter 1.)
6. Boyd, R. H. and Phillips, P. J. (1993). *The Science of Polymer Molecules*. Cambridge University Press, Cambridge. A more detailed presentation of polymer chemistry and physics, including polymer solutions. (Relevant to Chapters 1, 2, and 3.)
7. Broek, D. (1982). *Elementary Engineering Fracture Mechanics* (3rd edn). Martinus Nijhoff, The Hague. A well-presented and readable account of the subject. (Relevant to Chapter 5.)
8. Brown, R. P. (ed.) (1982). *Handbook of Plastics Test Methods*. George Godwin, London. A comprehensive assembly of test methods with a large bibliography: a reference book. (Relevant to Chapters 4, 5, and 7.)
9. Brydson, J. A. (1981). *Flow Properties of Polymer Melts*. George Godwin Ltd, London. A useful account with many research references. (Relevant to Chapter 7.)

10. Bucknall, C. B. (1977). *Toughened Plastics*. Applied Science, London. A special-ized text on rubber toughened plastics, but including more general discussions of crazing, yielding, and fracture in plastics. (Relevant to Chapter 5.)

11. Campbell, I. M. (1994). *Introduction to Synthetic Polymers*. Oxford University Press, Oxford. Nicely complements the present book, from the chemical perspective. Includes more detailed discussions of polymerization processes, and of techniques for chemical structure determination. May be of particular interest to students of chemical engineering. (Relevant to Chapters 1 and 2.)

12. Cogswell, F. N. (1981). *Polymer Melt Rheology*. George Godwin Ltd, London. A highly individualistic, interesting account; very readable and with a minimum of mathematics. (Relevant to Chapter 7.)

13. Cowie, J. M. G. (1991). *Polymers: Chemistry and Physics of Modern Materials* (2nd edn). Blackie, Glasgow/London. A good general coverage of synthesis, characterization and properties of polymers, including chapters on polymer liquid crystals and on polymers for the electronics industry.

14. Crawford, R. J. (1987). *Plastics Engineering* (2nd edn). Pergamon Press, Oxford. A general account of polymer engineering, with emphasis on thermoplastics. Many end-of-chapter problems. (Relevant to Chapters 4–8.)

15. Donald, A. M. and Windle, A. H. (1992). *Liquid Crystalline Polymers*. Cambridge University Press, Cambridge. An introduction to one of the most rapidly growing areas of polymer science, including a chapter on applications.

16. DuPont. *Design Handbooks* E. I. DuPont de Nemours, Wilmington, Delaware.
    (a) *Delrin Acetal Resins*.
    (b) *Zytel Nylon Resins*.
    Not only provides data on the products being offered by the manufacturers, but is also essentially a textbook for the practising designer, with design formulae and worked examples. (Relevant to Chapter 8.)

17. Ehrenstein, G. W. and Erhard, G. (1984). *Designing with Plastics*. Hanser, Munich. A practically oriented book on the application of design principles to polymeric components. (Relevant to Chapter 8.)

18. Fenner, R. T. (1979). *Principles of Polymer Processing*. Macmillan, London. A good account of thermoplastic processing suitable for graduate student or undergraduate specialist courses. (Relevant to Chapter 7.)

19. Ferry, J. D. (1980). *Viscoelastic Properties of Polymers*. (3rd edn) John Wiley, New York. An authoritative treatise covering all aspects of linear and non-linear viscoelastic behaviour in polymers. (Relevant to Chapter 4.)

20. Flory, P. J. (1971). *Principles of Polymer Chemistry*. Cornell University Press, Ithaca, N. Y., Eighth Printing. A comprehensive account, somewhat out of date but a classic in the field and beautifully written. Contains an elegant chapter on rubber elasticity. (Relevant to Chapters 1 and 3.)

21. General Electric Company. *Lexan, Noryl, Valox for Functional and Economical Design*. General Electric Plastics BV, Bergen op Zoom, Netherlands. Not only provides data on the products being offered by the manufacturers, but is also

essentially a textbook for the practising designer, with design formulae and worked examples. (Lexan is PC, Noryl is MPPO, and Valox is PBT.) (Relevant to Chapter 8.)

22. Gibson, R. F. (1994). *Principles of Composite Material Mechanics*. McGraw-Hill, New York. A particularly thorough, but also clearly presented, account of the mechanics and micromechanics of composites including laminates. There are many worked examples and end-of-chapter problems. (Relevant to Chapter 6.)

23. Hertzberg, R. W. and Manson, J. A. (1980). *Fatigue of Engineering Plastics*, Academic Press, New York. A specialized text aimed at the postgraduate, covering all relevant literature up to the date of publication. (Relevant to Chapter 5.)

24. Hull, D. (1981). *An Introduction to Composite Materials*. Cambridge University Press. A well written textbook aimed at the final year undergraduate and graduate student. It also concentrates upon stiffness and strength. (Relevant to Chapters 6 and 8.)

25. ICI (1973). *Presentation and use of data on the mechanical properties of thermoplastics* (2nd edn). Technical Service Note G123, ICI, Welwyn Garden City. (Relevant to Chapter 8.)

26. ICI (1978). *The Principles of Injection Moulding* (3rd edn). Technical Services Note G103, ICI, Welwyn Garden City. (Relevant to Chapter 7.)

27. ICI (1978). *Victrex PES—Data for Design*. Technical Service Note VX101, ICI, Welwyn Garden City. Data on stiffness, strength, and other properties presented mainly in graphical form. (Quoted in Chapter 8.)

28. Kinloch, A. J. and Young, R. J. (1983). *Fracture Behaviour of Polymers*. Applied Science, London. A good readable review of fracture in polymers and the underlying science. (Relevant to Chapter 5.)

29. Lindley, P. B. (1974). *Engineering Design with Natural Rubber* (4th edn). MRPRA, Hertford, England. A short but very informative and authoritative handbook on the subject, especially covering rubbers in springs and bearings. (Relevant to Chapter 8.)

30. McCrum, N. G., Read, B. E., and Williams, G. (1967). *Anelastic and Dielectric Effects in Polymeric Solids*. John Wiley, New York. A specialist text; will be useful as a reference. (Relevant to Chapter 4.)

31. Mandelkern, L. (1964). *Crystallization of Polymers*. McGraw-Hill, New York. A specialist account. (Relevant to Chapter 2.)

32. Mathews, F. L. and Rawlings, R. D. (1994). *Composite Materials: Engineering and Science*. Chapter and Hall, London. A wide-ranging introduction to composite materials in their various forms, including composites with metallic or ceramic matrices. Many (non-numerical) self-assessment questions in addition to some numerical end-of-chapter problems. (Relevant to Chapter 6.)

33. Middleman, S. (1977). *Fundamentals of Polymer Processing*. McGraw-Hill, New York. A thorough and quantitative account of the subject, covering heat transfer and melt flow in detail. (Relevant to Chapter 7.)

34. Mills, N. J. (1993). *Plastics: Microstructure and Engineering Applications* (2nd edn). Edward Arnold, London. A highly readable account, with the focus on engineering properties (including electrical and transport properties) and their microstructural origins. Includes design applications with the aid of three case studies. (Some relevance to all chapters.)

35. *Modern Plastics Encyclopedia* (published annually). McGraw-Hill, New York. An up-to-date review of all the significant plastics and resins in current use and of processing technology. Aimed at a general readership and, therefore, avoiding advanced concepts. Useful for materials selection, with tables of data on each polymer. Includes lists of applications. (Relevant to Chapter 8.)

36. Osswald, T. A. and Menges, G. (1996). *Materials Science of Polymers for Engineers*. Carl Hanser Verlag, Munich. Thorough coverage of the engineering properties of polymers, including melt properties and thermal, electrical optical, and transport properties in the solid state. An especially useful feature for reference purposes is a large Table giving over 30 property values for 51 polymers. (Relevant to Chapters 3–8.)

37. Parker, D. B. V. (1974). *Polymer Chemistry*. Applied Science Publishers, Barking, Essex, UK. Fairly detailed but readable account of polymer chemistry. (Relevant to Chapter 1.)

38. Piggott, M. R. (1980). *Load Bearing Fibre Composites*. Pergamon Press, Oxford. A good basic text aimed at undergraduate and graduate students in Materials Science and Engineering. The emphasis is on clear presentation of quantitative principles. (Relevant to Chapter 6.)

39. Powell, P. C. (1983). *Engineering with Polymers*. Chapman & Hall, London. A good account of polymer engineering; should be supplemented with introductory material on polymer science. (Relevant to Chapters 5, 6, 7, and 8.)

40. Pye, R. G. W. (1983). *Injection Mould Design*. George Godwin Ltd, London. An authoritative and detailed account; a reference book. (Relevant to Chapters 7 and 8.)

41. Richardson, P. N. (1974). *Introduction to Extrusion*. Society of Plastics Engineers Inc., Brookfield Centre, Conn. A refreshingly short but detailed and non-mathematical account of extrusion; a very interesting style with excellent diagrams. (Relevant to Chapter 7.)

42. Suh, Nam P. and Sung, Nak Ho (eds). (1979). *Science and Technology of Polymer Processing*, MIT Press, Cambridge, Mass. A wide-ranging review covering the current state of understanding. Includes contributions from a distinguished group of authors, mainly from industry. (Relevant to Chapter 7.)

43. Tadmor, Z. and Gogos, C. G. (1979). *Principles of Polymer Processing*. John Wiley, New York. A post-graduate text book, highly mathematical and very comprehensive. (Relevant to Chapter 7.)

44. Treloar, L. R. G. (1975). *The Physics of Rubber Elasticity* (3rd edn). Oxford University Press. An authoritative text on the stress–strain behaviour of rubbers. (Relevant to Chapters 2 and 3.)

45. Turner, S. (1983). *Mechanical Testing of Plastics*. George Godwin, London. An interesting and unique account of the field by one of the pioneers. (Relevant to Chapters 4, 5, and 8.)

46. Ward, I. M. (1983). *Mechanical Properties of Solid Polymers* (2nd edn). John Wiley, New York. An advanced textbook on the mechanical properties of plastics and rubbers. Includes a detailed discussion of the treatment of anisotropy, and the relevant mathematics. (Relevant to Chapters 2, 3, 4, and 5.)

47. Ward, I. M. and Hadley, D. W. (1993). *An Introduction to the Mechanical Properties of Solid Polymers*. John Wiley and Sons, Chichester. The third edition of the book above, but the coverage is slightly less comprehensive, begins at a less advanced level, and now includes some Problems and outline solutions. (Relevant to Chapters 2, 3, 4, and 5.)

48. Weir, C. I. (1975). *Introduction to Injection Moulding*. Society of Plastics Engineers Inc., Brookfield Center. A good authoritative and non-mathematical account of injection moulding; excellent diagrams. (Relevant to Chapter 7.)

49. Williams, J. G. (1984). *Fracture Mechanics of Polymers*. Ellis Horwood. Chichester, England. An advanced treatise aimed at the postgraduate worker. (Relevant to Chapter 5.)

50. Wunderlich, B. (1973–80). *Macromolecular Physics*. Academic Press, New York.

Vol. I. Crystal Structure, Morphology, Defects	(1973).
Vol. II. Crystal Nucleation, Growth, Annealing	(1976).
Vol. III. Crystal Morphology	(1980).

    A treatise covering all aspects of crystallinity in polymers (Relevant to Chapter 2.)

51. Young, R. J. and Lovell, P. A. (1991). *Introduction to Polymers* (2nd edn). Chapman and Hall, London. A general coverage of polymer science, with particular emphasis on synthesis and physical chemical characterization. (Relevant to Chapters 1–5.)

Further Reading no.	\multicolumn{8}{c}{Chapter no.}							
	1	2	3	4	5	6	7	8
1						×		×
2		×						
3	×							
4								×
5	×							
6	×	×	×					
7					×			
8				×	×		×	
9							×	
10					×			
11	×	×						
12							×	
13	×	×						
14				×	×	×	×	×
15		×						
16								×
17								×
18							×	
19				×				
20	×		×					
21								×
22						×		
23					×			
24						×		×
25								×
26							×	
27								×
28					×			
29								×
30				×				
31		×						
32						×		
33							×	
34	×	×	×	×	×	×	×	×
35								×
36			×	×	×	×	×	×
37	×							
38						×		
39					×	×	×	×
40							×	×
41							×	
42							×	
43							×	
44			×					
45				×	×			×
46		×	×	×	×			
47		×	×	×	×			
48							×	
49					×			
50		×						
51	×	×	×	×	×			

Answers to problems

Chapter 1

1.1 (1) 56 000 (2) 46 000 (3) 97 000
    (4) 56 000 (5) 44 000 (6) 86 000
1.2 (1), (2), (5), and (6).
1.3 (1)

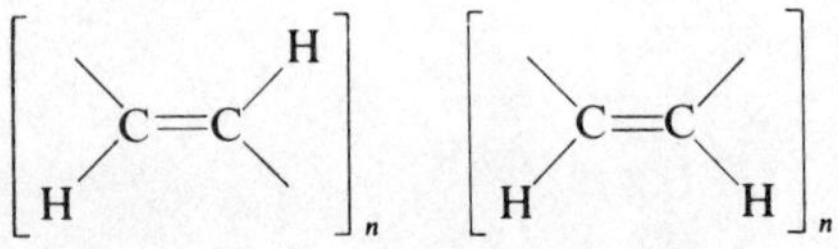

polyacetylene

(2)

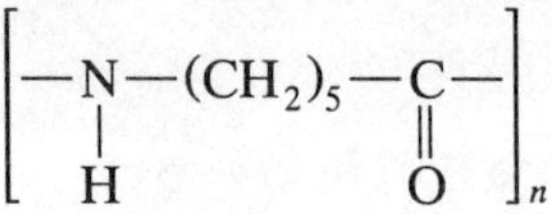

nylon 6

(3) $[-CH_2CH_2O-]_n$ poly(ethylene oxide)

(d)

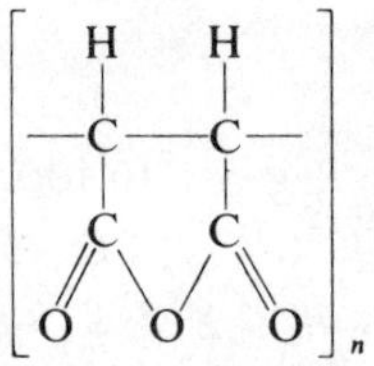

poly(maleic anhydride)

(1) and (4) form stereoisomers.
1.4 *Cis* and *trans* isomers formed by addition at carbons 1, 4. Also 1, 2 and 3, 4 addition products, both of which exhibit stereoisomerism.
1.5 Reduced crystallinity due to (C_4H_9) side groups will lower modulus and density.
1.6 (1) and (2) linear polyethylene; (3) branched polyethylene.
1.7 Cl—C—Cl

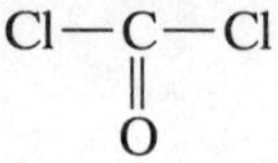

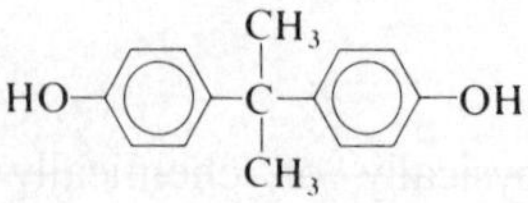

1.8   Polysulfone (PSF) (see 8.N.1).
1.9   (1) 14.75   (2) 64.7
      (3) 37.6   (4) 40.4.
1.10 Crosslinking through double bonds.
1.11 (1) $\overline{M}_n = 1982$.   (2) $\overline{M}_n = 9,662$.
      (3) No limits.
1.12 22574.
1.14 0.224; 332 chains.
1.15 $\overline{M}_n = 38\,687$; $\overline{M}_w = 56\,630$.
1.16 $\overline{M}_n = 13\,595$; $\overline{M}_w = 56\,349$.

Chapter 2

2.1   (1) 1011;   (2) 1477;   (3) 1429;
      (4) 2085;   (5) 1010 (kg m^{-3}).
2.2   (1) 943;   (2) 1231 (kg m^{-3}).
2.3   (1) 1238;   (2) 1451 (kg m^{-3}).
2.4   (1) 10.9°;   (2) 12.1°.
2.5   Yes.
2.6   143.5°C.
2.7   (1) 46.8%, 56.2%, 66.0%.   (2) 66.8%.
2.8   164.5°C.
2.9   121.7°C.
2.10 11.5%, 50%.
2.11 −82°C.
2.14 (1) 762 nm.   (2) 17 nm, 2.84 nm.
2.15 (1) C_4H_{10}: t, g^+, g^- (3 forms).
          C_5H_{12}: tt; tg^+; tg^-; g^+g^+; g^+g^-; g^-g^- (6 forms).
          C_6H_{14}: ttt; ttg^+; ttg^-; tg^+t; tg^+g^+; tg^+g^-; tg^-t;
              tg^-g^+; tg^-g^-; g^+tg^+; g^+tg^-; $g^+g^+g^+$; $g^+g^+g^-$;
              $g^+g^-g^+$; $g^+g^-g^-$; g^-tg^-; $g^-g^+g^-$; $g^-g^-g^-$ (18 forms)
      (2) 3^{4997}.
2.16 52.4%.
2.17 22.1%, 41.1%, 26.2%, 8.7%, 1.6%.
2.18 26.6 nm.
2.19 183 GPa.
2.21 $\Delta n_c^0 = 0.028$; $\Delta n_a^0 = 0.063$.
2.23 0.637.

Chapter 3

3.1 $T > T_g$. Physically or chemically crosslinked. Amorphous phase
      present.
3.5   0.95 K.

3.7 5.37×10^{25} m^{-3}; 150.

3.8 9.26×10^{25} m^{-3}.

3.9 $G_s/G_u = \phi^{\frac{1}{3}}$; 0.172 MPa.

3.10 Rubbers are very nearly incompressible; i.e. $G \ll K$ (where K is the bulk modulus).

3.11 1.3%.

3.13 $E = 1.35$ MPa; $G = 0.449$ MPa; $F_x = -4.61$ kN;
$Y = Z = 115.5$ mm; $F_x = F_y = -15.5$ kN;
$Z = 178$ mm; $W = 288$ J.

3.14 (1) 20 μm; (2) 25 MPa; (3) 4 kPa.

3.15 $F_x = 1.94$ kN; $F_y = 1.43$ kN.

3.16 34.4 mm; 2.77 mm.

3.18 $\sigma = 1.23$ MPa; 14.3% error.

3.19 (1) $C_1 = 93.5$ kPa, $C_2 = 122$ kPa, $G = 431$ kPa; (2) 75.5; (3) 64, 4.7.

Chapter 4

4.1 2.04 mm.

4.2 (1) 0.00398; (2) 55.9 N.

4.3 0.283%; 0.0177% (taking hoop stress $= 0.4 \times 37/8$ MPa).

4.4 0.280%; 0.011%.

4.7 18.8 mK s^{-1}; 23.28°C (taking $\Lambda = 0.315$; and $G' = 1.37$ GPa).

4.8 $\sigma = \gamma/J + \eta(d\gamma/dt)$; $d\gamma/dt = (d\sigma/dt)/G + \sigma/\eta$.

4.9 Kelvin: 10^{-9} $[1 - \exp(-t/1000)]$ Pa^{-1}.
Maxwell: $10^{-9}[1 + t/1000]$ Pa^{-1}.

4.14 (1) 5.23 J_0; (2) $(4.88 - 1.98i)J_0$.

4.15 7.09 GPa^{-1}. All relaxation times have same shift factor a_T: J_U and J_R are independent of temperature; no ageing or delayed crystallization occur during creep.

4.16 (1) 17.4°; (2) 27.7°; (3) 20.9°.

4.17 +2.22 MPa.

4.19 3.42 m.

4.20 0.029 %.

4.21 0.77%; 79 N.

4.22 $\sigma = 500\varepsilon + 5[1 - \exp(-200\varepsilon)]$ MPa.

Chapter 5

5.2 (1) 23.6 and 20.1 MPa. (2) 8×10^{-2} and 8×10^{-5} s^{-1}.

5.3 3.16×10^{-3} m^3 mol^{-1}.

5.4 165 kJ mol^{-1}; 2.44×10^{-3} m^3 mol^{-1}; 63.1 MPa.

5.5 3.14, 4.61, and 5.34×10^{-3} mol^{-1}.

5.7   51.0 and 56.6 MPa.

5.9   393 mm^{-1}; 17.5%.

5.10 (1) 2.4 MPa m$^{0.5}$;   (2) 8.56 MPa.

5.11 3.66 MPa m$^{0.5}$; 856 N.

5.12 (1) 54.3 N;   (2) 0.36 mm.

5.13 0.60, 2.31, 3.12, 7.7, 2.2, 1.21, 5.72, 1.10, 1.21, MPa m$^{0.5}$; All valid except saturated PA6.6 and POM.

5.14 0.886 mm.

5.15 0.306 J.

5.16 6.2 kJ m^{-2}.

5.17 (1) 1.12,   (2) 2.433.

5.18 1.42 Ms.

5.19 15.8 MPa; 0.2 Ms.

5.20 (1) 5.79 MPa, yield;   (2) 156 MPa;   (3) 75.6 MPa.

5.21 (1) 0.69, 0.78, 1.00, 2.24, 4.58, 6.23 mm
      (2) $-50°C$, $-28°C$, $-6°C$.

5.22 77.4 MPa.
      Increasing temperature or decreasing strain-rate has the effect of reducing yield stress. In sharply notched fracture mechanics specimens, yielding is confined to a single craze at the crack tip. The resulting crack opening displacement for the craze is not particularly sensitive to temperature or strain-rate, and the observed effects therefore reflect the changes in craze-surface (fibril-drawing) stress.

5.23 188 500 cycles.

5.24 1.09 MPa.

5.25 1.46 MPa.

5.27 $-39.3°C$.

Chapter 6

6.1 $c > a > b > d > e$.

6.3   (1) 13.7%;   (2) 21.4%;   (3) 29.8%.

6.4   1392 kg m^{-3}; 23.3%.

6.5   39.2, 205, 75.4 GPa; 20.4, 125, 54.2 MPa m^3 kg^{-1}.

6.6   (1) 4.65;   (2) 3.96;   (3) 3.49 GPa.

6.8   (1) 1.18, -0.0158 mm;   (2) 53 μm;   (3) 5.23°C.

6.9   6.0 GPa.

6.10 3.93×10^{-4}; 4.22×10^{-3}, 11.7×10^{-3}.

6.11 16.3 mm; -0.352 mm; 8.77°.

6.14 (1) 7.66, 52.5;   (2) -0.777, 48.4;   (3) -1.36, 66.5 ($\times 10^{-6}$ K^{-1}).

6.15 (1) 0.336;   (2) 0.393.

6.16 6.13, 34.2 ($\times 10^{-6}$ K^{-1}); 1.93°; 1.21 mm; 60μm.

6.17   (1) 928;   (2) 1507;   (3) 1702 (MPa).
6.19   (1) 189;   (2) 900 (MPa).
6.20   80 MPa transverse tensile.
6.21   Axial tensile 0–2.48°, shear 2.48–29.7°, transverse tensile 29.7–90°.
6.22   (1) 23.5;   (2) 31.2 (GPa).
6.23   (1) 52.6;   (2) 31.0 (MPa).
6.24   (1) 439 MPa;   (2) 560 MPa.

Chapter 7

7.4 $94\,300$ Pa s.
7.5   (2) $\mu v w l / h$; $w l \tau_0^{1-n} \mu_{a0}^n (v/h)^n$.
7.6   328, 229, 160 Pa s.
7.7   0.483; 323, 646, 1291 s^{-1}; 259, 181, 126 Pa s.
7.8   (1) -13.3;   (2) -4.95 (% K^{-1}).
7.10   (1) $\tau = [(h/2) - y](\Delta P/\Delta L)$.
       (2) $v = (1/2\mu)(hy - y^2)(\Delta P/\Delta L)$.
7.11   (1) As 7.10(1)
       (2) $v = (\Delta P/K\Delta L)^{1/n}(h/2)^{1+(1/n)}(1 - |1 - 2y/h|^{1+(1/n)})n/(1+n)$.
7.12   70 MPa m^{-1} (taking $\mu = 700$ Pa s); 61.5 mm; 3.0 mm; 18.1 mm s^{-1}
       (1.08 m min^{-1}).
7.13 6.43×10^{-6} m^3 s^{-1}.
7.16 1.13×10^{-4} m^3 s^{-1}, 5.21 MPa; 1.15×10^{-4} m^3 s^{-1}, 1.65 MPa.
7.17   494 kg h^{-1}; 20.4 MPa (taking $\mu = 794$ Pa s at 58 s^{-1}).
7.18   (1) $Q = 3.56 \times 10^{-21}(\Delta P)^2$ m^3 s^{-1}; 3.98×10^{-7} m^3 s^{-1},
       127 mm s^{-1}.
7.19   21.5°.
7.20   (3) 10%.
7.21 $\ln(v/v_0) = (FRT^*/Q\lambda^*\Delta Ha)[1 - \exp(-\Delta Haz/RT^*)]$, 143 mm s^{-1},
       0.394 mm.
7.22   1.37 N, 116 Pa.
7.23   (1) 3.21 mm.
7.24   (2) (i) £1595, (ii) £5660, (iii) £21 889.
7.25   0.61%.
7.26   (1) 4.60%;   (2) 108 MPa.
7.27   4.2%, 203 mm, 40.6 mm, 3.04 mm.
7.29   22.5 MPa, 120 tonnes.

Chapter 8

8.7   14 100.
8.8   (4) £6.15 for GF-PP, £6.67 for GF-PET.

8.12 Forms 4 welds with Y 24 mm from corners.

8.13 (1) 90° (2) 47.9°.

8.14 90°, 62.7°.

8.16 0.19 mm; 0.25 mm.

8.18 20.9 mm.

8.19 If seat uncovered, suggest 14 ribs 30 mm high, 3 mm wide, with side supports. If chairs covered, use fewer ribs of width 4.5 mm (causing sink marks but reducing distortion due to differential cooling rates).

8.20 24.6, 4.0, 23.4 mm.

8.21 0.79%.

8.22 (1) 6.28 mm; (2) 6.41×10^{-3}.

8.23 (1) 5 webs.

8.24 16.9 mm.

8.25 4 bolts.

8.26 Four pads $160 \times 160 \times 26$ mm.

8.27 (1) 17.6 GPa; (2) 80.6, 56.3 GPa.

8.28 0.17 : 1.

8.30 41 mm; $a = 12.3$ mm; 619 m.

8.31 6.21 mm (taking tensile strength $\sigma^* = 1.05$ GPa).

Index

ABS 18
  applications 199, 374, 414
  crazing and toughening 199, 227–8
  processing 348–50, 362
  properties 373
acrylic polymers 37, 334, 348
acrylonitrile 43, 102, 110
acrylonitrile–butadiene–styrene
    copolymer, *see* ABS
activation enthalpy 150, 175–7, 192–4
activation volume 190–4
addition polymers 21–3
additives 372
affine deformation 92
ageing
  chemical 229, 405
  physical 153, 158–61
aircraft 199, 414
alloys (metal) 241
amorphous fraction 45
anisotropy 49, 196, 223, 264, 265, 271
apparent viscosity 305, 306, 311, 313,
    314
applications of polymers
  aircraft 5, 199, 414
  boats 239, 256
  building and construction 6, 319
  cables 38
  cars and vehicles 1, 7–11, 14, 105,
    370–2
  chemical 369
  domestic 38, 76, 77
  electrical 6, 38, 376
  fibres 76, 77, 251
  films 38, 77, 357
  gears 77
  medical 1, 38
  optical 9, 71, 411–13
  packaging 7, 38, 77, 357
  pipes 38, 170, 199
  sports goods 239–41
  tyres 102, 240, 241, 283

aramid fibre (Kevlar[R]) 240, 247,
    250–2, 291
Arrhenius equation 150, 175, 313
aspect ratio of filler 242, 257, 275–8
atactic polymer 29
automobiles, *see* cars
Avogadro's number 77, 111

Baekeland 416
Bagley correction 311, 312
Bakelite 416
barrier flight extruder screw 327
biaxial orientation 72, 196, 223, 350
Biot number 317
birefringence 70, 71, 83
blends 32–4
block copolymer 26, 104, 172, 173
blow moulding 77, 349, 350
Boltzmann equation 88
Boltzmann superposition principle
    (BSP) 164–70, 183
branching 20, 22, 23
brittle fracture 184, 200–13,
    217–23, 407
Brownian motion 117
bulk deformation of melt 299
butadiene 41
butyl rubber 376

capillary flow 306–12
car tyres (reinforcement) 240, 241, 283
carbon black in rubbers 101, 240, 283–5
carbon fibre
  applications 239, 408, 409, 414
  composites 244, 247–52, 266–73
  manufacture from PAN 110
  structure and properties 239, 240,
    250, 251, 291

vacuum forming 346, 348
van der Waals bonds 71, 109
velocity profile 308
vented screw in extruder 323–7
vinyl polymer 22
viscoelasticity 117–8
    linear 119, 120, 127, 170, 185
    non-linear 120, 127, 162, 185
    of polymer liquids 313
viscosity, apparent 305, 306, 311
    definition 305
    determination 306–12
    RMM dependence 313, 353–5
    stress dependence 313
    temperature dependence 313, 353
voids 52, 227, 245, 319, 333, 383, 404
von Mises yield criterion 194
vulcanization 4, 28, 84, 101, 155

wall shear rate 308, 311, 330
wall thickness of mouldings 383, 384
warping 319
water
    absorption 40, 157, 196
    effects on polymer properties 40, 157,
        158, 196, 375, 392
    formation in polymerization 24, 25
weathering (outdoor ageing) 405

weight average molar mass 33–6
weld lines 338, 379, 381, 382
welding 385
wire coating 331
WLF theory 238, 313, 353
woven fabric 256, 274

X-ray diffraction 45, 50

Y (geometrical factor in fracture
        mechanics) 210, 211, 217,
        233, 235
yielding 185–97
    under multiaxial stress 194–8
    of net section in cracked bar 218
yield stress 184–97
    data for plastics 373
    lowering, by rubber particles 197
    pressure dependence 194–7
    strain-rate dependence 191–2
    temperature dependence 191–3
yield zone at crack tip 194, 211–17, 228

Zener model 140–51, 167, 174, 180, 181
Ziegler-Natta catalysts 23, 29, 42